The North Korean Nuclear Program

*To Joel Wit,
the man who negotiated
an end to the North Korean
nuclear program,
With best wishes,
Alexandre Mansourov
12/07/99*

More praise for
The North Korean Nuclear Program

"This is the comprehensive briefing book that American negotiators needed to have prior to tension filled negotiations over the DPRK's threats to withdraw from the NPT in 1993. It provides many of the missing pieces as part of the international effort to interpret the motives, determination, and level of know-how that have driven North Korea's past efforts to acquire nuclear capabilities."
—Scott Snyder, Program Officer,
Research and Studies Program,
United States Institute of Peace

"I applaud the editors' efforts in putting together such a comprehensive volume. This provides an important piece in ongoing efforts to understand the puzzle of the DPRK nuclear program, and perhaps more importantly, the North Korean regime."
—L. Gordon Flake, Executive Director,
The Mansfield Center for Pacific Affairs

"This book offers many new insights into the opaque North Korean decision-making process and casts light on why North Korea may be seeking nuclear weapons. American policy makers who fail to learn from the lessons gained by reading *The North Korean Nuclear Program* put both the United States and the East Asian region at risk."
—Peter Hayes, Co-Director,
Nautilus Institute for Security and
Sustainable Development

The North Korean Nuclear Program

Security, Strategy, and
New Perspectives from Russia

James Clay Moltz
and Alexandre Y. Mansourov,
Editors

Routledge
New York and London

Published in 2000 by
Routledge
29 West 35th Street
New York, NY 10001

Published in Great Britain by
Routledge
11 New Fetter Lane
London EC4P 4EE

Copyright © 2000 by Routledge

Printed in the United States of America on acid-free paper.

All rights reserved. No part of this book may be reprinted or reproduced or utilized in any form or by any electronic, mechanical or other means, now known or hereafter invented, including photocopying and recording or in any information storage or retrieval system, without permission in writing from the publishers.

Library of Congress Cataloging-in-Publication Data

The North Korean nuclear program : security, strategy, and new perspectives from Russia / edited by James Clay Moltz and Alexandre Y. Mansourov.
    p. cm.
   ISBN 0-415-92369-7 (H : alk. paper). — ISBN 0-415-92370-0 (P : alk. paper)
   1. Nuclear weapons—Korea (North) 2. Nuclear industry—Korea (North)
3. Korea (North)—Military relations—Russia (Federation) 4. Korea (North)—Foreign relations—Korea (North) 5. Russia (Federation)—Foreign relations—Korea (North) 6. Russia (Federation)—Foreign relations—Korea (North) I. Moltz, James Clay. II. Mansourov, Alexandre Y.
U264.5.K7N67 2000
355.02'17'095193—dc21   99-20510
                                                                                                 CIP

Contents

Acknowledgments   viii

Foreword   ix
    Mitchell B. Reiss

Introduction

  1 Russia, North Korea, and U.S. Policy toward the Nuclear Crisis   1
      James Clay Moltz

Part I
The History of the North Korean Nuclear Program

  2 A Technical History of Soviet–North Korean Nuclear Relations   15
      Georgiy Kaurov

  3 Nuclear Institutions and Organizations in North Korea   21
      Valery I. Denisov

  4 A Political History of Soviet–North Korean Nuclear Cooperation   27
      Alexander Zhebin

Part II
The Economic Context of the North Korean Nuclear Program

  5 Economic Aspects of the North Korean Nuclear Program   41
      Vladimir D. Andrianov

6 The North Korean Energy Sector   51
   Valentin I. Moiseyev

7 Economic Factors and the Stability of the North Korean Regime   60
   Natalya Bazhanova

8 The Natural Disasters of the Mid-1990s and Their
   Impact on the Implementation of the Agreed Framework   76
   Alexandre Y. Mansourov

Part III
Political and Military Factors behind the North Korean Nuclear Program

9 Nuclear Blackmail and North Korea's Search for a Place in the Sun   93
   Alexander Platkovskiy

10 Military-Strategic Aspects of the North Korean Nuclear Program   101
   Evgeniy P. Bazhanov

11 Leadership Politics in North Korea and the Nuclear Program   110
   Roald V. Savel'yev

Part IV
The International Context of the North Korean Nuclear Program

12 North Korea's Decision to Develop an Independent
   Nuclear Program   127
   Natalya Bazhanova

13 North Korea and the Nuclear Nonproliferation Regime   138
   Vladimir F. Li

14 North Korea's Negotiations with the Korean Peninsula
   Energy Development Organization (KEDO)   156
   Alexandre Y. Mansourov

15 China and the Korean Peninsula: Managing an Unstable Triangle   171
   Evgeniy P. Bazhanov and James Clay Moltz

16 The Korean Peninsula and the Security of Russia's
   Primorskiy Kray (Maritime Province)   179
   Larisa V. Zabrovskaya

Contents

Part V
Unsettled Problems and Future Issues

17 The Renewal of Russian–North Korean Relations   197
    James Clay Moltz

18 The Korean Peninsula: From Inter-Korean Confrontation to a System of Cooperative Security   210
    Alexander Zarubin

19 Russian Views of the Agreed Framework and the Four-Party Talks   219
    Evgeniy P. Bazhanov

20 Pyongyang's Stake in the Agreed Framework   236
    Alexandre Y. Mansourov

Notes   245

Contributors   269

Index   271

Acknowledgments

The editors would like to express their gratitude to Dr. Evgeniy Bazhanov of the Russian Diplomatic Academy and the Institute for Contemporary International Studies (ICIS) in Moscow for his considerable assistance in various stages of this book's development. Dr. Pyotr Razvin of the ICIS also provided valuable advice as well as considerable technical assistance in transferring the initial draft chapters from Moscow to Monterey. Other experts who helped in particular aspects of the book's final preparation include Dr. Vladimir Li (Diplomatic Academy) and Dr. Mitchell Reiss (KEDO). In addition, a number of individuals from the Center for Nonproliferation Studies at the Monterey Institute of International Studies deserve thanks for their help at various stages in the preparation of the manuscript: Stanley Shepard, Pat Levac, Filip Stabrowski, Kevin Orfall, Sarah Diehl, and Russell Willeford. The editors also are grateful to Routledge Press and, in particular, to associate politics editor Amy Shipper for her enthusiasm and their professionalism in handling the manuscript. Finally, the editors express their special thanks to Dr. Thomas Graham and the Rockefeller Foundation for their support of three workshops in Moscow organized by the CNS and ICIS that originally brought together the Russian authors featured here.

Foreword

Mitchell B. Reiss

Few countries have posed more vexing problems for the international community in recent years than the Democratic People's Republic of Korea (DPRK). Whether projecting strength with its ongoing nuclear weapons program, ballistic missile tests, and provocative military actions toward the Republic of Korea (ROK), or signaling weakness by appealing for humanitarian assistance to save its own starving population, the DPRK has presented a picture of a regime at once proud and insecure, defiant and vulnerable. Above all, it has eluded easy definition, manipulating uncertainty over its intentions, motives, and decision making to try to win tactical and strategic advantage. Indeed, the puzzling nature of the regime recalls Winston Churchill's famous 1939 description of Stalinist Russia: "I cannot forecast to you the action of Russia. It is a riddle wrapped in a mystery inside an enigma." More recently, former U.S. Ambassador to the ROK Donald P. Gregg has called the DPRK "the longest-running intelligence failure in U.S. history."

We are especially fortunate, then, that James Clay Moltz and Alexandre Y. Mansourov have labored long and hard to assemble this volume by Russian diplomats and scholars on North Korea and its nuclear program. It is a tonic to the paucity of hard facts surrounding the secretive North Korean regime. Moscow can offer firsthand experience in dealing with Pyongyang since the end of the Second World War. Soviet officials, diplomats, and scholars can provide both context and insight into the DPRK nuclear weapons program, correctly viewing it as a product of the country's history, ideology, politics, and security environment. They offer a perspective not often heard (and perhaps not even considered) among American policy officials and academics when considering Northeast Asian security issues. At a time when the United States and its allies are spending hundreds of millions of dollars trying to divine the internal workings of the Pyongyang regime, here is a book written by those who have actually lived, studied, worked, argued, and negotiated with North Koreans, in some cases over many years.

Not surprisingly, then, this volume contains some valuable assessments and sound judgments. Particularly noteworthy are two chapters by Natalya Bazhanova, a member of the Russian Academy of Sciences: one on the DPRK's decision to build the bomb and the other on the likelihood of economic reform taking root. On economic reform, she argues that "reforms are likely to create inflation, unemployment, and social instability . . . one can expect the Pyongyang-style 'economic adjustment policy' to continue to be characterized by internal contradictions that only make sense within Pyongyang's stilted worldview: efforts to attract foreign capital while strengthening political and ideological control over the population."

In another chapter, Evgeniy Bazhanov, a veteran Soviet diplomat, confirms the suspicions of many by informing us that the DPRK "has made serious efforts to create a missile force and to equip it with nuclear warheads." And in the most thought-provoking contribution to this volume, General Alexander Zarubin, a staff member of the Russian Security Council, charts a course for direct military-to-military contacts between the DPRK and ROK commands as "the way out of the current impasse" on the Korean Peninsula.

Many of these views also deserve careful review and attention because Russia may play a larger role in future developments on the Korean Peninsula than it has in the 1990s. As James Clay Moltz reminds us, "underlying historical, strategic, and regional economic forces are pushing the two sides [Russia and the DPRK] back together again." A 1998 Japanese initiative to hold a six-party discussion group on Korean security issues, consisting of Russia, Japan, China, the United States, and the two Koreas, confirms that Russia may once again assert its influence.

But this begs a larger question, namely, "What can Russia contribute to help resolve many of the problems that currently afflict the Korean Peninsula?" And it is on this point that the editors have performed an especially valuable service, because the Russian authors in this volume provide us with almost as much information about current Russian attitudes toward recent developments on the Korean Peninsula as they do about the nature of the regime in Pyongyang.

It is clear that Russia can bring to the table its experience and deep knowledge of the DPRK, its priorities, policies, preferences, and personalities. But it is also clear that at least some circles in Moscow harbor a lingering sense of bitterness, anger, and disappointment at how Russia has been "squeezed out" of the DPRK nuclear program, and, more generally, at how it has lost influence and status in Korea. (Indeed, the very fact that Moscow now emphasizes the need for a multinational conference on Korean security issues underscores how far Russian leverage has waned in the context of the current four-power negotiations.) This attitude evokes the image of the aging movie queen in *Sunset Boulevard*, Norma Desmond, declaiming, contrary to all evidence, that "I am big. It's the pictures that got small."

Significantly, this stance has colored perspectives among Russian diplomats and scholars regarding the Agreed Framework and its nuclear deal, as well as the ongoing activities of the Korean Peninsula Energy Development Organization (KEDO). In contrast to these views, Russia has always been welcome to join the organization. KEDO officials have, in fact, met on several occasions with members of the Russian government in Moscow and New York, and KEDO remains willing to discuss the light-water reactor project and its policies with Russian officials, parliamentarians, and academics in hopes of attracting Moscow as a member or as a contributing nonmember.

But whether or not Russia joins KEDO is not the main issue here. Rather, as the United States and its allies attempt to promote peace and security on the Korean Peninsula, the pertinent issue is whether Russia will continue to harbor resentment for past slights, real or imagined, or instead move ahead to embrace a new, more forward-looking policy. Certainly, Moscow could play the role of spoiler, trying to further complicate the already tenuous political and security environment in the region. But Moscow could also choose to play a constructive role in working with the United States, South Korea, and Japan to promote a more stable and secure Northeast Asia.

This is the key question facing Russia, and not only in Northeast Asia. Can it put aside its vestigial notions of Cold War grandeur and work constructively with the United States and its allies on a range of international security issues stretching from Kosovo to the Koreas? Expressed differently, can Moscow articulate a concept of its national interest that does not define itself by what it is against, but by what it is for? A concept that is not reflexively at odds with the West? This will be the great challenge for Russian foreign policy in the years to come. Whether it uses its unique position to help craft innovative approaches to ongoing regional conflicts will shape the nature of international relations in the next century.

1

Russia, North Korea, and U.S. Policy toward the Nuclear Crisis

James Clay Moltz

Since the North Korean nuclear crisis first erupted on the international stage with the March 1993 North Korean announcement that it was withdrawing from the Treaty on the Non-Proliferation of Nuclear Weapons (NPT), Western analysts have been trying to crack into the black box of North Korean decision making and understand its perspectives on nuclear technology. Ironically, however, at the very same time that the United States was trying to learn more about North Korea, the single best source of information on this highly secretive communist recluse—Russia—was being held at arm's length by Western negotiators. Indeed, the eventual signing of the October 1994 Agreed Framework between Pyongyang and Washington froze out the Russians from their previous position in North Korea, causing considerable frustration and irritation in Moscow. However, as time has passed, it has become clear that the United States still knows very little and perhaps understands even less about the forces that propelled Pyongyang to pursue a nuclear weapon and to risk alienating itself on the world scene even further by threatening to withdraw from the NPT.

The purpose of this book is to fill this knowledge gap, by drawing on the one source of information that is both the most rich historically and the most neglected of late: Russian experts who have lived in, worked in, or studied North Korea. This collection of specially commissioned papers represents the first Western publication to draw on the unique source of personal recollections, interview data, and archival information compiled by Russian experts on North Korea. Few Western experts can claim any experiences in the Democratic

People's Republic of Korea that even begin to approximate this level of expertise. This book should therefore provide a wealth of information for scholars, journalists, and government officials that has never before been available.

In the rest of this chapter, I first discuss other recent books on the North Korean nuclear program. Overall, these books have provided considerable new information on Western (and particularly South Korean, Japanese, and U.S.) perspectives and decision making on the crisis, which culminated in the October 1994 Agreed Framework and the 1995 formation of the Korean Peninsula Energy Development Organization (KEDO) in New York, in charge of building the two light-water reactors in North Korea. However, these books have not provided a great deal of insight into North Korean perspectives, strategy, and thinking about its nuclear relations with the West. In fairness, it is very difficult for Western observers to enter the North Korean mindset and understand the complex political, historical, and social constraints that shape leadership thinking on these issues. The authors of the present book are in a far better position to understand these factors, having been raised within the Soviet Union under similar constraints and often having spent many years within North Korea itself, exposed to the daily propaganda, regimentation of life, distrust of the West, and deprivation from outside sources of information. Thus, these authors are able to shed light on following questions: How does North Korean nuclear strategy fit within the dominant *juch'e* (self-reliance) ideology in Pyongyang? How does the regime differ from the "rational actor" model Western social scientists and policy analysts are familiar with? Within this North Korean "worldview," what is the role of the nuclear program in its relations with the West, and how is the program likely to be used in the future to shape (or leverage) those relations?

Given the recent East Asian economic crisis, pressure on the Agreed Framework is increasing, as funds for its implementation are becoming hard to come by. Indeed, it remains a serious question whether the unique experiment in collective security that KEDO and the Agreed Framework represent will ultimately fulfill its original functions, or whether the terms of the deal will have to be adjusted, with unpredictable and possibly dangerous consequences in Pyongyang. In the recent past, North Korea has threatened to restart its nuclear reactor and has engaged in a number of provocative missile sales that are likely to further jeopardize its relationship with the West. At the same time, it has blamed these actions on the slowness of the West to implement elements of the fuel oil deal that was part of the Agreed Framework, thus providing KEDO with at least a plausible means of remedying the situation and putting the responsibility for current problems in the relationship back in Pyongyang's court.

After a brief review of the recent literature on North Korea, showing what arguments have been made by some recent books on the nuclear crisis and what their studies add to our general knowledge of the problems at hand, this chapter then offers a substantive overview of the structure and main arguments presented in the book, highlighting new information uncovered by its Russian authors as well as the unique sources they have drawn upon. The chapter concludes with some tentative speculations about current trends in the implementation of the Agreed Framework and the future course of international negotiations with North Korea over its nuclear program. These can be viewed by the reader as themes to consider throughout the book and to apply to remaining questions facing decision makers today. Since 1993, Pyongyang has remained in formal violation of its nuclear safeguards agreement with the International Atomic Energy Agency (IAEA). Therefore, even if the current crisis in the international settlement over the reactor deal is solved, future nuclear developments and the unaccounted nuclear material will remain a matter of ongoing concern for some time to come.

Existing Studies of the North Korean Nuclear Program

Among recent studies of the diplomacy surrounding the North Korean nuclear crisis, several books stand out for contributing new information and insights into various aspects of the Agreed Framework and the cooperative efforts to freeze North Korea's nuclear program. Perhaps the most detailed account of these events from a U.S. perspective is that of former *New York Times*'s editorial board member Leon V. Sigal, whose book *Disarming Strangers: Nuclear Diplomacy with North Korea* (Princeton University Press, 1997) offers considerable new information not found in the press about the various actors on the U.S. side, including those outside of the government. He discusses, for example, the early North Korean contacts of individuals like University of California at Berkeley Professor Robert Scalapino, Nautilus Institute director and Australian analyst Peter Hayes, former Asia Society specialist Tony Namkung, and the Carnegie Endowment's Selig Harrison. He also provides fascinating details on the role of the Rockefeller Brothers Fund and W. Alton Jones Foundation in providing funding for these early ventures. He further delineates the Rockefeller Foundation's role in facilitating the visit in June 1994 by former U.S. President Jimmy Carter to speak with North Korean leader Kim Il Sung. This visit ultimately broke the stalemate in the nuclear negotiations by turning Pyongyang's attention to the *positive* effects of cooperation with the West, rather than on the sanctions that would be continued if he maintained a hard line. Sigal focuses his attention on the various governmental and nongovernmental efforts that led to the creation of the Agreed Framework and KEDO, noting that cooperative disarmament "is much less costly than coercion, and it works."[1]

The contribution of Michael J. Mazarr's recent book, *North Korea and the Bomb: A Case Study in Nonproliferation* (St. Martin's Press, 1995), is precisely the focus he places on the relationship of the crisis to the broader nonproliferation regime. While not as detailed as Sigal's book on some of the background factors in the evolving U.S. position, Mazarr covers the politics well and highlights the significance (and problems) of the eventual settlement both in terms of U.S. security and nonproliferation goals. Even from the perspective of 1995, Mazarr is more skeptical than Sigal of the value of the Agreed Framework, noting the risks of a flare-up of the nuclear crisis and his view that the deal eventually may dissolve. Thus, his prescription for ending the crisis involves "a rapid move to enmesh North Korea in a web of economic and political contacts."[2] However, North Korea's own behavior since 1995—in the continued stoking of tensions with South Korea—have helped put off this broader rapprochement by hardening existing opponents of such a warming in Seoul, Tokyo, and Washington. Similarly, North Korea's alleged exports of missile technologies to Pakistan and perhaps Iran have caused new concerns, even among supporters of a rapprochement, about Pyongyang's willingness to "behave" according to norms acceptable to the United States. Thus, despite the fact that the Agreed Framework is now teetering on the brink of collapse due to lack of funding and waning political support in the various KEDO capitals, none of Mazarr's larger web of suggested contacts with the North have been implemented. If Mazarr were writing today, he would be certainly express concern about the failure of diplomacy to move beyond the 1994 agreements.

While the two books already mentioned focus largely on the U.S. decision-making context, a third recent book of note—journalist Don Oberdorfer's *The Two Koreas: A Contemporary History* (Addison-Wesley, 1997)—provides a useful *regional* context to the North Korean nuclear crisis. Oberdorfer's history, which concludes with a long section on the Agreed Framework, provides considerable useful details on the tensions, conflicts, and failed diplomatic efforts that preceded the nuclear crisis of the early 1990s. Oberdorfer concludes by reiterating the importance of Korea, indeed its centrality, to hopes of long-term peace and stability in East Asia. But he notes that the United States is perhaps no longer in a position to affect its fate, citing the "gradual reduction of its leverage" and "the absence of clear U.S. goals."[3] Unfortunately, the economic news of late 1997 and 1998 did not improve the hopes of dynamic leadership and new diplomacy emerging from within East Asia itself, suggesting that further tensions may still be in store for the Korean Peninsula.

While a number of other books have been published on the more general topic of Northeast Asia and security in the two Koreas,[4] few have provided the level of detail offered by Sigal, Mazarr, and Oberdorfer on questions of relevance to the present volume. Yet, despite the significant contributions of these studies, the present book offers a focus and level of detail not possible without truly inside knowledge of North Korean politics, economics, and security. This is the knowledge that former Soviet and now Russian analysts and government officials can provide. Thus, this book seeks to add to this existing literature by providing new information to illuminate previous "black holes" in the North Korean debate. Short of an eventual opening of North Korean society, which is likely to take a considerable number of years, Russian experts are likely to remain the most informed experts on the North Korean situation for some time to come.

Contributions to the Present Volume

The book's twenty chapters are divided into five parts: Part I (The History of the North Korean Nuclear Program); Part II (The Economic Context of the North Korean Nuclear Program); Part III (Political and Military Factors behind the North Korean Nuclear Program); Part IV (The International Context of the North Korean Nuclear Program); and Part V (Unsettled Problems and Future Issues). The aim of the book is to provide a well-rounded analysis of both the history and current developments in North Korea regarding the nuclear program and the broader crisis it has set off in Northeast Asia. Included in the history is information from previously unpublished Russian Foreign Ministry archives and from the files of the Ministry of Atomic Energy. These and other chapters also draw on in-depth personal knowledge of individuals and issues in North Korea. Similarly, the chapters on current developments draw considerably on interviews with North Korean and senior Russian officials, providing new information and perspectives on previously "closed" topics, including the role of the North Korean military, economic conditions, details of North Korea's relations with its neighbors, and current political developments in North Korea's inner circles. The chapters on KEDO draw upon in-depth interviews by U.S.-based author Alexandre Mansourov with KEDO and U.S. officials containing considerable unpublished information.

Part I of the book covers in careful detail the history of the North Korean nuclear program. It consists of three chapters written by experienced experts on the North Korean nuclear scene. Chapter 2, written by Georgiy Kaurov, a leading Soviet official who worked formerly at the Russian Ministry of Atomic Energy, provides the most concise and detailed description available to date of Soviet–North Korean nuclear cooperation, beginning with joint mining activities in the 1940s, to the training program set up at the Soviets' Dubna facility in the 1950s, to the provision of a research reactor in the early 1960s, to the Soviet decision to provide a major power reactor in 1985, to the virtual cancellation of cooperative projects in the early 1990s due to funding problems. Dr. Kaurov's detailed recounting of this relationship provides, for the first time, actual figures for the value of Soviet technological assistance provided during those years and its contribution to the upgrading of North Korean technical capabilities. He concludes with some suggestions on how Russia might be brought into a new nuclear construction program to assist North Korea, should the KEDO organization fail.

While Dr. Kaurov's perspective is clearly supportive of the Russian Ministry of Atomic Energy's generally negative view of KEDO, it does offer an alternative means for completing an international project aimed at increasing North Korea's electrical capacity.

In Chapter 3, current Russian Ambassador to North Korea, Valery Denisov—a highly experienced diplomat with several prior tours of duty in Pyongyang—provides a review of the structure of North Korea's nuclear organization and decision making, as well as a discussion of the broader political and developmental goals of the program. His chapter offers a unique insight into organizations not well known, much less analyzed, in the United States. Ambassador Denisov's account draws on his more than a decade of work in Pyongyang and his lifetime of contacts with North Korean and Soviet/Russian officials working on cooperative projects between the two countries.

Chapter 4, written by former *Izvestiya* journalist Alexander Zhebin, traces the history of Soviet–North Korean nuclear cooperation from its *political* context. He examines Soviet motives in providing this assistance as well as North Korea's often quite different goals in receiving and applying it for particular national aims. Zhebin's expert analysis of North Korean strategy draws on his many years of reporting in Pyongyang and his intimate knowledge of North Korean officials, their thought processes, and their style of behavior. In addition, Zhebin's account is filled with rare interview information from Soviet military and intelligence officials with access to classified Soviet documents on the North Korean bomb program. He concludes that North Korea may have deployed a single nuclear device, or certainly may have come very close. However, he rejects the notion that Soviet scientists assisted in these programs in any way, noting that many key facilities (according to his interviews) were located in secret mountainous sites, which no foreigners (including Russians) had access to.

In Part II, the analysis turns to the economic side of the equation. In Chapter 5, well-known Russian economist Vladimir Andrianov analyzes some of the serious gaps in the North Korean economy, and how Pyongyang vainly hoped to use military means—both nuclear and missile—to make up for its many weaknesses compared with South Korea. He also considers how North Korea's economic problems provided an underlying rationale for the risky policy of nuclear confrontation with the West, as Pyongyang sought to negotiate economic concessions from the United States, Japan, and South Korea, as well as international aid agencies, by using the threat of nuclear weapons. In Chapter 6, Valentin Moiseyev, former head of Korean policy in the Russian Foreign Ministry, provides an extremely detailed history of the North Korean electric power section, tracing the development of hydroelectric plants as well as the failure of North Korea to develop quickly enough to keep up with electricity demand. He highlights gaps in fossil fuel supplies and the decision by Pyongyang in the early 1980s to turn aggressively toward nuclear power as a long-term energy solution. These points raise a still-puzzling question for analysts of the nuclear weapons program: Was North Korea's move toward an expanded nuclear program driven mainly by its energy needs or by the perceived security threats it faced?

Chapter 7, by Natalya Bazhanova of the Institute of Oriental Studies in Moscow, provides a broader picture of the North Korean economic scene by analyzing the structural problems in the existing system and the country's failure to introduce adequate incentives into the workplace. Dr. Bazhanova also traces the nearly total reliance of the country on foreign trade, despite the edicts of Kim Il Sung's *juch'e* ideology.

However, as much as the North Korean system can be blamed for the country's current economic difficulties, any balanced analysis of existing conditions has to take into account the disastrous economic impacts of the flood of 1995 and the drought of 1996. These twin catastrophes, coming in successive years, greatly exacerbated Pyongyang's already considerable economic dilemmas: dealing with a loss of foreign trade and investment capital from Russia and a concomitant energy crisis these losses caused for domestic industry. As pointed out by Alexandre Y. Mansourov, now a fellow at Harvard University's Korean Institute and previously a young Soviet diplomat in Pyongyang, these natural disasters crushed any hopes of near-term recovery in North Korea and forced it to rely on a risky policy (short of conducting broad-reaching economic reforms) of military threats followed by appeals for foreign aid.

Part III of the book moves from these economic factors to an analysis of the contributing political and military factors that led North Korea to engage in nuclear weapons research. In Chapter 9, Russian journalist Alexander Platkovskiy (formerly the *Komsomolskaya Pravda* correspondent in Pyongyang) traces an ironic and contradictory past in North Korean politics. In 1989, heir apparent Kim Jong Il sponsored a Youth Festival in Pyongyang to raise his profile, which resulted in the first meetings between North Korean and foreign youths and the sowing of new ideas on the possibility of political reforms in the North. But the rapid collapse of socialism in Eastern Europe in the fall of 1989 and then in the Soviet Union in 1991 caused the regime to squelch all plans for reform and embark instead on a harder political line at home, while using an elevated foreign threat and its nuclear diplomacy to mask its own uncertainties and failure to deliver economically to its own population. But, like the Khrushchev era in Russia, the seeds planted by the Youth Festival in 1989 still exist in a number of rising North Korean leaders, and possibly in Kim Jong Il himself.

Evgeniy Bazhanov—a long-time analyst of East Asian developments, a former diplomat, and now deputy rector of the Russian Diplomatic Academy—provides an account in Chapter 10 of two contradictory tracks in North Korean military policy during the 1990s. On the one hand, North Korea engaged in a massive military build-up. On the other, it also engaged in a number of confidence-building measures with the South (including the December 1991 denuclearization agreement) and for the first time accepted IAEA inspections of its nuclear facilities. While the latter agreements were temporarily repudiated, none of these measures have been officially rejected by the North. Thus, Dr. Bazhanov remains cautiously optimistic that if the security concerns of the North are addressed in future international negotiations, Pyongyang may move more quickly than some expect to broader acceptance of its international nonproliferation commitments.

In Chapter 11, long-time Soviet diplomat in Pyongyang Roald Savel'yev analyzes the Agreed Framework and the long-term nuclear deal with KEDO from the perspective of domestic politics. Drawing on his detailed, personal knowledge of North Korean political leaders, Savel'yev suggests that while certain aspects of the deal appear to provide considerable benefits to the North, the opening of the country through the course of the deal is also viewed as a serious threat to the regime itself. Military leaders sense a similar threat to their ability to control both national resources and the population at large. Savel'yev argues that these fears ensure that Pyongyang will take a very cautious (and sometimes hostile) approach to the implementation of the Agreed Framework—in order to maintain its control.

In Part IV, five Russian analysts discuss the international context of the North Korean nuclear crisis. Natalya Bazhanova looks in Chapter 12 at the international pressures facing Pyongyang in the late 1980s as a possible reason for the crisis itself. Among the key factors she highlights are the reform processes in the Soviet Union and China, as well as the increase in U.S. military activities. Dr. Bazhanova also identifies a crucial element of national pride in the decision by Kim Il Sung to pursue a nuclear weapon, which she likens to the French nuclear decision under Charles De Gaulle. She argues that the nuclear weapons program was no bluff, but a serious effort by the top leadership both to secure its own sovereignty and to show the rest of the world that it was not a two-bit country that could be bullied into submission by the great powers or its Southern neighbor.

Vladimir Li, director of the Russian Diplomatic Academy's Center for East Asian Studies, picks up some of the same themes in Chapter 13, which analyzes North Korea's relationship with the IAEA and the Permanent Members of the U.N. Security Council regarding its compliance with its NPT safeguards obligations. Dr. Li also offers a detailed discussion of the failure of China to serve as a reliable replacement for Russia in securing North Korea's safety in the region as a possible motivation for the bomb. However, in examining the evidence for and against an actual bomb, Dr. Li suggests that Russian intelligence reports (from which he quotes) about an actual device may have been the result of North Korean deception. For this information, Dr. Li draws on KGB memoirs not published in the West, which admit that such disinformation regarding East Asia was routinely passed to higher-ups in the Soviet government in order to provide interesting intelligence that agents believed their superiors wanted to hear (whether substantiated or not). Still, Dr. Li treats the plans for a North Korean bomb seriously and argues that the Agreed Framework alone will not solve the nuclear crisis.

Chapter 14 provides the best account to date on the recent history of North Korea's relations with KEDO. Drawing on extensive interviews with KEDO and U.S. government officials, as well as his own personal knowledge of North Korean politics and interviews with Russian diplomats, Alexandre Mansourov pieces together a highly detailed account of the disputes and accomplishments that have littered the trail of KEDO–North Korean relations. He tracks several years of negotiations between the two sides from 1994 to the present, and provides the first truly authoritative account of these negotiations from the perspective of North Korea's negotiating behavior. Dr. Mansourov expects that North Korea's strategy of "implementing while negotiating" will continue through the life of the reactor deal, meaning that the West should fully expect Pyongyang to seek new concessions and benefits at each juncture along the way.

Russian Foreign Ministry analyst Evgeniy Bazhanov and I discuss in Chapter 15 the intricate relationship between North Korea and its last communist ally, China. We make the case that despite opening lucrative trade ties with Seoul, China has gone to considerable efforts to maintain close and friendly relations with Pyongyang. However, we emphasize that Beijing's top priority in relations with Pyongyang is that the North Korean nuclear crisis not get so out of hand that it might threaten China's own developmental or security interests. For this reason, we argue that there are definite limits to Beijing's support for Pyongyang, as expressed by Chinese officials in informal conversations with Russian experts. Indeed, we show that some Chinese are openly contemptuous of North Korea's "arrogance," its lack of gratitude for China's Korean War assistance, and its apparent willingness to attempt to drag China into new confrontations with the West. This undercurrent of Chinese discontent with the relationship complicates the situation facing North Korean leaders. Despite these trends, however, the evidence suggests that China may still lean toward Pyongyang if pressed by the West. Specifically, a future expansion of U.S. military deployments in East Asia, together with increasing evidence of military preparations by Tokyo and Seoul, could lead China to move closer to North Korea, thus creating a potentially dangerous wild card for the twenty-first century.

Part IV concludes with a fascinating study by Vladivostok scholar Larisa Zabrovskaya, which traces the little-known details of Russian–North Korean regional relations, including their historical context. Zabrovskaya shows why Russia is likely to retain a significant interest in developments in North Korea, both for economic and security reasons, given its shared border and the ongoing integration of various aspects of their economies. For these reasons, the nuclear crisis will continue to involve Russia, which she argues cannot be viewed—as it often is in Washington—as a disinterested player in U.S.–North Korean negotiations.

Finally, Part V of the book looks ahead to the future of the North Korean nuclear crisis from a number of perspectives. In Chapter 17, I examine the ongoing Russian–North Korean rapprochement, describing the fundamental change in Russian policy that has taken place from the alienation of Pyongyang pursued by the early Yeltsin administration to the virtual courting of North Korea being undertaken by certain parts of the Russian government today. These trends, consistent with the findings of Zabrovskaya, suggest that Russia's influence in the eventual settlement of the Korean crisis is likely to be felt in more ways than is apparent from the current terms of the Agreed Framework and despite Russia's lack of membership in KEDO.

In this vein, Alexander Zarubin of the Russian Security Council proposes an alternative security regime in Chapter 18 for beginning multilateral confidence-building measures through the creation of a new military structure on the peninsula. General Zarubin sketches out a plan for a step-by-step series of exchanges among regional military forces that would eventually lead to an integration of activities by the North and South Korean armies.

In Chapter 19, Dr. Bazhanov provides details of Russian government perspectives on means for solving the Korean crisis. In particular, he shows how a multilateral conference among interested parties could lead to new mechanisms for solving North-South issues that have festered for decades in the poisoned environment of bilateral or trilateral talks that have failed to include other interested powers, some which command greater trust in Pyongyang. He also suggests some practical steps for implementing a denuclearization of the Korean Peninsula and for considerable demilitarization as well as pledges of noninterference by outside powers. Given the stalemate in the existing four-party talks (between North Korea, South Korea, China, and the United States), Dr. Bazhanov's ideas may provide a possible avenue for making practical steps toward progress through the use of a new venue.

Finally, Alexandre Mansourov discusses the future implementation of the Agreed Framework from the perspective of the North Korean leadership. He addresses the vexing question of how Kim Jong Il and his inner circle are likely to behave in future negotiations on the implementation of the reactor deal and the broader security discussions with South Korea (and the United States). Dr. Mansourov concludes on an optimistic note, arguing that while Kim Jong Il will push to the limits to achieve as many benefits as possible for his beleaguered regime, he will in the end step back from ultimate military confrontation because he knows that this is a battle he cannot win. Nevertheless, this does not rule out significant possible problems for the West along the way, as well as continued difficulty in handling the confusing short-term bargaining strategies, bluffs, and threats that the North may use in the pursuit of its ultimate long-term goals: regime survival and Korean reunification (however contradictory these may seem from a Western perspective).

Recent Developments in the Implementation of the Agreed Framework

The difficulty of presenting any book on an issue of ongoing international importance is that events change daily. For example, in the past three years alone there have been several crises in the North Korean situation having to do with international food aid, the incursion of North Korean submarines into South Korean waters, the discovery of new North Korean tunnels under the DMZ, the defection of a high-level party official from the North in Beijing, KEDO financial problems, the East Asian economic crisis, North Korean missile tests, alleged underground nuclear facilities, and many other more minor events. In April 1998, for example, in response to Pyongyang's perception that KEDO had failed to deliver heavy fuel oil according to the agreed schedule, North Korea delivered a particularly serious threat to the international community: declaring that it was halting its compliance with the terms of the Agreed Framework and might restart its plutonium-producing research reactor at Yongbyon. It also forced IAEA inspectors to leave their posts, thus providing the potential that North Korean scientists might reenter "frozen" facilities, remove or alter fissile materials under control, or tamper with existing monitoring devices.

At the same time, however, there have been glimmers of hope during the past few years in the groundbreaking ceremony for the reactor project at Sinp'o in August 1997, as well as in the starting of the four-party peace talks and the initial stages of South Korean President Kim Dae-jung's "sunshine diplomacy" with the North in spring 1998. Thus, efforts to move beyond the current nuclear crisis and the broader political/strategic crisis that surrounds it have proceeded in fits and starts.

This book represents an effort to put a solid floor of information under our existing, sketchy knowledge of the North Korean nuclear program: its history, the decisions that led to its acceleration in the late 1980s, and the various domestic and international political issues it has raised. It also seeks to provide new information on North Korea's relations with its neighbors, as well as the international organizations responsible both for implementing the Agreed Framework and ensuring its compliance with its NPT obligations: KEDO, the United Nations, and the IAEA. Hopefully, this book will stimulate readers to identify new questions and possible new solutions regarding the current North Korean situation.

From the perspective of this writing, hopes for a near-term solution to the North Korean nuclear crisis still appear very dim. The Agreed Framework is facing serious challenges, despite the fact that the bulk of its funding has not yet come due and the politically difficult turnover of significant nuclear technology to run the two planned reactors is yet to be undertaken. Pyongyang's threatened military tests and probes into South Korea and even the territorial waters of Japan continue to raise doubts about the agreement's long-term viability. Meanwhile, given the serious divisions extant in the U.S. Congress on foreign assistance programs, particularly to countries with dubious human rights, anti-terrorism, and nonproliferation records like North Korea,[5] supporters have been wary of opening the Agreed Framework to amendment, while opponents have offered no credible alternatives. Seoul and Tokyo remain preoccupied with their domestic economic crises and fearful of a new round of North Korean missile tests. Both capitals appear reluctant to either pay for or withdraw from the deal. Under these conditions, Pyongyang may take drastic action to return itself to the focus of world attention for the purposes of bargaining. Alternatively, it could seek quietly to subvert international controls by making new advances in its nuclear and missile capabilities in the expectation of the eventual collapse of the Agreed Framework. A third option would be a more accommodative policy, but this direction has not been the normal stock-in-trade of the North Korean regime. Thus, short of a sudden (and unexpected) North Korean collapse or the adoption of innovative new approaches to the current crisis by one side—and their recognition as such by the other—it appears that the problems that have plagued the peninsula since the early 1990s are likely to continue well into the next century.

One of the key premises of this book is that Russia has a stake in the Korean Peninsula, due both to its intimate knowledge of the North and its direct border with the peninsula on the southern tip of its Far Eastern region. For economic, political, and strategic reasons, therefore, Moscow will maintain an interest in any future settlement. As the chapters in this book show, moreover, Russian analysts have a significant amount of information, understanding, and new ideas to contribute to the solution of the North Korean crisis. South Korean President Kim Dae-jung stated recently: "Russia is truly important for us, and its cooperation is essential to peace on the Korean Peninsula."[6] With Russia as part of the solution, North Korea would likely have a greater incentive to yield to the desires of the rest of the international community than to continue to oppose them. It is our hope in organizing this volume that the book will assist in the identification of an eventual solution to the crisis, or at least highlight a new political process that might lead in that direction.

Part I

The History of the North Korean Nuclear Program

2

A Technical History of Soviet–North Korean Nuclear Relations

Georgiy Kaurov

Cooperation in the peaceful use of nuclear energy between the Soviet Union and North Korea was one element in the overall plan for promoting socialist economic integration in the Far East. Within this cooperative framework, Soviet goals were to raise the standards of North Korea's scientific and technological potential while also promoting the country's industrialization. The two sides based their cooperation on the following principles: (1) its voluntary nature; (2) mutual respect for state sovereignty and noninterference in the internal affairs of the other side; and (3) friendly mutual assistance. This chapter reviews the history of Soviet–North Korean nuclear cooperation focusing on several key elements: the technology provided, the facilities built, and the financial arrangements associated with these deals. It also discusses the collapse of this cooperation, due to financial reasons, in the early 1990s. The information provided here draws on Ministry of Atomic Energy (Minatom) documents and reports not previously available to Western scholars or officials. The chapter concludes with some observations on the current Korean Peninsula Energy Development Organization (KEDO) deal and Russia's interests in renewing its place on the peninsula.

First Steps

An intergovernmental agreement on cooperation in the field of atomic energy, signed in 1959, laid the foundation for joint nuclear activities between the Soviet Union and North Korea. On the basis of this agreement, the two countries signed a number of so-called Series 9559 contracts, which concerned such areas of bilateral cooperation as the conduct of geological studies, the construction of a nuclear research center (called a "Furniture Factory" by the Korean side), and the training of Korean specialists.

As a result of joint geological studies conducted by Soviet and North Korean specialists, Korean officials chose a site for the construction of a nuclear research center 92 kilometers (km) from Pyongyang on the right bank of the Kuryong River, 8 km from the district center of Yongbyon. The site was situated amid hills 40–60 meters high in a newly cultivated area previously occupied by rice fields. During the site-selection process, however, the planners made a mistake in failing to take into account the high-water mark of the Kuryong River.

The normal level of the Kuryong River in the summertime is 28 meters. But, during the July–August flood season, the level rises to the 33-meter mark usually once every five years and, once every fifty years, to the 34-meter mark. During the floods of 1964, the river level rose to the 33-meter mark, and its waters penetrated inside the basement rooms of the center. This forced the evacuation of the basement rooms and the relocation of some nuclear-related equipment from the radiochemical laboratory, including filters and an evaporate installation for the IRT-2000 nuclear research reactor. Flooding of the basement rooms occurred repeatedly in subsequent years as well.

The Yongbyon Scientific-Research Center (alias Object 9559 or the Furniture Factory) was set up with the assistance of Soviet specialists and consisted of the following facilities: an IRT-2000 nuclear research reactor, a radiochemical laboratory, a K-60,000 cobalt installation, and a B-25 betatron.[1] In addition, the following auxiliary technical facilities were built according to Soviet blueprints, with the equipment having been partially supplied by the Soviet Union: a set of UDS-10 decontamination drains, a waste storage site, a special laundry facility (for decontaminating protective clothing and undergarments of site personnel), and a boiler plant generating 40 tons of steam per hour.[2] Thirty Soviet specialists participated in the construction of these installations and in preparations to put them into operation.

The IRT-2000 nuclear research reactor is the main facility at the center. This is the reactor from which International Atomic Energy Agency (IAEA) officials believe that plutonium may have been diverted during secret, unsupervised refueling operations. The power of the original Soviet-supplied reactor was 2 megawatts (MW). Later, relying on their own efforts, North Korean nuclear specialists managed to increase its capacity first to 5 MW and then to 7 MW. The start-up of the IRT-2000 in 1965 allowed North Korean specialists to study physical and chemical processes that occur under the impact of ionizing irradiation; to research the effect of radiation on solid substances, including semiconductors; to conduct an activation analysis; to measure neutron sections of atomic nuclei; to conduct gamma-spectroscopic and radiation research; and to carry ⌊ analysis of the biological impact of ionizing radiation.³

The reactor made it possible to produce Sodium-24 (1.10^4 curie/year), Sulfur-35 (1.10^4 curie/year), Phosphorus-32 (6.10^4 curie/year), and Cobalt-60 (1.10^4 curie/year). The highest activity was Sodium-24 (7 curie/gram), Phosphorus-32 (30 curie/gram), Cobalt-60 (3 curie/gram), and Gold-198 (100 curie/gram).⁴

The radiochemical laboratory originally met the most modern technological standards and consisted of twenty glove boxes and twenty hot cells. The laboratory made it possible to extract radionucleids (with an activity equivalent to 5 kilograms [kg] of radium) from the irradiated fuel assemblies used in the reactor, as well as to conduct radiochemical research at the highest level. The B-25 betatron permitted study of the internal structures of materials with the help of beta-radiation at an energy level of 25 mega-electron volts. The K-60,000 cobalt installation allowed various materials to be exposed to gamma radiation of the Cobalt-60 isotope with a power equivalent to 60 kg of radium. The total start-up costs for the Yongbyon Scientific-Research Center are estimated to have been about $500 million (in 1962 prices).⁵

During the whole period of cooperation, more than 300 North Korean nuclear specialists of various qualifications were trained at various Soviet institutes of higher education. These facilities included the Moscow Engineering Physics Institute (MEPhI), the Bauman Higher Technical School (*VTU imeni Baumana*), the Moscow Energy Institute (MEI), and other higher educational establishments. Some North Korean nuclear specialists also worked at the nuclear scientific research complexes in the cities of Dubna and Obninsk.

In 1965, after having completed most of the construction work at the Yongbyon Scientific-Research Center, Soviet specialists departed from the Democratic People's Republic of Korea (DPRK). But the cooperation continued in the form of authoritative supervision over the exploitation of the betatron and cobalt installations, as well as in the form of Soviet provision of nuclear fuel supplies for the reactor.

The nuclear fuel supplies to the DPRK (specifically, to the "Yonhab" firm) were carried out by the "Tekhsnabexport" Foreign Trade Association (FTA) in the form of IRT-type fuel assemblies. In those years, the Soviet Union also supplied nuclear fuel for a critical assembly that was built independently by the North Korean side.⁶ Activities at these installations are now conducted under IAEA supervision in accordance with the DPRK-IAEA Agreement on Nuclear Safeguards signed in 1977 (an IAEA Information Circular [INFCIRC] 66 facility-based safeguards agreement). Each contract for the shipment of nuclear fuel stipulates that the North Korean side first issues assurances that the fuel will not be used for the purposes violating the Treaty on the Non-Proliferation of Nuclear

Weapons (NPT). The volume of the DPRK-bound supplies of four-tube IRT-24 type research thermal fuel assemblies (Uranium-235 at a 36 percent enrichment level) for the 1986–90 period are as follows: 1986 (10); 1988 (10); 1987 (10); and 1990 (10). The total cost was 80 million rubles (paid through a clearing arrangement).[7]

In 1991, with the consent of the Soviet Ministry of Foreign Affairs, the Tekhsnabexport FTA signed a contract with the Yonhab firm regarding supply of 10 IRT-2M fuel assemblies at a cost of $185,000 for delivery in 1991.[8] However, due to the inability of the Yonhab firm to make payments, the fuel was not supplied either in 1991 or in 1992.

Early in 1993, the "Ryuksolbi" firm (successor of the Yonhab firm) made another request that the Tekhsnabexport FTA supply the above-mentioned fuel. However, due to the fact that in March 1993 the DPRK announced its intention to withdraw from the NPT, Russia's Minatom notified the Ryuksolbi firm that Russia would not supply the requested nuclear fuel.[9]

Soveity–North Korean cooperation expanded on December 26, 1985, at the initiative of Pyongyang, when the governments of the Soviet Union and the DPRK signed an "Agreement on Economic and Technical Cooperation in the Construction of a Nuclear Power Plant in the Democratic People's Republic of Korea." The agreement provided for cooperation on the design, construction, commissioning, and use of a nuclear power plant in the DPRK consisting of four energy blocks with surface reactor installations of the VVER-440 type. The Soviet government undertook to open a credit line to the DPRK government in order to enable it to pay for the costs to be incurred by various Soviet organizations in connection with the supply of technological equipment, instruments, and materials, as well as the development of blueprints for the nuclear plant's construction. The amount of the loan and its terms were to be determined by the two sides after the development and approval of the technical project. Prior to the formal approval of the project, the various costs incurred by Soviet organizations were to be reimbursed by the North Korean side in accordance with the terms of the Soviet–North Korean agreement on mutual supplies of goods and mutual settlements for the 1986–1990 period.

Among its obligations, the Soviet side undertook:

(1) to render technical assistance,
   (a) in selecting a nuclear power plant construction site, where the maximum calculated earthquake force was not to exceed eight points on the Richter scale; and
   (b) in conducting the full scope of preparatory work at the site chosen;
(2) to supply the DPRK with the equipment and instruments for conducting the survey work;

(3) to develop a feasibility study for the power plant project and hand it over to the DPRK for approval;

(4) to carry out technical work under the nuclear power plant project at the agreed level; and

(5) to accept North Korean specialists in the USSR for on-site technical training.

Soviet organizations were committed to provide nuclear fuel supplies to the DPRK in the form of fuel assemblies for the whole period of the nuclear plant's exploitation. The North Korean side guaranteed that nuclear materials, nuclear equipment, and installations to be imported from the Soviet Union and materials and installations produced from them or using them would neither be diverted for nuclear weapons production nor facilitate the attainment of any other military goal, and that they would be kept under IAEA controls and would be provided with physical protection measures in accordance with North Korean legislation.[10] In early 1992, at the request of North Korea, Russia and the DPRK signed an agreement to substitute four VVER-440s with three newly designed MP-640–type reactors with even greater safety features. By early 1992, within the framework of the 1985 agreement, Russian specialists had carried out work on the selection of the construction site for the nuclear plant (in the vicinity of the city of Sinp'o) and completed work on the technical design of the project. However, the DPRK failed to complete its payments for these efforts. As a result, in May 1992, work on the implementation of the 1985 agreement was halted. The DPRK declared that, with the disintegration of the Soviet Union, the Russian Federation had become another country that was not bound by the USSR's obligations regarding the financing of the nuclear power plant, and, therefore, the DPRK was not obligated to pay for the work completed.

In April 1993, the DPRK announced its plan to withdraw from the NPT. Consequently, on April 19, 1993, the President of the Russian Federation issued Order #249/RP, which halted all work under the 1985 agreement. As of this writing, the North Korean debt on this project to Minatom totals $1.72 million. Later, the DPRK government held negotiations with the U.S. government regarding the dismantlement of the IRT-2000 reactor, shutting down the radiochemical laboratory, and liquidating a number of other installations built by North Korean specialists, as well as plans for dealing with North Korea's spent nuclear fuel. At U.S. initiative, the United States, the Republic of Korea, and Japan set up KEDO consortium, which other parties—including the European Union and nine other states—have since joined.

However, these U.S. actions led to the abrogation of the 1985 Soviet-DPRK Agreement by Pyongyang, which has caused Russia economic and political damage, while also showing Russia considerable disrespect. U.S. Ambassador-at-Large

Robert Gallucci invited Russia to participate in KEDO only as a second-rate member. Moscow rejected this invitation emphatically. Subsequently, the DPRK agreed to the supply of two light-water reactors (LWRs) from South Korea. Nevertheless, taking into account the DPRK's renewed membership in the NPT, Minatom considers it expedient to reaffirm its readiness to develop and implement in full the nuclear plant project with the VVER reactor of medium (640 MW) and high (1,000 MW) capacity in the DPRK.

Conclusion

Minatom is aware that the DPRK's nuclear and political establishments are divided between those who favor the resumption of Russian–North Korean cooperation and those who advocate further rapprochement with KEDO and the West. It fully realizes that the absence of political support and economic assistance from Russia to the DPRK creates obstacles to the realization of cooperation in the nuclear field, including in the construction of nuclear power plants. Nonetheless, Minatom believes that the resumption of full-scale collaboration in the field of science and engineering—including in the peaceful use of nuclear energy—could be of mutual benefit to both states.

In general, one can imagine three possible ways for Russia to participate in the resolution of the nuclear problem on the Korean Peninsula. The first approach is based on its traditional practice with North Korea: that is, exclusive bilateralism. The second approach is to offer KEDO Russian technology, equipment, and engineering expertise in the construction of the LWRs at Sinp'o, to the benefit of all parties concerned. Finally, in June 1998, Russia's new Minister of Atomic Energy Evgeniy Adamov proposed yet another alternative to the KEDO project. He suggested that the construction of new nuclear power plants in the Russian Far East near the North Korean border and their provision of electricity to North Korea might greatly reduce KEDO's construction costs, while, at the same time, allowing Russia to play a meaningful role in the project.[11] Given KEDO's financial difficulties, it seems expedient for the international community to consider this option as it seeks to find a reliable means of fulfilling its new obligations to the North Korean side. In the meantime, bearing in mind Russia's strategic interests on the Korean Peninsula, Moscow should—for the time being—refrain from any active assistance to KEDO's efforts to further the implementation of the Agreed Framework with the DPRK.

3

Nuclear Institutions and Organizations in North Korea

Valery I. Denisov

This chapter analyzes the development of North Korea's nuclear laboratories and related research organizations. It discusses a variety of institutes and their technical capabilities, as well as the organs of political oversight assigned to them. The information presented here is drawn from studies by Russian specialists, North Korean publications and reports, and the author's own knowledge of North Korea from several tours of duty in the Soviet embassy in Pyongyang. The chapter concludes by arguing that while the Agreed Framework has contributed to the solution of some of the underlying problems that led to the recent nuclear crisis, it has not *solved* them. Thus, we can expect that the crisis is far from over.

Nuclear Organizations and Soviet Training

The history of nuclear development in the Democratic People's Republic of Korea (DPRK) is more than forty years old. North Korean scientists began theoretical study on nuclear questions in the mid-1950s, when they started their practical training at the United Institute of Nuclear Research in Dubna, located in the Soviet Union. They were interested in studying electronic physics, radiochemistry, high-energy physics, and other subjects. From the start, the training of North Korean specialists in the Soviet Union was carried out solely in the interests of the peaceful use of atomic energy. Soviet–North Korean agreements signed in this connection specifically emphasized the peaceful nature of bilateral cooperation in the nuclear sphere. Other North Korean scientists received their education in Japan, East Germany, and West Germany, and some underwent practical training at Chinese nuclear centers as well.

The scientific and experimental infrastructure in the nuclear field was also built with Soviet technical assistance. Soviet specialists took part in the construction of the nuclear center in Yongbyon (92 kilometers north of Pyongyang).[1]

Expansion of the Program

In the late 1960s, the North Korean leadership made a decision to accelerate the development of nuclear science and technology. New research institutes, laboratories, and chairs were established in the country. The initial goal of this decision was to create the basis for the development of a nuclear energy sector. The DPRK had always suffered from acute electric power shortages. In 1970, at the Fifth Congress of the Workers' Party of Korea (WPK) and also at the Sixth Congress in 1980, delegates again stressed the necessity of constructing "nuclear power plants on a large scale in order to sharply increase the generation of electrical power."[2] At the same time, the North Korean leadership took into account such factors as the absence of explored oil deposits and the impossibility of compensating for electric power shortages by means of hydro- and thermo-electric power plants.

The North Korean leadership decided to develop a nuclear energy sector on the basis of gas-graphite reactors (which can be run on unenriched uranium) because the country possessed sufficient deposits of natural uranium. These deposits have been estimated to amount to 26 million tons of ore. There are also substantial graphite deposits in North Korea.[3] These reserves can be processed to a grade applicable for use in gas-graphite reactors, where it serves as a moderator.

In order to accelerate scientific and technological developments, including those in the nuclear field, the March 1988 Plenum of the Central Committee of the WPK made a decision to elaborate a Three-Year Plan (1988–90) for the development of North Korea's science and technology, which identified four main directions: electronics, thermo-technology, chemistry, and metallurgy. This plan paid special attention to the development of electronics, particularly large integrated circuits, computer technologies, new materials, robotics, digital program controls, and others.[4]

The plan envisaged some increase in financing for science and technology programs. In 1990, allocations for science constituted 3.8 percent of national income, according to North Korean figures. Before the plan was adopted, these allocations did not exceed 2 percent. However, those funds were not enough, causing the construction of many projects, including nuclear, to be disrupted. The North Korean leadership, nevertheless, spent considerable sums of money to import foreign equipment necessary for the development of the nuclear sector.[5]

The Nuclear Weapons Program

In the 1970s, Kim Il Sung made a decision to begin work on the development of a domestic nuclear weapons capability. By all judgments, he was motivated primarily by the fact that by that time Pyongyang had lost its economic competition with South Korea and did not see any hope of winning a conflict on nonmilitary terms. Besides, by now the North Korean leadership regarded the policies of the Soviet Union and China with growing suspicion, knowing that in the near future these major allies of the DPRK could "abandon" or "betray" Pyongyang and establish diplomatic relations with the Republic of Korea. Hence, Kim Il Sung concluded that the development of a nuclear deterrent was the only means of ensuring the regime's survival.[6]

After Pyongyang made a political decision to start working on the development of nuclear weapons, it adopted a number of practical steps aimed at expanding the network of research institutions dealing with developments in the field of nuclear physics, energy, radiochemistry, and others. The Nuclear Research Center in Yongbyon, the Nuclear Energy Research Institute, and the Radiological Institute were some of the organizations established at this time. In addition, a Department of Nuclear Physics was opened at the Pyongyang State University and a nuclear reactor technology chair was opened at the Kimch'aek Polytechnic University. In terms of new equipment, a Soviet-made research cyclotron was installed at Kim Il Sung State University in Pyongyang and an industrial cyclotron was installed in one of Pyongyang's suburbs.[7]

Key Nuclear Organizations

Meanwhile, a reorganization of scientific research activities was also carried out in the 1970s. A majority of the nuclear research institutes were transferred from Pyongyang to the city of Pyonsong (50 kilometers from the capital) and combined into a scientific center. At the Pyonsong scientific center there are now seventeen research institutes and one experimental test facility servicing them all.

The institutes of this center are part of the research network of the North Korean Academy of Sciences. But they receive administrative guidance regarding methods, control over the implementation of scientific and technical tasks, and funding from the State Committee on Science and Technology, as well as from the Ministry of Finance.

Institutions studying nuclear problems are included in this research. Among them there are several worth noting.

The Institute of Physics was founded in 1952. At present, 250 persons work at the institute, including staff members employed in experimental production. Its director is Dr. Cho Chen Nam, who is a specialist in laser physics. Its deputy director is Dr. Ryo Yin Gan, whose specialty is signeto-electronic materials. Five doctors of science and twenty-five candidates of science work at the institute.

The institute conducts research on the following problems: (1) lasers and optics; (2) the physics of solids (magnetic materials, amorphous materials, monocrystalline growth, signeto-electric materials, and superconductive materials);

(3) changes in the properties of materials under extreme conditions—high pressures (40,000 atmospheres), low (helium) temperatures, vacuums (down to 10^{-10} torr[8]), and the creation of diamonds through explosions and via static pressure (in research conducted jointly with the Institute of Machine Studies); and (4) acoustics and surface waves (by means of the frequency-resonance method). In addition, the institute has a theoretical physics group doing research on plasma physics, energy-induced states, and other questions.

At the institute, specialists are conducting research aimed at developing powerful lasers (up to 1 kilowatt). Their efforts have succeeded in creating a carbon dioxide longitudinal transmission laser with the power of continuous emanation of up to 600 watts.

The various departments originally created within the Institute of Physics have served as the basis for the creation of several independent research centers, including the Institute of Atomic Physics, the Institute of Semiconductors, and the Institute of Mathematics.

Among other organizations, the Institute of Mathematics has been involved as a contributor to the nuclear program. At this facility, there are laboratories focusing on programming, mathematical physics, differential equations, fundamental mathematics, quantitative modeling, the mechanics of liquids and gases, and computational mathematics.

The institute's director is Professor Ho Gon, a specialist in programming. Also working there are Academician Lee Cha Gon (head of the Fundamental Mathematics Laboratory and editor in chief of the journal *Mathematics*), five teaching professors, and four doctors of science. Academician Lee and Kin Soh Yin are specialists in algebra, Professor Lee Jong Rin is a specialist in mathematical statistics, and Dr. Song Chang Ho is a specialist in computational and applied mathematics.

There is an Academic Computing Center under the auspices of the Institute of Mathematics. It has an old French IRIS-50 computer, which was received in 1967. Its speed is only 200,000 operations per second and its operative memory is 256 kilobytes; its external memory includes Bulgarian discs of 25 megabytes each. The center also has Japanese personal computers. A now-dated, Polish-made ES-1034 computer was also installed at the institute.[9]

This institute cooperates with theoretical physicist Dr. Nam Hong Woo, who is an employee of the Institute of Physics. Dr. Nam is a top-level scientist specializing in nuclear physics and elementary particle physics. He studies the problems of focusing laser beams in the atmosphere and beam proliferation instability in the atmosphere. He also conducts research on a solution to the Bolzmann Equation for the description of Uranium-235 and Uranium-238 fission processes.

A third major organization active in the North Korean nuclear program is the Institute of Electronic Control Machines. This institute was established in 1984.

Its main task is the development of flexible production systems. Two hundred specialists, among them two doctors of science and twenty-six candidates of science, are employed at the institute. Research is conducted on six themes and includes laboratories on controls, servers, mathematical support, digital program controls, robotics, and sensor technologies. The institute also includes a design bureau employing twenty people and an experimental workshop used for the production of printing plates.

A fourth center that plays a role in North Korean nuclear activities is the Institute of Electronics. The main direction of the institute's research is the development and practical introduction into production of large integrated circuits for computers, as well as the development of elements and devices based on medium- and small-size integrated circuits.

The institute employs 402 specialists, including five doctors of science and fifty candidates of science. The institute consists of several laboratories in the following areas: (1) the design and development of large integrated circuits (chips); (2) the design and development of medium and small integrated circuits; (3) hybrid semiconductors; (4) fiber-optic communications; and (5) elements and circuits research. Each lab employs twenty-five to thirty specialists. The institute also has an experimental plant for integrated circuits production. This plant is equipped with some sophisticated West European and American machinery, as well as some Japanese equipment.

Besides these four research institutes, there are other indisputably nuclear-related facilities in North Korea as well. They include a fuel rod fabrication plant, a 5 megawatt (MW) nuclear research reactor, and a radiochemical laboratory (all in Yongbyon), as well as uranium mines in Pakch'on and Pyongsan and two uranium enrichment plants. There are also an uncompleted 50 MW nuclear power reactor in Yongbyon (whose dual purpose was to generate electrical power and to produce weapons-grade plutonium) and an uncompleted gas-graphite 200 MW nuclear reactor utilizing natural uranium in Taech'on, as well as the current KEDO construction site for the new light-water reactors at Sinp'o. Most of these facilities are supposed to have been closed or their construction frozen according to the terms of the U.S.-DPRK Agreed Framework (except at Sinp'a).

Control over the development of the nuclear-energy sector is exercised by the Ministry of Atomic Energy (headed by Minister Choe Hak Kyun, an alternate member of the Central Committee of the Workers' Party of Korea [WPK] and a Supreme Peoples' Assembly [SPA] deputy). The alleged military nuclear program must have been supervised by the Ministry of the People's Armed Forces, which was headed by the then–Defense Minister O Jin-u, a member of the Politburo of the WPK Central Committee and a KPA marshal and SPA member. Nuclear research institutions are supervised by the State Committee on Science and Technology, which is chaired by Choi Hee Cheng, an alternate member of the

Central Committee of the WPK and an SPA deputy. General guidance and oversight of nuclear activities is implemented by Chon Byon Ho, a member of the Politburo of the WPK Central Committee and a secretary of the WPK Central Committee. The development of the nuclear energy sector and of its alleged military component is "personally" controlled by the DPRK leader Kim Jong Il.

In the estimation of certain Russian nuclear specialists, a full nuclear fuel cycle exists in the DPRK, and the nuclear infrastructure can be used for weapons-grade plutonium production.[10] Theoretically, 8,000 fuel rods can yield an amount of Plutonium-239 sufficient to produce one or two nuclear bombs. But the availability of weapons-grade plutonium in itself does not determine a state's real capability for developing a nuclear weapon.[11] Indeed, most Russian specialists believe that the current scientific-technical level of the DPRK's nuclear program does not allow the North Koreans to build a nuclear explosive device.[12]

Conclusion

The implementation of the U.S.-DPRK agreements reached in Geneva and Kuala Lumpur is likely to go a long way in providing a solution for the DPRK's nuclear problem. Nevertheless, one should not overlook the fact that this process unfolds along a bumpy road in a controversial way. One should not rule out that at a certain stage the North Korean leadership might resort to its "nuclear card" to reach certain goals again.

The recent allegations in the press about North Korean construction of new facilities in the mountains outside of Pyongyang and the U.S. recriminations that have surrounded them[13] show that considerable mistrust remains between the two sides. There are also problems related to the lack of transparency and the uncertainty of information. In order to keep these tensions from boiling over, both sides need to pursue diplomacy as a first step, and not succumb to the temptation for overreaction and mutual hostility. If these guidelines are kept in mind, both sides are likely to maintain their reciprocal commitments to keep the deal alive. However, if the U.S. administration allows its policy to rely on unsubstantiated allegations, which may serve its domestic political agenda, it will ultimately bring about the destruction of the Agreed Framework.

4

A Political History of Soviet–North Korean Nuclear Cooperation

Alexander Zhebin

In the opinion of many experts, the nuclear problem on the Korean Peninsula—or what Western analysts prefer to call the "North Korean nuclear problem"—is one of the most serious, long-lasting, and still-unresolved crises that the international community has faced since the end of the Cold War. This is why it is natural for specialists studying this problem to be interested in exploring the reasons that originally led the Democratic Peoples' Republic of Korea (DPRK) to undertake the development of a nuclear program of an applied military nature in parallel with the country's well-known activities in the peaceful use of nuclear energy.

Undoubtedly, analyzing the scale and nature of both national activities and international cooperation of any specific country in the field of the peaceful use of nuclear energy might permit one, to a certain degree, to determine that state's capability to use the material, scientific, technological, and human resources accumulated in this sphere for military purposes and the level of advancement of such programs. At the same time, international experience testifies to the fact that the character and scale of secret nuclear programs of a military nature are rarely related to the official activities conducted by nonnuclear states in the civilian field (although they may, of course, draw on them).

This chapter discusses the role of Soviet–North Korean bilateral cooperation on the inter-governmental level in creating and developing the material and technological basis for both scientific research in the nuclear field as well as for the use of nuclear energy in the economic and social spheres (energy production, health care, etc.). As part of this historical analysis, it focuses on the training of personnel. It draws on evaluations made by a number of Soviet and Russian specialists who participated in the implementation of the cooperative programs of both a bilateral and multilateral nature in the field of the peaceful use of nuclear energy in the DPRK and who are therefore qualified to make a significant contribution to our understanding of the DPRK's activities in the nuclear field. The chapter also surveys some of North Korea's efforts to facilitate both legal and semilegal "brain drain" from Russia and other former Soviet republics, as well as of Pyongyang's illegal operations aimed at obtaining access to nuclear secrets and materials through intelligence gathering and smuggling.

Cooperation in Training North Korean Nuclear Specialists through Exchanges

Soviet assistance to North Korea in training specialists in the nuclear research field was the first practical step in the implementation of the two countries' agreements on cooperation in the peaceful use of nuclear energy.

The general principles for training nuclear specialists in the Soviet Union were laid out in the "Agreement between the Government of the USSR and the Government of the DPRK on the Education of Citizens of the DPRK in the USSR Civil Higher Education Establishments," signed in Moscow on May 6, 1952. According to this document, the Soviet government agreed to admit undergraduate and graduate students from the DPRK to allow them to complete their educations in Soviet universities and institutes.[1]

General principles for cooperation between the two countries in science and technology were expressed in a five-year "Agreement on Science and Technology Cooperation between the USSR and the DPRK," signed on February 5, 1955, in Moscow and in the "Agreement on Scientific Cooperation between the Academy of Sciences of the USSR and the Academy of Sciences of the DPRK," signed on October 11, 1957.[2]

According to the first of these two documents on bilateral cooperation, the two sides expressed their intention to exchange experience "in all fields of the people's economy," to transfer technical documentation to each other free of charge, and to exchange relevant information and specialists for rendering technical assistance. Within the framework of academic cooperation, they planned to exchange overall scientific research plans, specific information on scientific research conducted and results achieved, as well as to coordinate research and to implement various research projects by joint efforts through the exchange of scientists. They also planned to render each other assistance in obtaining materials and equipment for scientific purposes.[3] Later on, agreements of this kind were periodically renewed and revalidated on average every five years.

In 1956, the United Institute for Nuclear Research (UINR) was established in the city of Dubna (near Moscow) to serve as an international science and research center for the socialist countries. The DPRK was among the institute's original members and signed its founding agreement and charter. North Korea has continued to participate in its work since then. Some of the main directions of the UINR activities include the "conduct of theoretical and experimental research in the field of elementary particles physics" and of other fundamental properties of matter and the "utilization of scientific research results of an applied character in the development of industrial, medical, and other technologies." Its founding documents emphasized that "scientific research results achieved at the institute may be used only for peaceful purposes for the benefit of all mankind."[4]

Among other entities, the UINR's organizational chart includes laboratories (in fact, large scientific research institutes) of high-energy physics, nuclear problems, theoretical physics, neutron physics, and others.[5]

It is worth noting that the institute constitutes an international intergovernmental organization located on Russian territory. This status was reaffirmed in the "Agreement between the RF Government and the UINR Administration," signed in 1995 by then Russian Prime Minister Viktor Chernomyrdin and UINR representatives. The document defines the institute's legal status and the privileges and rights of the institute and its staff.[6]

A Committee of Plenipotentiary Representatives of the Governments of the Institute Member-States is the highest administrative body of the organization. Choe Hak Kyun, the DPRK Atomic Energy Minister, is the North Korean representative. The institute's work is funded by dues paid by the member states, which jointly manage the institute's activities. At present, there are eighteen institute member states, including the DPRK.[7]

In accordance with the "Agreement on the Establishment of the Institute and Its Charter," DPRK representatives began to arrive for training and joint work in Dubna in 1956. Since then, about 250 North Korean scientists and specialists have taken part in the UINR's work. Among the two main directions of research—theoretical and experimental—they have displayed greater interest in the latter direction. About 80 percent of DPRK representatives have worked in this sphere, generally, in the Laboratories of Nuclear Problems (LNP), Nuclear Reactions (LNR), and Neutron Physics (LNP); whereas about 20 percent of the North Korean scientists working at the institute dealt with theoretical problems.

Several North Korean scientific centers—the Institute of Nuclear Physics, the Institute of Atomic Energy, and the Kimch'aek Polytechnical Institute—have participated in joint projects with the UINR on a number of topics, including studies of the fundamental properties of neutrinos (specifically in disintegration processes), the properties of condensed environments, the development of equipment for physical research, and the automation of processes for obtaining and processing experimental data.

But only slightly more than 10 percent of the North Korean specialists working at the UINR defended their dissertations there, and only two of them defended doctorate-level theses. Nevertheless, some DPRK specialists, who worked at the UINR, have succeeded in achieving top positions in their national nuclear research program. Dr. Taek Kwan-oh, for example, rose to become the president of the Scientific-Research Center on Atomic Energy in Yongbyon.

Recently, the DPRK's participation in the activities of the UINR has decreased. Whereas, in 1992, sixteen North Korean scientists worked there, by 1993 there were just eight. At present, there are only three DPRK representatives. This is less than one percent of the total number of the UINR scientific employees.

Already in the late 1950s, Soviet scientists and specialists began to visit the DPRK for lecturing and consulting, they also organized practical training in the nuclear research field at DPRK Academy of Sciences institutions dealing with physics, mathematics, and engineering problems, as well as at the nuclear physics chairs and faculties established in the DPRK, including those at Kim Il Sung University and the Kimch'aek Polytechnical Institute (now also a university).[8]

Thus, in 1957, at North Korea's request, the USSR Academy of Sciences dispatched researcher I. M. Gramenitsky to the Physics and Mathematics Institute of the DPRK Academy of Sciences to assist in organizing research on nuclear interactions by means of thick-layered emulsions. While working in North Korea, Gramenitsky held ten seminars and delivered several lectures.[9]

In parallel with the decrease in North Korea's participation in UINR activities, the number of specialists participating in exchanges also declined. For example, Soviet nuclear scientists lectured for the last time in the DPRK in 1991.

Soviet Assistance in the Establishment of the DPRK's Scientific Base

In September 1959, the Soviet Union and the DPRK signed an agreement on Soviet technical assistance to the DPRK in setting up a nuclear research center and developing cooperation in the peaceful use of nuclear energy.[10]

In light of Kim Il Sung's guidelines for nuclear research development, which were outlined in the Report of the Central Committee of the Workers' Party of Korea (WPK) to the Fourth Party Congress in September 1961, scientists and specialists were assigned the task of "carrying out research work on the peaceful use of nuclear energy, making wide use of radioactive isotopes, and producing various isotopes and measuring devices...."[11]

However, in the opinion of some Soviet specialists who have studied the development of science and education in the DPRK, the first stage of research in this field was not conducted fast enough and lacked coordination. Only by the mid-1960s, after scientific forces had been consolidated and a sufficiently strong material and technological base had been developed, did conditions emerge for more fruitful research.[12]

The situation changed in 1965 after a 2 megawatt (MW) IRT-2000 nuclear research reactor built with Soviet assistance was put into operation at the nuclear center in Yongbyon. China, which had also provided training to some North Korean scientists and nuclear specialists, participated in the construction of this center as well.[13]

Later on, the North Koreans—relying exclusively on their own efforts—modernized this reactor without the participation of Soviet experts. They increased its power-generating capacity first to 4 MW and then to 8 MW. This information was revealed by DPRK Atomic Energy Deputy Minister Park Hyon-gyu in an interview with the author.[14]

This reactor was safeguarded under the International Atomic Energy Agency's (IAEA) facility control regime in accordance with an agreement signed between the agency and the DPRK on July 20, 1977.

The Soviet Union supplied fuel assemblies for this reactor. Throughout the entire period of its operation, a total of about 40 kilograms of nuclear fuel, calculated in terms of the principal fissionable material (Uranium-235), was supplied to the DPRK. According to data from the Russian Ministry of Atomic Energy (Minatom), the supply volume of this fuel was in compliance with the requirements of a reactor of this type (i.e., there was no oversupply of fuel elements).[15]

By 1970, North Korean nuclear scientists and engineers had made some progress in their efforts, which enabled the Fifth Congress of the WPK to set forth general guidelines for work on peaceful uses of nuclear energy in its directive "On the Six-Year Plan for the Development of the People's Economy of the DPRK" (adopted in November 1970). These guidelines stated: "Broadening the achievements made in research on the use of radioactive isotopes and radiation in various branches of the people's economy, we must step up research work on the development of a nuclear industry on the basis of our own raw materials and technology," adding that the program must solve scientific and technical problems on the basis of "the maximum effective utilization of nuclear equipment and raw materials."[16] The WPK Central Committee also set forth the tasks of liquidating some lags and lapses in the field of theoretical research and of initiating research in the field of thermonuclear reactions. These tasks were also reiterated at the Sixth Congress of the WPK in 1980.

In this connection, it is noteworthy that beginning in the early 1970s, the North Korean leadership—consistent with its *juch'e* ideology (or policy of self-reliance)—all but ceased to invite Soviet scientists for practical participation in projects involving the development of the material and technological bases for research in the nuclear field. An indirect indication of the fact that North Korea was pursuing such projects independently, and on a rather large scale, is the purchase of large quantities of instruments and other equipment applicable both to theoretical and experimental nuclear research made by the DPRK in the Soviet Union in the 1970s.

In 1979, the DPRK turned to the Soviet Union with a request to purchase fuel for an experimental nuclear installation built solely by North Korean specialists at Kim Il Sung University in Pyongyang. The Soviet Union rejected this request on the grounds of its long-standing policy to sell nuclear fuel only for Soviet-made equipment.[17]

With Soviet assistance, two more facilities were built in the DPRK: a radio-chemical laboratory for the production of radioactive isotopes, and a cyclotron at the Atomic Energy Institute in Pyongyang (the latter built within the framework of an IAEA technical assistance program and put into operation in 1992).

In 1993, the DPRK sought to invite Russian specialists to participate in the exploitation of this cyclotron and also made a request that Russia resume supplies of fuel assemblies for the nuclear research reactor. But these North Korean requests were denied following the March 1992 Order of the Russian Federation President concerning exports of nuclear materials to nonnuclear states whose nuclear facilities lacked IAEA safeguards and the April 1993 Order of the Russian Federation President concerning the DPRK's announcement of its intention to withdraw from the Treaty on the Non-Proliferation of Nuclear Weapons (NPT).

The Soviet Union and the Industrial Use of Nuclear Power in the DPRK

In the mid-1980s, the DPRK leadership began to place the issue of construction of nuclear power plants in the DPRK on the policy agenda. This shift can be explained both by prestige considerations (its rival, South Korea, had already put three nuclear power plants into operation and had six more under construction) and by economic factors (the possibilities of extensive growth had almost been exhausted by the mid-1980s).

In his speech to the senior staff of the WPK Central Committee on August 3, 1985, Kim Jong Il pointed out the necessity "to solve the scientific and technical issues of constructing ... nuclear power plants appropriate to the conditions of our country."[18] In the same speech, Kim Jong Il set forth the task of elevating "the science and technology of the Fatherland up to the world level" and, to this end, "to intensify even more scientific research aimed at the development ... of the physics of ultrahigh pressures and ultralow temperatures, to solve actively the problems connected with mastering laser technology, plasma research, utilization of atomic and solar energy, and the practical application of these achievements into the people's economy."[19]

In 1984, during a state visit to the Soviet Union, Kim Il Sung raised in general terms the issue of the Soviet assistance to the DPRK in the construction of a

A Political History of Soviet–North Korean Nuclear Cooperation 33

nuclear power plant. The Soviet side explained to the North Koreans that, in compliance with the obligations undertaken by the Soviet Union under the NPT, such a project could be implemented only after the DPRK joined the treaty.

After the DPRK made the required steps in this direction, the Soviet and DPRK governments signed an agreement in Moscow on December 25, 1985, on the provision of technical assistance to North Korea in the construction of a nuclear power plant. It is worthwhile to review several clauses of this document.

First, according to the agreement, the North Korean side undertook the obligation to guarantee that nuclear materials, equipment, and devices imported from the Soviet Union or made on the basis of those devices "shall not be used for the production of nuclear weapons or nuclear explosive devices, nor shall they be used to attain any military goal," and that "they will be safeguarded by the IAEA." Second, the DPRK was obligated not to transfer to third countries the documentation, equipment, or materials received from the Soviet Union for implementation of this agreement. Third, the DPRK was to ensure physical protection of the materials and equipment mentioned above at a level not lower than that required by IAEA regulations. Fourth, the agreement did not provide for any restrictions on equipment supplies or the invitation of specialists from third countries. Fifth, the agreement did not include a timetable for the construction of the nuclear power plant. Moreover, the question of allocating a state loan to the DPRK for the purpose of financing this construction was slated to be considered only after the elaboration and approval of the technical project and the conclusion of contracts on equipment supplies for the project.[20]

Initially, the DPRK rejected the assistance of Soviet specialists in choosing a site for the nuclear power plant. However, after the Soviet side ruled out all seven alternative sites proposed by North Korean experts, the latter invited a group of Soviet specialists to start working on site selection in 1986. This fieldwork had been largely completed by the beginning of 1993, but the North Koreans had refused to pay for the research done by various Soviet/Russian governmental and private enterprises. Russian experts estimate these debt obligations at between $1.7 and $4.7 million.[21] Nevertheless, the initiation of the preparatory work for the power plant's construction allowed the DPRK leadership to include this task in its economic development plans.

In a report delivered by Premier of the DPRK Administrative Council Lee Gun-mo to the Second Session of the Eighth Supreme People's Assembly of the DPRK on April 21, 1987, which approved the tasks set forth in North Korea's Third Seven-Year (1987–93) Plan for economic development, and in the Law on the Third Seven-Year Plan signed by Kim Il Sung on April 23, 1987, direct guidelines were set forth for building several nuclear power plants during these years, although their number was not specified.[22]

One should point out that these goals were clearly unrealistic and were dictated by propagandistic considerations rather than by economic calculations. In accordance with some of the preliminary estimates, and taking into account the necessary preparatory and design work, the first block of any nuclear power plant could hardly have been commissioned before 2005.

After the DPRK's decision to suspend its membership in the NPT in April 1993, the implementation of the 1985 Soviet-DPRK reactor agreement was frozen, in compliance with the April 1993 Order of the Russian Federation President (mentioned above).

The Russian government believed that nonproliferation issues should not be used to gain economic benefits. It was also assumed that construction of a modern nuclear power plant under the supervision of Russian specialists would provide a certain guarantee of North Korean compliance with the safety requirements for the construction and consequent exploitation of nuclear reactors. This was a very important factor, considering the proximity of the Russian border and the low technological level of nuclear power plant experience in the DPRK.

The production of radioisotopes for various branches of the DPRK economy was somewhat more successful. According to data from the DPRK Ministry of Atomic Energy, more than twenty types of isotopes, including Iodine-131, Phosphorus-32, and Selenium-75, have been produced in North Korea since the late 1980s.[23]

About 200 North Korean enterprises, organizations, and establishments have used these radioisotopes and related equipment. Among them are 130 enterprises in industry and construction, twenty-five in medicine, and others in agriculture and other fields.[24]

Reprocessing or storage of the spent nuclear fuel could become yet another direction of possible cooperation between Russia and the DPRK. Russia was supposed to take the spent nuclear fuel for temporary storage on a commercial basis and to complete the construction of a reprocessing plant in Krasnoyarsk with the funds thus earned. According to news reports quoting Minatom sources, Russia discussed such arrangements with Switzerland, Taiwan, the DPRK, and a number of other countries. These unnamed sources maintained that "these were all preliminary discussions that were never formalized by any obligations or agreements." These sources also excluded the possibility that Russia could return the plutonium produced as a result of reprocessing to the countries that exported their spent fuel for reprocessing.[25]

Russia was assigned a similar role for spent fuel within the framework of its proposed participation in the Korean Peninsula Energy Development Organization (KEDO). Russia viewed this role as "third rate" and considered its approach to KEDO with extreme reservation. As a result, no Russian-KEDO agreements have been reached to this date.

Russian Assessments of the North Korean Nuclear Program: From Bluff to Real Threat

In Russia, governmental officials and academic analysts hold varied and sometimes mutually contradictory assessments regarding North Korea's level of advancement in its military nuclear program, including on the central issue of whether or not the DPRK already possesses nuclear weapons.

The most authoritative assessment given before the disintegration of the Soviet Union was contained in a secret note sent by the Soviet KGB to the Central Committee of the Communist Party on February 22, 1990. This document maintains that "scientific research and development work on the creation of nuclear weapons is underway in the DPRK," and that this program is "personally" controlled by Kim Jong Il. Moreover, the note maintains that at the Nuclear Research Center in Yongbyon, North Korean scientists have "completed the development of [their] first nuclear explosive device." But, according to the report, the DPRK government had decided not to test it "in order to hide the fact of the creation of a nuclear weapon from the world community."[26]

The Russian news media, however, noticed that later assessments of the level of progress of the DPRK nuclear program by a number of governmental bodies and officials became more reserved. For example, while a report prepared by the Russian Foreign Intelligence Service (FIS) ("The New Post–Cold War Challenge: Proliferation of Weapons of Mass Destruction") maintains that the DPRK's applied military nuclear program is at "an advanced stage," it also expresses "serious doubts" that the DPRK has made "any breakthrough" in developing its own nuclear weapons yet.[27]

In 1992, Viktor N. Mikhailov, former chief of the Soviet nuclear weapons program and then head of Minatom, stated: "As a specialist, I believe that, considering the level of development of North Korean industry, the creation of the A-bomb there is impossible, at least in the near future, due to difficulties of a technological nature." He has not changed his opinion since then.[28]

In the opinion of Gennady Yevstafyev, head of the FIS Department in Charge of Control Over Weapons of Mass Destruction, North Korea was on the threshold of creating a nuclear weapon when it had to abandon the program due to considerable pressure from the world community and its tremendous cost.[29]

Vladimir Belous, a retired major-general and a well-known Russian specialist in the field of arms control and disarmament, also expresses his doubts about the ability of North Korean specialists to produce a combat-ready nuclear weapon.[30]

Another FIS report published in March 1995 maintains that "the present scientific and technical level and the technological equipment of nuclear facilities in the DPRK do not allow North Korean specialists to create a nuclear explosive device applicable for field tests, even less so to model a cold test of a plutonium-type military-purpose charge under laboratory conditions."[31]

The head of an enterprise within Minatom voices the opinion that, theoretically, there are other methods of producing the necessary quantity of weapons-grade material. For instance, chemical enrichment of uranium ores could be undertaken. That is why, he argues, one should seek to inspect the sites where natural uranium is extracted in the vicinity of Sunch'on and Pyongsan. It is no wonder, he points out, that access to these sites was always prohibited to all foreigners, including Soviet specialists. According to data from one Russian expert, who asked to remain unidentified, the extraction of uranium ore in these areas has been underway since the 1960s.[32]

One former Soviet military official, who worked in Korea for many years and prefers to remain anonymous, reports that all North Korean facilities earnestly working for defense purposes were hidden within mountains in areas where foreigners are not allowed. Thus, if North Korea had plants producing components for nuclear weapons, no specialist would be able to pinpoint their location. The same source believes that some time ago the North Koreans possibly did have plans to develop their own nuclear weapons, but, in the face of insurmountable difficulties, they decided to put an emphasis instead on their missile program and the development of a modern air force on the basis of Soviet MiG-29 planes.[33]

In the author's own view, the 1990 version of events presented by the former Soviet KGB is nearest to the truth, because it was meant to be secret and was not dictated or influenced by any political considerations.

Illegal Channels of North Korean Access to Nuclear Secrets

Statements by Russian officials and news reports testify to the fact that the DPRK government and North Korean intelligence services are undertaking attempts to gain access to Russian secrets in the field of nuclear weapons and their means of delivery by using both unofficial and outright illegal channels.

In December 1992, Russian Security Minister Victor Barannikov reported in a speech before the Seventh Congress of People's Deputies that his agents had blocked the departure of sixty-four Russian missile specialists to a third country, which, with their assistance, intended to build military-purpose missile complexes capable of delivering nuclear weapons.[34] Although the minister did not specify which country these specialists had planned to go to, journalists quite quickly located some members of this expedition and learned that it was North Korea.[35]

In 1994, two DPRK citizens were detained in the Primorskiy Kray region. They had tried to sell 8 kilograms (kg) of heroin in order to raise the money to buy Russian military secrets mainly in the nuclear field. In particular, they had intended to buy technologies related to the dismantlement of nuclear reactors at one of Russia's shipyards, as well as the patrol schedule of Russian nuclear submarines.[36]

The newspaper *Kommersant*, an influential voice of the Russian business community, reported with a reference to its own sources that in just two months in 1992, 56 kg of plutonium had been illegally transported to North Korea from the newly independent states (NIS) of the former Soviet Union. Radioactive substances are usually exported from Russia by railway in carriages carrying scrap metal via the route going through Vladivostok and China.[37] However, one should note that these facts of alleged smuggling of radioactive materials from the NIS to the DPRK have not been confirmed yet by other sources.

Conclusion

The above analysis of the activities of Soviet nuclear science centers and education establishments where DPRK representatives studied or worked in the past leads to the conclusion that there is no doubt that these specialists dealt only with programs and research related to the peaceful use of nuclear energy.

Former Minatom Director Viktor Mikhailov stated unequivocally in 1992, "The nuclear weapons complex of the USSR has never had anything to do with any possible nuclear weapons program in the DPRK. We do not control their potential work in this field."[38]

In this connection, the allegations by some foreign news media that the Soviet Union provided North Korea with technology for nuclear weapons development during the Brezhnev, Andropov, and Chernenko years are groundless and have not been substantiated by any documents.

This negative assessment is also supported by the very fact of the existence of the above-mentioned secret KGB note to the Central Committee. It would not have made any sense for the director of the KGB to inform the top Soviet leadership about a program carried out with the Kremlin's alleged knowledge and support.

However, this is not to say that nuclear research, together with scientific and experimental work in other fields of knowledge (such as chemistry or biology), could not have been reoriented by the North Koreans to develop these types of weapons. But the responsibility for such policy changes would lie entirely with the North Korean government, which would then have abused the noble goals of international science and educational exchanges and have diverted enormous material, financial, and human resources from the solution of peaceful tasks.

As far as the Soviet Union was concerned, being a great power, on equal footing with the U.N. Security Council's other permanent members, it obviously had no interest in seeing nuclear weapons acquired by such an unpredictable and unreliable partner as the North Korean regime, especially from the early 1960s on. To argue to the contrary is to ignore not only elementary logic but the above-mentioned historical facts as well.

Part II

The Economic Context of the North Korean Nuclear Program

5

Economic Aspects of the North Korean Nuclear Program

Vladimir D. Andrianov

In order to define the place and importance of the nuclear energy sector in the general structure of the North Korean economy, it is important to recognize that all economic activity in the North takes place within the confines of an all-out competition with South Korea. For this reason, the nuclear factor must be put both into the context of its role in the North Korean economy and in the ongoing competition with South Korea.

This chapter begins by reviewing the economy of North Korea, which was in steep decline relative to that of South Korea in the early 1990s, and the net effect was to exacerbate feelings of vulnerability in Pyongyang. It then analyzes the economic infrastructure in the North, specifically transport and communications, highlighting a number of weaknesses. Finally, it provides an overview of conditions in the energy sector in North Korea. Given the ongoing conflict with the South, the chapter stresses the important national security significance attached to the nuclear program and how calculations of relative power must be factored into any informed analysis of the roots of the nuclear crisis instigated by the North in the early 1990s.

North Korea's Comparative Decline versus South Korea

In recent years, the Democratic Peoples' Republic of Korea (DPRK) economy has been undergoing the deepest economic crisis in its history. Beginning in the late 1980s, North Korea's gross domestic product (GDP) began to fall: –3.0 percent in 1991, –7.6 percent in 1992, –5.4 percent in 1993, –1.7 percent in 1994, –2.0 percent in 1995, –2.1 percent in 1996, and –6.8 percent in 1997.[1]

The crisis situation in the North Korean economy is preconditioned by a number of factors: chronic deficits of energy resources, fuel, and raw materials; an abrupt decline in the volume of trade and the level of economic cooperation with foreign countries; exceedingly high military spending; and an inefficient utilization of financial and material resources.

Industrial production makes up 50 percent of North Korea's GDP, with up to 40 percent of the country's population being employed in this sector. Recent years have witnessed production drops in practically all leading industrial sectors, including both primary and export-oriented areas of industry. In particular, there have been severe declines in the production of steel and steel-based products, fabrics and textiles, foodstuffs, and consumer goods.[2]

In contrast, South Korea, one of the newly industrialized countries, has registered some of the highest economic growth rates in the Asian-Pacific region, despite its recent troubles. The annual GDP growth rates in South Korea have been as follows: 9.1 percent in 1991, 5.1 percent in 1992, 5.8 percent in 1993, 8.4 percent in 1994, 8.2 percent in 1995, 7.5 percent in 1996, and 5.5 percent in 1997.[3] Industrial production in South Korea is developing at even faster rates, putting its dynamics in sharp contrast with the industrial stagnation seen in North Korea. At present, South Korea occupies one of the leading places in the world in terms of the volume of production of a whole range of industrial goods. In the mid-1990s, South Korea became the world's second-largest producer of ships, television sets, transistors, and synthetic fibers and fabrics; the fourth-largest producer of electronics and electric equipment; the fifth-largest producer of ethylene and naphtha, steel and steel items; and the sixth-largest car manufacturer.[4] While its economy may register modest declines in the late 1990s due to the East Asian financial crisis, the South Korean economy remains generally sound and has the productive potential to rebound quickly.

By contrast, practically all sectors of the DPRK's industry are at the technological level already reached by South Korean industry in the late 1970s and early 1980s. A serious lag is evident in the production of consumer electronics and semiconductors, as well as in the petrochemical and oil refinery sectors. The DPRK has not yet mastered the production of nylon and polyetheric fibers. The production of communications equipment and computers is limited to the assembly process only.

There is a significant gap between the two countries in the production of such consumer goods as textiles made of synthetic fabrics and foodstuffs. The DPRK produces cement and glass, but the quality is low due to outdated equipment.

The situation in the DPRK's metallurgy and metal-processing industry is comparatively better, primarily due to the introduction of relatively modern technology for producing nonferrous metals, which is largely oriented to the production of materials for defense industries and is comparable to that of South Korea.

Economic Aspects of the North Korean Nuclear Program 43

Agriculture was once one of the DPRK's strongest sectors, but, by the late 1980s, grain production had declined, forcing the country to import 10 percent of its demand.[5] In 1989, grain production grew to 5.48 million tons, but the next year it fell dramatically.[6] This downward trend has continued. Recent low grain output may stem in part from the decline in the production of chemical fertilizers and insecticides, induced by the oil shortage, or from the long-term overuse of chemical fertilizers, which has exhausted the naturally poor soil. Of course, aging agricultural machinery and the shortage of fuel have also played their role. Economically, the decline in grain output is a cause for concern because often it is the agricultural sector that determines market growth. With agricultural profits rising, the market for other goods grows as well. If agriculture is stagnant overall, markets are likely to be constrained too.

One third of the DPRK's arable land is utilized for rice cultivation, with the rest devoted to a variety of grains, vegetables, and fruits, including corn, barley, wheat, millet, oats, potatoes, and soybeans. Due to a combination of economic inefficiencies and unusually cold weather, North Korea closed the early 1990s with a grain output of just 3.9 million tons in 1993, which put it in a poor situation for the crisis during the 1995 and 1996 floods.[7] Even by 1997, grain output had reached only 3.49 million tons. These figures compare negatively with average annual yields of 4.2 million tons before the onset on the food crisis.[8] In the early 1990s, North Korea undertook to import as much as 20 percent of its food demand from countries such as Australia, China, and Turkey to meet its growing deficits. Today, however, it lacks the resources even for these purchases, and must rely largely on foreign aid.

In 1990, the DPRK's farm population constituted of 37.8 percent of the total population, whereas the rural population in the Republic of Korea (ROK) comprises only 15.1 percent of its total population.[9] Today, North Korea's farming population is almost the same as that of South Korea in 1978, which, at that time, constituted 35.3 percent of the ROK's total population.[10] In 1997, the DPRK's total revenues from agriculture, forestry, and fisheries constituted 28.9 percent of its GNP, as compared with 5.0 percent in the ROK's GNP.[11]

As far as forestry is concerned, about 75 percent of the DPRK's territory is covered with forests and woodlands, but top-quality timber is rare. Moreover, in order to develop the timber industry further, the DPRK needs to import modern equipment and to implement a large-scale reforestation plan. Extensive use of wood for fuel and widespread flooding during 1995, 1996, and, to a lesser degree, 1998 have exacerbated these conditions.

Regarding fisheries, the DPRK's Third Seven-Year Plan (1987–93) ambitiously set the target for the output of marine products at 11 million tons, including 3 million tons of fish, but estimated coastal and deep-sea fishing yielded only 1.7 million tons of fish in 1993.[12] The state invested heavily in the fisheries sector, but higher yields are hampered by limitations in ship maintenance, shipbuilding, and fishing technologies.

In the field of livestock production, the Third Seven-Year Plan envisages an increase in the annual egg production up to 7 billion eggs and that of meat up to 1.7 million tons with the aim of improving the food supply for the population. Unfortunately, by the early 1990s, declining rates for meat production, including poultry products, indicated that these targets would not be met.[13]

The mining sector in North Korea accounts for 6.7 percent of its GNP, although the production of anthracite coal, iron ore, and nonferrous metals is down. In South Korea, the mining sector provides just 0.3 percent of its GNP.[14] The DPRK's natural endowments of minerals and metals enable it to preserve its advantage over the ROK in per capita output of iron ore and coal, with ratios of 11:1 and 1.5:1, respectively.[15]

Economic Infrastructure

The DPRK's transportation system depends heavily on railways, which carry 90 percent of total freight. The remaining capacity is covered by road transport (7 percent) and marine transport (3 percent). The transport situation in South Korea is more balanced, with about half of freights being transported by road.[16] North Korea's transport infrastructure is comparatively underdeveloped. The share of paved roads is estimated at only 8.5 percent, with a mere 2.2 percent constituting serviceable highways.[17] In the DPRK, there are about 5,000 kilometers of railroads, of which 98 percent are single track and 60 percent electrified.[18] Most seaports are in need of improvement and expansion. At present, the ports of Chongjin, Rajin, and Sonbong are undergoing modernization of their piers and of loading and unloading equipment, and as well as of their warehouses. The estimated capacity of all North Korean ports is 35 million tons, which is only 14 percent of the total capacity of the South Korean ports. The DPRK's car and shipbuilding production is low, which further limits transportation services. North Korea's acute energy shortage presents a serious obstacle to future infrastructure development as well.

Services and Construction

North Korea's service sector is small but growing, although largely due to the relative decline of mining and manufacturing. In 1997, it accounted for 35 percent of the GNP, of which 25.1 percent represented government-provided services. By comparison, in 1997, the South Korean service sector accounted for 51.3 percent of the ROK's GNP, with only 8.3 percent represented by government-provided services.[19] The share of the DPRK's construction sector in the GNP declined by 9.9 percent in 1997, adding to the two previous years' drops of −11.8 percent in 1996 and −3.2 percent in 1995. This sector now represents only 6.3 percent of GNP, compared with 14.6 percent in the South.[20]

Significant differences in the material production sphere of North and South Korea are reflected in the character of their foreign economic ties. While South Korea relied on an export-oriented model of development, North Korea has preached self-reliance, consistent with its *juch'e* ideology. South Korea is widely recognized as one of the world's greatest trading powers by the volume of its export-import operations: $280 billion in 1996. By contrast, North Korea, with its foreign trade volume of only $2.5 billion, occupies one of the last places in the world.[21]

In recent years, though, certain changes have been noted in the sphere of the DPRK's economic and foreign economic policy. These primarily concern new initiatives aimed at attracting foreign capital, forming joint-stock ventures, and establishing free economic zones. The North Korean government made a decision in 1990 to set up the Rajin-Sonbong Free Economic Zone (FEZ). It is situated in the northeast of the DPRK, with an area of 746 square kilometers and a population of 130,000 (in 1995). Within the zone, the DPRK is experimenting with new forms and methods of cooperation with foreign investors.

The FEZ is being organized in order to capitalize upon the natural and geographic advantages of the so-called Golden Triangle and to create a center for transit trade and international business activities in Northeast Asia. The zone is adjacent to Russia and China. Goods from northeastern China are expected to be transported through the zone to North Korean seaports. There are two ports in the zone—Sonbong and Rajin—and to the south is the port of Ch'ongjin. All three ports are ice-free.

The DPRK plans to develop the zone over a period of eighteen years in three stages: 1993–95, 1996–2000, and 2001–10. In the first stage, a modern infrastructure of roads, railways, and ports was to have been developed. In the second stage, attention is supposed to be paid to the further development of ports and to the construction (or modernization) of industrial enterprises oriented at foreign markets. During the third stage, the government plans to develop all sectors of the economy of the zone as well as tourism. By 2010, the ports are expected to process up to 100 million tons of cargo, and its population is expected to reach 1 million people.[22]

Any foreign company is allowed to invest capital in the development of the zone. The DPRK has announced its readiness to accept investments from both state and private companies, financial institutions, international organizations and individual investors, including those from Japan, the United States, and other countries.[23]

The zone allows enterprises with 100 percent foreign ownership to be formed. Investments in any form are to be admitted, including equipment, materials, and capital. Funds can be invested into infrastructure development, hotel construction, service-sector enterprises, and, most desirably, construction of heavy- and consumer-goods manufacturing industries.

The program for the development of the zone is being managed by a committee headed by the premier of the DPRK's Administrative Council. The legislative basis regulating the activities of foreign companies in the zone has already been established: more than twenty normative acts were adopted in the past several years, including laws on the zone's legal status, on foreign enterprises, and other matters.[24]

Substantial privileges are granted to those companies that invest their capital into export-oriented enterprises, the processing of imported raw materials, and the production of finished goods. Foreign investments into high-tech industries are supposed to be exempt from taxes for the first three years. The tax on profits is 14 percent. Foreign businessmen in the zone may rent land and real estate for a period of up to fifty years. The government guarantees a stable rate of exchange of foreign currencies to the DPRK's won. No visa arrangements for entry or exit are to be implemented in the zone.[25] But only some of these measures have been put into practice to date.

The zone's establishment is to be an integral part of the Tumen River Basin Development Program. Many foreign companies have expressed their interest in the FEZ's development. For instance, contracts amounting to $65 million are reported to have been concluded with the Peoples' Republic of China's companies investing in the construction of roads and railways and reconstruction of port facilities. There are proposals from leading South Korean companies as well.[26] According to North Korean sources, there are a total of 111 projects in the zone, with an estimated value of $750 million. However, actual investments to date have been made in only 77 of these ventures, with total investment capital reaching only $57.92 million.[27]

The North Korean Energy Sector

The DPRK's energy-intensive industries are considerably dependent on domestic energy sources and are thus extremely vulnerable to power shortages. Though the fast-flowing rivers of North Korea constitute a stable source of hydroelectric power, coal is the major domestic source of energy. Most power plants run on coal and a majority of manufacturing plants use coal as their fuel source. Oil is a precious import commodity used for transport vehicles and as a raw material for the chemical industry only.

According to 1996 data, North Korea's main energy sources are coal (59 percent), wood and biomass (31 percent), crude oil (4 percent), and hydropower (4 percent).[28] The energy sources in the ROK are considerably more diverse: oil (54.8 percent), coal (26.3 percent), nuclear power (14.9 percent), gas (3.2 percent) and hydropower (0.6 percent).[29] Because both countries lack explored sources of crude oil and gas, energy shortages are determined by the level of domestically produced energy and by the availability of oil and gas imports.

Thus, the DPRK's current energy shortages have resulted from a decline in domestic coal production and a dramatic decrease in oil imports since 1988. In that year, crude-oil imports were 3.16 million tons; whereas by 1993 crude-oil imports had declined to 1.3 million tons due to the loss of former Soviet subsidies. In 1995, according to the terms of the Agreed Framework, the Korean Peninsula Energy Development Organization (KEDO) shipped 150,000 tons of heavy fuel oil to North Korea, and, since October 1995, has provided annual shipments of 500,000 tons per year.[30] Previously, China was North Korea's largest and most reliable oil supplier. The DPRK received about 1.2 million tons of crude oil from China in the late 1980s, but these supplies gradually declined to 0.2 million tons in 1994. Iran provided 1.0 million tons in 1990, but in 1992 its imports declined to 0.2 million tons. In the early 1990s, Russia also exported crude oil to the DPRK, but the 0.2 million tons it supplied by 1993 were just a shadow of the guaranteed 0.85 million tons of crude oil that had been supplied annually by the Soviet Union before 1987. This abrupt decline in oil imports was due in large part to the policy changes of the DPRK's major trading partners. Prior to 1990, North Korea could buy oil from China at a "friendly" price of about $60 per ton (i.e., approximately half of the world market price). But both the former USSR (after 1990) and China (after 1991) demanded that the DPRK pay for oil supplies in hard currency at world market prices. North Korea's inability to pay for larger oil and gas imports on new terms, aggravated by the decline in domestic coal production, resulting in a significant energy deficit that hurt North Korean industrial and chemical production and those sectors of the economy that depended on transportation, such as fisheries and agriculture.

In sum, the DPRK's economy has always been dependent on domestic coal production and crude-oil imports. With imports dropping, North Korea has had to rely even more on coal. However, its coal production has gradually declined since 1989 when it totaled 43 million tons, falling to 32.4 million tons by 1995.[31] The decline in coal production can be explained by looking at the country's general economic crisis. The structural limitations of the DPRK's economy prevent the generation of capital investments needed to maintain or replace the obsolete and worn-out equipment and machinery or to purchase for hard currency enough coal and oil to compensate for the domestic energy deficit and the loss of Soviet oil imports.

Nuclear Energy Sector

The development of the nuclear energy sector in North Korea began rather early, approximately in the mid-1950s. Its development was based mainly on science and technology achievements of the Soviet Union and China, which provided assistance to the DPRK within the framework of science and technology cooperation agreements. As early as in 1956, an agreement on cooperation between the DPRK and the USSR in the field of peaceful use of nuclear energy had been signed. Dozens of specialists from North Korea underwent nuclear training in the Soviet Union and more practical training in China. With Soviet and Chinese assistance, a scientific-technical center was established in Yongbyon, where applied military nuclear research was conducted alongside work in the field of nuclear energy.

In 1965, a Soviet-supplied 5 megawatt (MW) research reactor was commissioned in Yongbyon. In the beginning of the 1980s, North Korean nuclear specialists began construction in Yongbyon of a 50 MW nuclear power reactor that was to be put into operation by the end of 1995. But it was never completed.

It is important to note that the lack of sufficient domestic energy sources and the DPRK's heavy dependence on energy imports predetermine the importance of the development of the atomic energy sector in North Korea. By the early 1980s, the DPRK's dependence on energy imports was about 98 percent, which was almost 5 percent higher than a similar South Korean indicator. North Korea, just like South Korea, is totally dependent on imports of oil and natural gas and partially so on anthracite coal imports. Thus, the development of nuclear energy in North Korea may be one way of solving its energy problem. In South Korea, which is also quite heavily dependent on energy imports, the share of nuclear energy in the overall output of electricity is 38.2 percent, with the nuclear energy sector providing 12.1 percent of the primary consumption of energy resources.[32]

But the DPRK's nuclear energy sector is still in its formative period compared with that of South Korea, where thirteen reactors are already operating as of fall 1998 and an additional eighteen reactors are under construction for completion by 2015.[33]

Among North Korean nuclear energy facilities there are two basic types: scientific-research centers—like the experimental nuclear physics laboratory at Pyongyang University and the laboratory of the Radiochemistry Institute in Yongbyon; and industrial facilities—like a fuel-rod fabrication plant, the uranium mines in Pakch'on and Pyongsan, and two uranium-enrichment plants. North Korea's natural deposits of uranium are estimated at 26 million tons of uranium ore containing more than 15,000 tons of uranium.[34]

Plans existed in the mid-1980s for the construction of a three-block Soviet nuclear reactor with a capacity of 635 MW each, but it did not reach fruition due to financing problems. Currently, North Korea expects to receive two 1,000 MW light-water reactors from KEDO by the year 2003, although, in reality, no one expects this target date to be met. In the meantime, therefore, North Korea will have to rely on its existing sources of energy, plus supplies from foreign donors and providers. South Korea's Hyundai Corporation plans to build a 100,000 kilowatt thermal power plant in Pyongyang in the coming years, if North Korea can come up with the financing.[35]

Overall, the current poor state of development in the North Korean energy sector testifies to a very low efficiency of capital investments and to the wasteful nature of past allocations of state funds for these purposes. While in South Korea nuclear energy plays an important role in the country's overall energy balance, in North Korea it is a matter for the distant future.

Possible Motivations behind the North Korean Nuclear Crisis of the 1990s

Study of the DPRK's nuclear program raises an inevitable question: what were the motives that led Pyongyang to get involved in such a provocative and dangerous undertaking?

Pyongyang's first goal, undoubtedly, was to use the nuclear issue as a means of redressing its obvious loss of the economic competition with Seoul by the early 1990s. Developing a nuclear weapon would have, in Pyongyang's thinking, greatly boosted its international prestige. This explains the DPRK's attempt to withdraw from the Treaty on the Non-Proliferation of Nuclear Weapons (NPT) and to break out from the International Atomic Energy Agency's (IAEA) controls in such a manner that could present the world with a fait accompli. Pyongyang was aware of the danger that if IAEA inspectors found out about its nuclear preparations, it might lead to a situation in which, even prior to its acquisition of new weapons, the DPRK would share the fate of Iraq and suffer a devastating foreign attack. One should not rule out the possibility that—after its initial production of weapons-grade plutonium and possibly even nuclear warheads—Pyongyang might have organized clandestine sales to its traditional buyers of its missiles: that is, other totalitarian regimes striving to acquire nuclear weapons.

Another probable motive behind Pyongyang's nuclear ambitions was its eventual intention to exchange its nuclear program for significant political and economic concessions from the United States, Japan, and South Korea at the highest rate possible. In this respect, the DPRK's leadership presumed that the farther its nuclear weapons efforts had advanced, the higher the price those countries would have to pay for the DPRK's abandonment of it. Among possible demands on the part of the DPRK could have been a reduction or even termination of the U.S. military presence on the peninsula, elimination of political and economic discrimination against it, and the provision of economic assistance. All of these concessions would serve to increase its relative power compared with its rival in the South.

North Korea's recent agreements with KEDO testify to the fact that the DPRK seems to have quite successfully played its nuclear card, achieving the goals it set. The question is whether it can continue to threaten the West and Seoul enough to keep the aid flowing. North Korea's April 1998 threat to withdraw from the Agreed Framework and to restart its 5 MW plutonium-producing reactor, as well as its August 1998 test of a medium-range *Taepodong*-1 missile, attest to the continuation of this strategy. In both instances, Pyongyang tried to use threats as a means of extracting concessions for food and other aid. North Korea has even offered to trade its missile program and halt its exports of these weapons for $500,000 per year in compensation from the United States and its allies.[36]

What remains in doubt, however, is whether this type of crisis strategy can save the North Korean economy from its underlying structural inefficiencies. The completion of the Agreed Framework and the start-up of the two nuclear reactors are a long way off. In the meantime, North Korea needs to engage in a thorough restructuring of its economy if it hopes to use these reactors to level the competitive playing field with South Korea. To date, however, there is little evidence of progress in this direction. Thus, the prognosis for the North Korean economy remains dim.

6

The North Korean Energy Sector

Valentin I. Moiseyev

The energy problem is one of the most serious structural dilemmas facing the North Korean economy. A lack of fuel and energy resources and the failure of the electric energy sector to meet national economic needs are among the main factors restricting adequate use of North Korea's existing industrial capacity and blocking the normal functioning of the economy. This chapter examines the current North Korean energy situation in all its aspects, providing a history of the development of the energy sector, political factors surrounding key decisions, and reasons for the existing inefficiencies. It concludes by arguing that the nuclear reactors constructed by the Korean Peninsula Energy Development Organization (KEDO) may provide additional output (if they are eventually completed), but they will not help North Korea overcome the essential structural problems of its energy sector, which requires much broader foreign assistance and provisions of technology.

The Foundations of the North Korean Energy Sector

The Democratic Peoples' Republic of Korea's (DPRK) total fuel and energy resources of combustible fossils, wood fuel, and hydropower, as calculated in uniform energy units (7,000 calories/kilogram), are estimated to be 7.672 billion tons. Of this total, coal makes up 6.617 billion tons; turf, 120 million tons; wood, 16 million tons; and hydropower, 9 million tons.[1] No industrially usable oil and gas deposits have yet been discovered in the country.

Among local resources, coal and hydropower are the sources of nearly all of the electricity generated. Coal deposits are estimated by various sources at between 2.5 and 6.6 billion tons, of which the share of lignite is about 20 percent.[2]

Hydropower resources amount to approximately 10 million kilowatts (kW).[3] Taking into account the instability of the existing infrastructure and the sharp seasonal dependence on the water level in rivers and lakes, their use requires increased capital investments.

In the future, the DPRK could also use sea tidal power—estimated at 4.6 million kW—to produce electric energy.[4] In the early 1980s, an experimental tidal power plant was constructed in the area of Namp'o, at which technology is still under development for industrial use of this type of power. There are also some small wind power plants in the mountains and other remote areas.

Approaches to the Expansion of North Korea's Electric Energy Sector

The theoretical foundation for the North Korean leadership's approach to the creation of an electric energy sector—similar to that in other areas of the national economy—can be found in the formulations of Kim Il Sung's works and speeches. Among these statements,[5] his speech before the staff of the Ministry of Electric Energy Industry on February 4, 1958 ("On Some Urgent Tasks for the Workers of the Electric Energy Sector") is considered particularly important:

> The significance of this work is first of all in the fact that ... it has outlined a clear path towards ensuring rapid growth in electric energy production, maximizing existing productive capacities, and opening the possibility of creating an independent electric energy foundation in the near-term with relatively small investments.[6]

Taking into account the postwar reconstruction-oriented nature of the economy of this period, Kim Il Sung's speech set the task of combining the restoration of destroyed power plants with the construction of new ones, promoting the production of necessary electric energy equipment domestically, and ensuring faster development of the electric energy sector as compared with the other sectors.

Within the framework of the earlier policy of "*juch'e*-ization" of the economy, plans were made to create large power plants with state capital investments and small and medium ones "on the basis of the all-people's movement"[7]: that is, through the use of local resources and domestically produced equipment. Since then, the electric energy sector has been developed in essentially the same way as other domestic industries.

For North Korean leaders, it was of primary importance to define their position concerning a combination of hydropower and thermal power resources. The problem was that until the late 1960s only hydropower plants (HPPs) and small thermal power plants (TPPs) attached to factories existed in the country. The inadequate development of the coal mining sector and a desire to develop irrigation systems simultaneously were the main reasons. At the same time, the instability of North Korea's river management regime has caused disruptions in electricity supplies to customers. A shortage of reserve capacities at HPPs has prevented electric power production from achieving a stable level.

At the Plenary Session of the Central Committee of the Workers' Party of Korea (WPK) that took place in June 1959, a special decision was adopted on ways of further developing the electric power production industry that envisaged a policy aimed at the parallel development of hydro and thermal power energy. At the Fourth Congress of the WPK, this policy was confirmed, and a goal was set to accelerate the rate of development of the electric energy sector.

Naturally, the implementation of this policy was linked to the attraction of foreign, primarily Soviet, assistance because the Koreans themselves had neither the funds nor the technical means required for the construction of hydropower and thermal power plants. Soviet specialists insistently recommended the creation of a thermal power energy sector on the basis of oil, pointing to the then-cheap price of oil as the major argument and promising supplies of necessary quantities from the Soviet Union, where some new promising oil fields had been discovered by that time. The pro–Soviet-minded North Korean technocrats also held the same view. However, considering North Korea's *juch'e* tenets and, fearing a raw material dependence on the Soviet Union (adding to the existing technological dependence), Kim Il Sung insisted on using domestic coal as fuel for the thermal power plants. Practice proved that he was right in these concerns. In fact, the Soviet Union not only never supplied the DPRK with the necessary quantities of oil, but it failed even to provide the quantities agreed upon in bilateral contracts—the one and only oil-fired thermal power plant, constructed in Ounggi (Sonbong), has never performed at its full capacity.

The last known directives of Kim Il Sung on energy sector development were given at the Sixth Congress of the WPK in 1980. He noted in the Summary Report of the Central Committee that:

> to increase the production of electric energy, it is necessary to create a wide network of hydropower plants (HPPs), ... and in the areas of the Western coast ... to build numerous tide-power plants.
>
> Together with the construction of HPPs it is necessary to widen the construction of thermal power plants (TPPs), ... to increase the capacities of the performing TPPs by means of upgrading of their equipment, ... to start a wide construction of TPPs performing on low-calorie coals, as well as that of medium and small TPPs using the radiating and excessive heat....
>
> In order to sharply increase electric energy production, it is necessary to build an atomic power plant and other plants using new energy resources.[8]

Creation of the Electric Energy Sector

During Japan's colonial domination, the total established capacity of the power plants in Korea (in 1944), of which 84 percent were located in the North, amounted to 2.94 million kW.[9] These plants produced 92 percent of all the electric power produced in the country. On the eve and during the military actions for Korea's liberation in August 1945, however, the Japanese intentionally demolished and dismantled about 60 percent of the electric energy production capacities.

In the process of implementing the first national economic plans of 1947 and 1948, as well as the two-year plan of 1949–50 for the restoration and development of the national economy, the electric energy production sector of North Korea was largely restored. The total power capacity of the electric power plants in 1949 constituted 1.55 million kW.[10]

The war of 1950–53 again dealt a severe blow to the electric power sector. Sixty-eight percent of all the power plants and up to 60 percent of transformers and distribution facilities were partially or totally incapacitated, with electric energy production capacities in 1953 reduced to 336,000 kW (or 4.6 times less than prewar levels).[11]

The subsequent restoration of the electric power sector took place in accordance with two central government economic planning periods stretching from 1954 to 1960. During the so-called Three-Year Plan (1954–56), the main task was to restore the productive capacity of the power sector, while the Five-Year Plan (1957–60, completed one year early) promoted a broad program of new capital construction, modernization, and technical reequipment of these enterprises according to the demands of industrialization. During these three- and five-year planning periods, about 12.6 percent of all capital investment into industry was allocated to electric power sector development, with the sum allocated during the Five-Year Plan period rising 1.6 times compared with the previous three-year period.[12]

As a result, all of the most important electric power plants of the DPRK were completely restored and reconstructed for greater power output, including the Sup'ung HPP with an output of 735,000 kW, the cascade-type HPPs on the Chanjin River (390,000 kW), the Hoch'on River (394,000 kW), and the Puch'on River (260,000 kW). A significant portion of the obsolete power production equipment at the largest HPPs was replaced with more up-to-date technology. For instance, at the Sup'ung HPP, six of the seven existing power units were replaced with new ones; at the cascade-type HPPs, seven new hydropower units were installed, while the thirty-seven existing ones were rebuilt based on new technology. As a result, in 1958 the power capacity of the Sup'ung HPP was 20 percent higher than in 1949, and that of the Chanjin River cascade-type HPP rose 22 percent.[13]

Alongside the restoration of the previously existing electric power plants, the

Five-Year Plan period saw the beginning of construction at several new sites: the Tonno HPP (90,000 kW), the Kangae HPP (246,000 kW), a cascade-type HPP on the Sodusu River (455,000 kW), and the Ounbong HPP (400,000 kW).

During that period, the restoration and construction of electric power plants (except the Tonno HPP) was carried out with technical and material assistance from several foreign countries, including the Soviet Union, Czechoslovakia, China, West Germany, and Austria (see Table 6.1).

In accordance with the new policy, smaller power plants were also built at the same time and numbered 1,149 by 1960. The total capacity of the electric power plants by the end of the Five-Year Plan period had reached 1.875 million kW, of which the share of HPPs was 1.805 million kW, while that of thermal power plants was only 70,200 kW.[14] The total capacity of the electric power plants had thus reached a level 1.2 times higher than that of the prewar period.

Substantial work was done in the field of reconstruction and expansion of the electricity transmission lines and the power transformation facilities. In particular, the Sup'ung-Pyongyang high-voltage electric transmission line was built. By 1960, the total length of electric transmission lines in the country had reached about 8,000 kilometers, with the length of the distribution network amounting to some 12,000 kilometers. A 3.5 million kW power transformation capacity had been rebuilt and put into service. By 1958, a unified energy grid had been created in the DPRK.[15]

The DPRK Electric Energy Sector in the 1960s and the 1970s

Qualitative changes took place in the North Korean electric energy sector in the 1960s, during the implementation of the first Seven-Year Plan. First, by this time, a domestic basis had already been laid for developing this sector through restoration of the destroyed capacities and the use of those under construction. Second, the creation of new electric power plants with the assistance of foreign countries had been carried out using new technologies that made their utilization coefficients higher. Third, and most important, the structure of electric power production had begun to change due to the creation of thermal power plants.

During the implementation of the Seven-Year Plan, investment in the electric energy industry grew noticeably in comparison with the previous period, with their annual amount averaging approximately 15 percent of all allocations to industry in general. Investment increased particularly in the late 1960s, when a resolution on a further build-up of the fuel and energy industry was adopted at the May 1965 Plenary Session of the WPK Central Committee.[16]

In 1967, Soviet and North Korean engineers completed the construction of a 500,000 kW central TPP in Pyongyang, the first and the largest in the country, and, by 1970, the construction of two units of the Pukchang TPP (200,000 kW) was completed. As a result of these developments, the decade saw the output of

TPPs increase by eleven times, and their share in the total volume of the established power capacities rise by 23 percent.

In the field of hydropower, the Kangae HPP and the Ounbong HPP were put into operation. Also, a significant quantity of locally used small- and medium-size power plants became operational in the 1960s. Their total number eventually exceeded 1,200.

The generation capacity of all power plants by 1970—as compared with 1960—had grown by 1.8 times and amounted to 3.375 million kW. Electric energy production reached 16.5 billion kilowatt-hours (compared with 9.1 billion kilowatt-hours in 1960).

Despite the obvious positive developments in the North Korean electric power sector, miscalculations in planning for the growth of its capacities also became readily obvious during this period. For instance, the rate of increase in electric energy production was 1.4 times smaller than the rate of general industrial growth during these years.[17] While in the 1956–70 period (the so-called industrialization period) electricity output increased 3.2 times, gross industrial production for the economy as a whole rose 11.6 times.[18] In addition, one should take into account the growth of energy intensive industries of substantial proportions: a chemical industry, nonferrous metallurgy, a construction materials industry, as well as the nearly complete electrification of the railway system.

During the 1970s, the electric energy sector fell further behind economic development needs. Nevertheless, from the point of view of the absolute build-up of new capacity, this period was quite favorable. During these years, an additional 1.97 million kW of capacity was put into operation, boosting total capacity to 5.345 million kW in 1980.

The largest increase took place in the thermal power sector (1.4 million kW), including expansion of the capacity of the Pukchang TPP to 1.6 million kW, the newly built 200,000 kW central TPP at Ounggi, and the Ch'onch'on TPP (200,000 kW). In 1975, the electric energy production at the TPPs for the first time exceeded one half of the total output.

At the Sodusu cascade HPP, two power plants were commissioned in the late 1970s with capacities of 180,000 and 230,000 kW, respectively. Moreover, a large number of small- and medium-size HPPs were built, achieving an aggregate capacity of 160,000 kW. Overall, electric energy output increased to 24 billion kilowatt-hours in 1980.

From the 1980s to the Current State of the Electric Power Sector

In the 1980s, the productive capacity of the electric power sector increased by approximately 1.4 million kW. Four additional power units were installed at the Pukchang TPP, whose new capacity reached 1.6 million kW. Upgrades were also installed at the Ch'onjin central TPP (150,000 kW), the Taedong HPP (200,000 kW), the T'aep'yonman HPP (190,000 kW), and the Wiwon HPP (390,000 kW). A number of small- and medium-size HPPs were also put into operation in the 1980s. The total established capacity of North Korea's electrical generation sector reached 6.725 million kW in 1990, and total electricity output climbed to some 35 billion kilowatt-hours.

The shortage of electric energy supply relative to the more rapid expansion of national energy demand, however, became increasingly acute. This situation led to the exploitation of equipment in excess of reasonable norms, causing disruptions of regular off-line repair schedules and eventual breakdowns. Neither the electric current frequency, which fell to 50–55 hertz, nor the voltage, which fell to 190–200 volts, was duly maintained. Household consumption of electric energy comprised only 3 percent of total production. Only Pyongyang was lighted in the evening and at night. In the 1990s, even the capital is not illuminated after dark. The shortage of coal for TPPs has also become a chronic problem recently.

Under these conditions, the state set a new goal to increase considerably the capacity of the electric power sector. The third Seven-Year Plan for national economic development for 1987–93, adopted at the Supreme People's Assembly session in April 1987, envisaged an increase in electric energy production to 100 billion kilowatt-hours by the end of the planned period. To these ends, plans were made to create new capacities of 4 million kW in the hydropower sector by constructing HPPs at T'aech'on, Kumganggang, Hich'on, Nam-gang, Kumyangang, and Oranch'on. The construction of TPPs using low-calorie coal were planned at the Anchu TPP and the Eastern Pyongyang Central TPP under an agreement that had already been reached with the Soviet Union in December 1985. Soviet loans had already been allocated for these projects. Also, construction of several large nuclear power plants was planned.

Significant assistance to the electric energy sector development was expected from foreign countries, primarily the Soviet Union. The WPK's proposals, elaborated during 1987–88, concerning long-term economic cooperation between the two countries requested construction assistance for the following projects:

—a nuclear power plant (NPP) with a 1,760-megawatt (MW) capacity in accordance with the December 1985 agreement with a target completion date of 1995;
—an NPP with a 4,000 MW capacity due to be completed in 2000;
—an Eastern Pyongyang Central TPP with a 200 MW capacity in accordance with the December 1985 agreement (with a prospective upgrade to 400 MW);
—the Anchu TPP, with a 1,200 MW capacity, in accordance with the December 1985 agreement; and

—a 500 kW electric power transmission line together with the supply of 32 transformers with a capacity of 333,000 kW.

Thus, according to the WPK's plans, Soviet assistance was expected to create an electric generation capacity of 7,360 MW by the year 2000, of which 3,160 MW had already been agreed to in December 1985. In fact, however, only two power units of 50 MW each were put into actual operation with Soviet assistance at the Eastern Pyongyang Central TPP in 1991. Overall, an additional capacity of only 980 MW was put into operation during the period following the adoption of the third Seven-Year Plan.

Clearly, the Seven-Year Plan's goals in the electric power sector (similar to its goals in other areas of the North Korean economy) were not achieved, as officials finally admitted at the December 1993 Plenary Session of the WPK Central Committee.

Since that time, the difficulties the country has been living through have continued to affect the electric power sector. Electricity production has fallen during the mid-1990s, amounting in 1995 to only approximately 20 billion kilowatt-hours (or less than half of the country's requirements).[19]

In 1995, the total established capacity of all North Korean power plants constituted 7.646 million kW, including 4.496 million kW from HPPs, 3.15 million kW from TPPs, and none from NPPs. But in terms of actual electric energy production, about 70 percent of electric energy production was provided by the TPPs.

Conclusions

The electric energy sector of the DPRK is in a deep crisis, which is of a structural nature. This problem, in turn, lies at the root of the general economic crisis in the country.

Foreign assistance to increase capacity and to modernize equipment in the country's TPPs is necessary for a normalization of the situation in the electric power sector because the DPRK lacks both the financial means and technical capability to accomplish these tasks. Also, funds and assistance to expand production in the coal-mining industry are needed.

In connection with the limited reserves of domestic energy resources, the use of nuclear energy to produce electricity for North Korea is therefore urgent. However, given the current progress of the KEDO project, it is unlikely that such capacity will be available to alleviate the current crisis anytime soon.

Table 6.1

List of the DPRK's Major Operating Power Plants as of 1995

Power plant name	Operation date	Restoration date	Established capacity (thousand kilowatts)	Country that gave technical assistance for construction (restoration)	Notes
HPPs					
1. Kumganggang cascade	1930	1958	13.5	Japan (CzSSR)	
2. Puren cascade	1932	–	28.5	Japan	
3. Puch'on-gang	1932	1956	260	Japan (CzSSR)	
4. Chanjin-gang	1936	1958	390	Japan (CzSSR)	
5. Sup'ung	1942	1958	735	Japan (CzSSR)	Two power units of total capacity of 210,000 kW produce electric energy for China
6. Hoch'on-gang	1942	1958	394	Japan (CZSSR)	
7. Tonno-gang	1959	–	90		
8. Kangae	1965	–	246	CzSSR	
9. Ounbong	1970		400	PRC	200,000 kW electric energy is produced for China
10. Sodusu-1	1974		180	FRG, Austria	four power units of 45,000 kW each
11. Sodusu-2	1978		230	FRG, Austria	
12. Sodusu-3	1982		45	–	
13. Mirim	1980		32	–	
14. Taedong-gang	1982		200	FRG, Austria	four power units of 50,000 kW Machinery section is underground
15. Ponhwa	1983		32		
16. Hwan-gang	1980's		20		
17. Tonhwa	1980's		20		
18. T'aep'yonman	1989		190		100,000 kW electric energy is produced for China
19. Wiwon	1989		390		190,000 kW electric energy is produced for China
20. Nam-gang	1994		200		
21. Other small and medium HPPs	1930–95		400		
	TOTAL		4,496, including 735 with Soviet assistance		
TPPs					
1. Pyongyang	1968		500	USSR, FRG	A 100,000 kW power unit was put into operation with the FRG technical assistance
2. Ounggi	1973		200	USSR, FRG	A 100,000 kW power unit was put into operation with the FRG and Austrian technical assistance
3. Ch'onch'on-gang	1978		200	Austria, PRC	
4. Pukchang	1985		1,600	USSR	
5. Ch'onjin	1985		150	USSR	
6. Sunc'hon	1994		200	PRC	
7. Eastern Pyongyang	1991		100	USSR	The first line capacity: 200,000 kW the second line capacity: 200,000 kW
8. Additional HPPs and other sources	1960–95		200		
	TOTAL		3,150, including 2,350 with Soviet assistance		
	GRAND TOTAL		7,646, including 3,085 with Soviet assistance		

7

Economic Factors and the Stability of the North Korean Regime

Natalya Bazhanova

The economy has always been the Achilles' heel of the North Korean regime. Even under Kim Il Sung, the regime suffered from a number of structural problems. But the regime survived and at times even prospered thanks to the hard work of the Korean people and generous economic and military aid from the Soviet Union and other countries. In the 1990s, these conditions have now changed, due to a variety of internal and external factors. In particular, North Korea has failed to adapt to changing conditions in the world economy, and its population has borne the brunt of this suffering.

This chapter analyzes the recent exacerbation of the country's economic problems from a social and political perspective, drawing on both Russian and North Korean sources. The chapter first considers the general political conditions prevailing in North Korea and the reasons why the top hierarchy remains loyal to Kim Jong Il, despite his failure to reverse the economic decline. It next examines the regime's current economic difficulties in greater detail and provides an analysis of their root causes. The data presented make the case that North Korea's economy is in a crisis, which is greatly increasing the pressures for change. But the question remains: Does the regime have the capability of recognizing the reasons for this crisis and the means for overcoming it? In this light, the chapter considers the conditions under which reformist policies might arise within the North Korean leadership, arguing that—despite popular dissatisfaction with the decline in living standards—they will likely be dictated from the top down, and not the reverse. The chapter concludes with a rather pessimistic forecast of the prospects for the Kim Jong Il regime in regard to the economy.

The Current Political Context

The death of Kim Il Sung in July 1994 gave rise to certain hopes of change in the Democratic Peoples' Republic of Korea (DPRK). They surfaced not only abroad but in North Korea itself. Citizens of the country—diplomats, scholars, journalists, and laymen—dropped such hints to foreigners, although in a careful, veiled form. They mentioned the possibility of a transition to a more collective leadership, with a certain liberalization of society.[1]

But five years have passed since the death of Kim Il Sung, and there have been no visible changes to date. Kim Jong Il has concentrated all powers in his own hands and, like his father, rules the country through the exercise of supreme authority over all decisions. One can detect many signs in propaganda and in practical activities of the state-party apparatus of the DPRK indicating that a unified "supreme will" determines everything, and, without question, it is the will of Kim junior. There are no direct manifestations of any struggle for power in Pyongyang. A split at the top is usually accompanied by dismissals of leadership members and top bureaucrats, at times even by all-out "purges" of the apparatus, zigzags in internal and external policies.[2] Nothing of this sort can be observed in the North. To the contrary, one can detect numerous manifestations of Kim Jong Il's seemingly unlimited power. For instance, there is a relentless worshipping of the supreme leader in the DPRK's official propaganda, with a paternalistic Korean flavor. Soviet propaganda used to refer to Joseph Stalin as "the father of all peoples" and compared Leonid Brezhnev with "a mountain eagle." In contrast, Kim junior is portrayed as a god on Earth possessing supernatural qualities.[3]

On the whole, it is evident that the essence of the North Korean regime remains unchanged. It is the same old *juch'e* model in ideology, politics, and economy, despite the fact that recently certain new peculiarities, even "riddles," have emerged in the way the system functions. Thus, only in October 1997, more than three years after his father's death, was Kim Jong Il elected the secretary general of the Workers' Party of Korea (WPK).[4] The North Korean leadership decided in summer 1998, finally, not to fill the position of the DPRK president, officially because of their desire to leave it in perpetuity to Kim Il Sung. The recent session of the Supreme People's Assembly, held in September 1998, elevated the status of the National Defense Commission, thereby stressing the increased role of the armed forces.[5] Overall, there is a noticeable increase in the number of generals taking over major positions as key advisors and aides of Kim junior.[6]

These peculiarities and riddles, however, can be explained. Kim Jong Il may have been slow to formalize his leadership position due to his desire to avoid having to take responsibility for North Korea's serious economic difficulties. Prolongation of the mourning period for Kim senior and the concentration of the population's attention on the personality of Kim Il Sung served as additional outlets for popular discontent, and continue to do so. Greater reliance on the armed forces is also understandable—now, as never before, the regime needs strict military discipline. It is equally important to show the generals that the leader cares about them personally so that they remain loyal to him. This does not mean that the regime is being transformed into a military junta. Kim junior is simply using, as Stalin did in his time, a whole range of instruments to maintain his hold on power: the armed forces, the party organs, and the state security apparatus. At this stage, the military's leverage is being used more energetically. Yet, at the same time, the armed forces are in fact controlled by party committees and official ideological constraints.[7]

One might ask whether there could be any challenges to the power of Kim Jong Il from inside his entourage. At present, there are no preconditions for such challenges. First, Kim junior has surrounded himself with loyal followers, including his close relatives. Second, the new leader follows a very clever line, displaying respect for his subordinates and richly rewarding those who have excelled in their work (by promotions, privileges, and decorations).[8] Third, the ruling class of the DPRK (together with Kim clan members), which constitutes no less than 7 percent of the population, has a stake in the leader, the "strong man" who can guarantee the preservation of their power.

These factors may outweigh Kim Jong Il's shortcomings—that is, his alleged lack of original ideas, charisma, and the capacity to make decisions.

Loyalty to the new "Great Leader" at the lower levels of North Korean society is achieved through massive ideological brainwashing, complete isolation of the population from independent sources of information, control over all facets of social life, and violent reprisals against those who disobey.[9]

Nevertheless, the system may still encounter problems. Among them, one can list a reported increase in crime, growing speculation and bribery, a rise in defections among North Koreans working abroad (including highly placed officials), and widespread apathy at work.[10] All these phenomena can be explained by the growing social differentiation between the elite and ordinary citizens, the increasing exposure of some North Koreans to life in other countries, the lack of material incentives to work hard, poor living conditions, and the irrationality of the system of economic management.

It is hard to say how widespread the negative sentiments in the DPRK are, or to what degree they have an antiregime character. There are no scientific instruments to measure correctly the "social temperature" in a totalitarian society. Nevertheless, it is evident that the dissatisfaction of the lower classes cannot undermine the foundations of the regime. Dissidents, whose existence is very

Economic Factors and the Stability of the North Korean Regime   63

doubtful, do not dare to speak out. As far as any organized opposition is concerned, its existence is out of the question.

To understand how groundless any hopes for the opposition movement in North Korea are, it is useful to recall the late history of the Soviet Union. Under Brezhnev's rule, Soviet citizens had much more freedom than North Koreans have today. Yet, despite their deep dissatisfaction with the situation in the country, the Soviet people kept silent and did not dare to challenge any actions of the authorities—from the invasion of Afghanistan to the appointment of the senile and hopeless Konstantin Chernenko as the new leader of the Soviet state.

However, it would be wrong to consider Kim Jong Il's regime unsinkable forever. As noted earlier, its Achilles' heel is the economy, which may at some point cause the regime to collapse out of an inability to support its population.

The Current Economic Situation

For the past several years, North Korea has registered consistent negative economic growth. Cumulatively from 1990–97, the economy has decreased by 42.2 percent.[11] Labor productivity has declined dramatically as well. Consequently, the 1994 Three-Year Economic Adjustment Plan, endorsed in 1994, failed, and is not even mentioned anymore.

Industrial production has played a major role in the current decline. In 1994, the output of the mining industry declined by 5.5 percent, that of heavy industry by 5.2 percent, and light industry by 4.2 percent. In 1995, imports of crude oil (which is not produced domestically) fell to 30 percent of the overall refining capacity of 3.5 million metric tons per year; coal output was only 40 percent of the annual requirement of 52 million metric tons. Electricity output was a mere 35 percent of total capacity of 7.1 million kilowatts.[12] In 1995, the DPRK's industrial capacity utilization ratio fell to a meager 30 percent.[13]

Concerning agriculture, the period 1995–98 has been replete with disasters. The countryside has experienced alternating torrential flooding followed by terrible droughts. Pyongyang claims that these calamities have caused more than $15 billion in economic damage, arguably equivalent to 70 percent of North Korea's GNP. Even though one may be suspicious of Pyongyang's attempts to attribute its recurring food shortages to natural disasters, it is clear that the DPRK's agriculture has suffered considerably due to these unforeseen problems.

In 1991–94, the shortage of grain was in the range of 1.7 to 2.0 million tons. In 1995, it exceeded 3 million tons. In 1995, total agricultural output declined by no less than 10 percent, adding to a cumulative decline of 13.8 percent in the previous three years.[14] In 1996 and 1997, the results were just slightly better than in 1995.

The decline in industrial and agricultural production also hurt state coffers. From 1994 to 1997, Pyongyang reportedly collected 20–40 percent less in governmental revenues than it had planned.[15] Capital spending was practically halted. Among fifty key enterprises that were to be built, construction of thirty plants was never even started.[16]

In addition, among other weak points of the North Korean economy today, one should mention growing transportation bottlenecks, tremendous sectoral disproportions, and an acute shortage of qualified labor. Material conditions of life for the population are close to unbearable, if judged by the modern Western standards. Rationing is all-encompassing, but even this bottom-line "no frills" safety net occasionally fails to reach those who suffer the most, namely children and the elderly. The situation has reached the point where authorities have been compelled to launch a nationwide campaign "to skip all meals one day a month" and "to eat twice a day only."[17]

Foreign trade has continued to decline through most of the 1990s, although not as sharply as in 1990–92 when its volume decreased by almost 40 percent directly after the collapse of Soviet subsidies. Nevertheless, in 1994, it declined by another 6 percent, in 1995 by 10 percent, and in 1996 by 7.8 percent. Some statistics note a possible increase in 1997–98, but largely due to North Korea's decision to export some of its gold reserves and to increase by five times its grain imports.[18] Not surprisingly, foreign debt has ballooned to $9.7 billion, with $3.6 billion owed to former communist countries and $6.1 billion owed to various Western nations. North Korea is notorious for refusing to honor its international debt obligations, and its international credit rating ranks 117th among 119 nations evaluated.[19]

The Root Causes of North Korea's Economic Difficulties

Beyond some of the cyclical factors noted above, the fundamental structural cause of the DPRK's economic difficulties is well known: the Stalinist model of socioeconomic development adopted by Pyongyang back in the 1940s. All means of production in North Korea are owned by the state. All economic activities, down to the smallest details, are planned through a highly centralized administrative mechanism. Everything is executed through the command type of management, by order from above.

There is no freedom of action for producers, no incentives for high-quality, consumer-oriented production or services. The economy is still geared to extensive development, to the augmentation of quantity at the expense of the quality of goods. This type of the economic system is not responsive to scientific and technological progress. Moreover, it cannot produce even enough to meet existing demand.[20]

In a country with a relatively small economy and underdeveloped industrial structure, centralized planning offers a number of advantages facilitating extensive growth. However, as the economy reaches a certain level of development, centralized planning is likely to impede the smooth transition to intensive growth and may even cause economic stagnation.

In the command economy, the law of value does not work, prices are distorted, and incentives for productive work are few. Moral incentives, coupled with coercion, provide positive results for some time. But the old system ultimately fails and cannot function properly any longer. This is what occurred in the former Soviet Union.

Today, the North Korean population is beginning to react against evidence of corruption, speculation, privileges for top officials, and the poor state of health care and educational facilities. Ideological slogans are no longer effective. There is a growing trend toward social apathy, thievery, and a decline in work discipline, productivity, and the overall quality of work.

Another factor adding to Pyongyang's woes is the modification of the standard Stalinist system into the *juch'e* system. For decades, North Korea relied mainly on its internal resources: it strived to develop all branches of its economy at once. The DPRK discounted the importance of external economic relations and instead was ruled by the whims of capricious dictators.[21]

Even when Pyongyang attempted to forge new ties with world markets, these attempts were not aimed at finding any long-term solutions to their economic plight. They were stop-gap measures to respond to the challenges of the moment. North Korean leaders were influenced to an excessive degree by distorted perceptions and inadequate awareness of the economic realities of their own country. As a result, their decisions were invariably based on faulty assumptions.

One can mention a few other factors that aggravate the situation in the North Korean economy even further. The DPRK spends enormous amounts of money on the military. The military syndrome, a result of the ideological animosity between the North and the South and its allies is a common feature of totalitarian regimes trying to control their own populations. The North Korean leadership keeps talking about the necessity of allocating more resources to national defense in order "to avert the growing threat of war on the Korean Peninsula."[22]

North Korea's economic troubles have worsened due to external factors as well. For a long time, the DPRK was able to satisfy its most pressing problems with the assistance of other communist states. Traditionally, the Soviet Union provided up to 50 percent of North Korea's foreign trade volume and China about 15 percent. Together, communist countries provided a substantial 70 percent of all the DPRK's foreign trade.[23] At present, China, Japan, and South Korea together account for more than 60 percent of the DPRK's external trade.[24]

Besides trade, there were other channels through which North Korea received assistance from its communist brethren. These states helped Pyongyang build factories, processing plants, and the country's energy and transportation infrastructure. In fact, most of the basic foundation of the DPRK's economy was created with "fraternal" aid. Fifty percent of industrial goods were produced in enterprises built by the Soviet Union.[25] Only with this kind of economic cooperation was the communist government in North Korea able to survive.

But the situation has changed drastically in the past few years. Its partners have lost both the incentive and capability to assist Pyongyang. Consequently, North Korea's volume of trade with Russia and Eastern Europe has decreased dramatically. Pyongyang's trade with its once most-important partner, Russia, has decreased tenfold. The DPRK can no longer hope for any credits, technical assistance, or forgiveness of debts by Russia. China, too, is no longer a supporting pillar for solving the urgent problems of the North Korean economy.[26]

Consequently, North Korea has lost key markets, financing, and sources of cheap energy supplies. This has led to reduced industrial output and further shrinkage of foreign trade. North Korea's attempts to find new partners have failed thus far. Its antiquated economic system, lack of sufficient export resources, and unpopular political practices are all factors that have turned the developed market economies away from investing in the DPRK.[27]

As far as the Third World nations are concerned, cooperation with them can help North Korea solve some of its problems. But these countries are not in a position to save Pyongyang from technical backwardness and the overall stagnation of its economy.[28]

The fundamental difficulties of North Korea in the world market stem from the isolation of its economy, its inflexible foreign trade mechanisms, and the North Korean leadership's emphasis on political rather than purely economic objectives in the conduct of its business with foreigners.

In fact, as in the past, the DPRK is not part of the global marketplace. Its per capita exports comprise only 5–15 percent of the levels found in the former Eastern bloc countries.[29] The share that exports provide to its national income is about 10 percent (which is three to four times less than the average national level).[30]

It is worth mentioning that even what is done with foreign trade earnings does not always benefit the economy. Pyongyang lavishly spends enormous amounts of money to buy foreign materials and equipment for the construction of lavish but useless edifices.

Among the causes of North Korea's difficulties, one should also mention limitations in the country's resources, as well as natural calamities that have seriously reduced food production, caused malnutrition, and prompted Pyongyang to take extraordinary measures.[31]

Overall, these problems can be summarized as follows. The DPRK is largely cut off from access to foreign technology and information networks. It has a grossly inefficient manufacturing sector still characterized by a highly centralized, obsolete command system of management, significant structural disbalances, rapidly aging and inefficient plants, and severe quality-control problems.

Pressures for Change

Pressures for change have been building up in the North Korean economy for quite a while. At this juncture, the pressure seems to be reaching a critical mass.[32]

Economic Factors and the Stability of the North Korean Regime 67

First, there is the military factor. The military balance on the peninsula has been changing in favor of the South for many years. The DPRK simply cannot afford the arms race any more. Attempts to overcome the widening gap with the "nuclear card" were checked and blocked. The continued arms build-up may ruin the economy completely. The economic burden of the arms race is less and less bearable given the worsening economic crisis. The obvious conclusion, from the military point of view, is that Pyongyang has to mend fences not only with the United States but also with South Korea.

Second, the situation with food supplies is disastrous. Food shortages are provoking widespread discontent. A growing number of North Koreans are fleeing into China. North Korean workers are reported to be escaping from logging sites in Russia's Far Eastern forests. It seems that the patience of the North Korean population may be running thin, and hungry throngs may eventually be forced to revolt.

There is no way to solve the food problem under Pyongyang's current economic and foreign policy strategies. The shock tactics aimed at increasing labor productivity have failed to bring about positive results. Price fixing and the allocation of goods by centralized decisions made by the government cannot induce creativity and zeal for production. Compulsion to attain officially set targets tends to undermine the quality of work and distort economic incentives.[33]

Labor productivity continues to fall across the entire span of North Korean industry and agriculture. Exports are stagnant and foreign debts are on the rise. As a result, the government is able to collect only a small portion of its projected total revenue targets. Pyongyang simply cannot resume investments in the economy.

Also, one should keep in mind the growing disparity between North and South Korea. The Republic of Korea's (ROK) GNP is already twenty-five times larger than that of the DPRK.[34] The gap is likely to continue to grow, creating mortal danger to the DPRK.

All of the above-mentioned factors make it imperative for North Korean leaders to move in the direction of change. They must institute internal economic reforms to give people an incentive to work. They must put in place some external economic reforms allowing the DPRK to gain access to capital, goods, and technology, which can be achieved only through accommodation with Seoul.

Is the System Capable of Reform?

The drastic reforms capable of curing the North Korean economy and putting it on the road toward healthy development must include the following measures: introduction of a market system, privatization of major sectors of the economy, integration of the North Korean economy with the world economy, curtailment of military expenditures, and economic cooperation with South Korea.

Pyongyang's ability to understand these imperatives can be judged by Kim

Jong Il's pronouncements and actions in the past, during Kim Il Sung's reign and at present. It is obvious that at first Kim Jong Il tended to view economics in ideological and political terms. But today his stance appears to be changing. He is becoming more of a pragmatist who acknowledges the logic of economics as a branch of science in its own right. Beginning in the 1980s, Kim Jong Il tried in practice to correct various imbalances in certain sectors of the North Korean economy. Despite his consistently negative attitude toward the issue of North Korea's opening and reform expressed in his writings, he has begun to implement a few elements of the Chinese-style "open-door" policy. In this connection, it is worth noting the Joint Venture Law of 1984, which was followed by the establishment of a free economic and trade zone in Rajin and Sonbong and other actions. Kim Jong Il began to emphasize that joint ventures had to be pursued with countries with advanced technology.[35]

In addition, Pyongyang has made some attempts to develop economic ties with the South Korean business circles. One can judge Kim Jong Il's realization of the dismal economic reality in the DPRK by his emphasis, as early as in 1989, on promoting agriculture and light industry, his encouragement of technological innovations, and his expansion of the underdeveloped infrastructure of the economy.

The Agreed Framework on the nuclear issue, signed between the DPRK and the United States in Geneva in October 1994, is another important sign of realism in Pyongyang's thinking. This move by the DPRK is obviously aimed at laying the groundwork for economic cooperation with the United States.[36] The ruling elite's awareness of the economic crisis facing the DPRK may have contributed to the decision to postpone Kim Jong Il's assumption of the head of the WPK until October 1997 and the new state leadership position until September 1998.

North Korea's response to the economic slowdown may be partly explained by reference to the concept of "limited rationality."[37] According to this concept, policy alternatives tend to represent stop-gap measures in the short term rather than a fundamental solution to the problem. Through this process, the vested interests of the ruling elite are not damaged.[38] There are special conditions in the North Korean case. First, there is the special status of Kim Jong Il. Second, there appear to be no clashes of interests among various classes and groups in the North; the entire society is characterized by conformity to the *juch'e* ideology.[39] Third, the North Korean political system is totally insulated from outside influences.

The role of the elite is defined by the goals and values of the system and institutions dictated by Kim junior.[40] In the top circles there is no one who can be described as an outspoken and independent-minded political figure capable or even willing to have his own point of view. At the top, the security of Kim Jong Il is ensured by his choice of personal aides. At the lower levels of society, thecompliance has thus far been facilitated by the complete isolation of the populace from outside influence, total indoctrination, intimidation, and reprisals against any hint of dissent.

Surely, in the DPRK there are those who despise the cult of personality and long for a more reasonable, humane, and prosperous life. There are even reasons to believe that the ruling class of North Korea is divided on purely economic matters into so-called moderates and hard-liners, with the moderates pushing for change. The moderates dominate among technocrats in the government who are responsible for the day-to-day management of the economy. They use every pretext and opportunity to generate openings to the outside world and to introduce reforms into the economy. However, the work of the technocrats as a whole is often sabotaged by the hard-liners, who do it not because they don't understand the near-catastrophic economic situation facing the country, but because they view any real reforms as the certain death warrant to the present regime in the DPRK.

Strategic and economic decisions are made by Kim Jong Il and a small circle of his confidants. Kim junior has now largely completed the legitimization of his dynastic succession. This was done on the basis of his allegiance to the cause of his deceased father and ongoing insistence on the necessity of finishing his revolution. To start risky reforms at present would be tantamount for Kim Jong Il to the further weakening of his fragile prestige among the elite of the country and to the undermining of his own political power base. If reforms failed to pay off right away (which they would), Kim junior's claims to power and leadership would suffer a serious blow.

In sum, one can state that North Korea is not ready yet for the introduction of drastic reforms and fears their potential consequences. The reforms are likely to create inflation, unemployment, and social instability. The exposure of the population to the realities of South Korea pose a special danger to North Korean authorities. It might cause traumatic shocks in the society, popular revolts, and a massive outflow of refugees.

Therefore, one can expect the Pyongyang-style "economic adjustment policy" to continue to be characterized by internal contradictions that make sense only within Pyongyang's stilted worldview: efforts to attract foreign capital while strengthening political and ideological control over the population. However, because the North Korean economy is faced with such a severe crisis, if the leadership continues to practice its conservative methods, it will be almost impossible for it to avoid economic collapse. Therefore, change in the political structure may take place after all. In fact, one cannot eliminate the possibility of a sudden, unforeseen collapse of the *juch'e* system.

Possible Methods of Reform

Among measures aimed at improving the economic situation in the DPRK, one should certainly include renewed attempts by North Korean authorities to attract foreign investment in the Rajin-Sonbong free zone, especially in its infrastructure. One can also expect the formation of additional economic zones in Sinuiju, Wonsan, and Nampo. They are certain to be used for joint-venture production of consumer goods for export.[41]

North Korean authorities may also try to turn the existing and future free-trade zones into the bases for high-tech industries, reprocessing of export goods, transit, and tourism. To achieve these ends, Pyongyang is likely to do its best to build telecommunication networks, hotels, bridges, roads, airports, and ports there. The North Korean leadership surely realizes that in order to attract foreign participation into the development of the infrastructure of its free-trade zones, it will have to continue the policy of easing restrictions on foreign investment. Since 1992, Pyongyang has already undertaken the following steps:

1. defined incentives for investment in more concrete terms;
2. allowed foreign investors to repatriate profits not only in goods, but also in hard currency;
3. made foreign nationals eligible for employment along with North Koreans, as well as allowed the use of imported raw materials;
4. accepted foreign insurance policies in joint ventures;
5. allowed the settlement of disputes not only through North Korean courts and agencies for arbitration, but also by foreign courts and arbitration authorities;
6. changed laws and regulations in such a way that they now clearly stipulate measures aimed at protecting the property rights of foreign investors, the assets of foreign-owned enterprises, their legal rights and interests, and their invested capital and earnings;
7. allowed the earned income of foreign-owned enterprises and the wages that foreign employees receive from them to be repatriated into foreign currency.

Some other measures have been undertaken as well, in order to raise the effectiveness of the legal system in attracting foreign capital. (See Table 7.1 for a list of all DPRK laws and regulations on foreign investment since 1984.)

However, one can predict that these efforts will not in themselves suffice because, in addition to new legal arrangements, general liberalization and transformation of political and social life must come first. Only then will the policy of gradual reforms be able, perhaps, to receive adequate political and societal support. At this juncture, Kim Jong Il is not ready for such measures.

There is the possibility that the DPRK might attempt to expand its economic ties with the ROK, especially by dealing directly with South Korean big business, bypassing the government in Seoul. These contacts could bring mutual benefits even under the present circumstances of North-South hostility. If the profits involved are sufficiently attractive, economic exchange could serve to reduce gradually the level of tension, whatever the state of other dimensions of the bilateral relationship between the North and the South might be at the time. In the short-to-medium term, the DPRK needs only to make small gains for a larger win overall.

Without harming their own political interests, the North Korean authorities could try to attract South Korean capital for the construction of a new economic infrastructure and for the purchase of advanced technology. Also, they could use it to initiate more constructive relations with international financial institutions and to increase the importation of food aid from abroad. The ROK is already the fourth-largest trading partner of the DPRK. Economic ties between the North and the South will definitely continue to develop and will eventually be formalized. Pyongyang is sure to continue its drive to obtain any kind of foreign aid from individual countries and international agencies, using all pretexts imaginable and forgetting about its own pride.

Meanwhile, the DPRK government is likely to deal with the problems of agriculture by attempting to improve farming skills, modernize its irrigation systems, and increase the production of chemical fertilizers and farm equipment. Private farming is unlikely to be introduced due to ideological, political, and even economic fears, given the unpreparedness of the rural population for this long-forgotten method of farming. The government is placing a renewed emphasis on light industry in order to provide food and clothing for the people, satisfy the requirements of the armed forces, and increase the DPRK's export potential. In Rajin-Sonbong, new elements of private enterprise have been introduced, including individual labor in family households.[42] Additional investments may be made in the construction of small factories. Furthermore, one can expect the DPRK government to make new efforts aimed at improving the quality of North Korean labor, in order to make it capable of dealing with advanced techniques in information gathering and processing. In October 1998, the Rajin Business Institute was opened to train North Korean students within the free economic zone in Western business techniques, thanks to the support of the U.N. Development Project and private Japanese foundations.[43] Finally, another possible objective of Pyongyang's efforts may be the promotion of foreign tourism as a source of hard currency and a method of improving the economic infrastructure. The prime target will be the tourist development of the scenic Mount Kumgang area.[44] Hyundai Corporation founder Chung Ju-yung recently toured the region and announced plans to invest up to $950 million for a fleet of tour ships, hotels, and related facilities for South Korean visitors.[45]

Conclusion: Prospects

Despite progress on the margins, the above-mentioned methods are likely to be ineffective in arresting the current unfavorable trends in the North Korean economy. The streamlining of North Korea's legislation on foreign investment is

Table 7.1

The DPRK Laws and Regulations on Foreign Investment and Businesses[50]

Law/Regulation	Date of Promulgation
Law on Joint Venture	September 8, 1984
Law on Income Tax on Joint Venture Firms and Specific Regulations on Income Tax on Joint Venture Firms	March 1985
Law on Income Tax on Aliens and Detailed Rules on Alien Income Tax	March 1985
Specific Rules on Enforcing the Joint Venture Law	March 1985
Law on Foreigners' Investment	October 5, 1992
Law on Foreign Enterprises	October 5, 1992
Law on Contractual Joint Venture	October 5, 1992
Amendments to Specific Rules on Enforcing the Joint Venture Law	October 5, 1992
Law on Taxation of Foreign Investment Enterprises and Foreign Persons	January 1993
Law on Foreign Exchange Administration	January 1993
Law on Free Economic and Trade Zones	January 1993
Law on Mineral Deposits	March 1993
Law on Leasing of Land	October 1993
Law on Foreign Investment Banks	November 1993
Specific Regulations on Entry and Exit of Foreigners in and from the DPRK's Free Economic and Trade Zone	November 29, 1993
Labor Regulations for Foreign-Invested Enterprises in Free Economic and Trade Zone	December 30, 1993
Amendments to the Joint Venture Law	January 1994
Regulations on Resident Representative Officials of Foreign Businesses in Free Economic and Trade Zone	February 21, 1994
Implementing Regulations for the Law on Taxation of Foreign-Invested Enterprises and Foreign Persons	February 21, 1994
Implementing Regulations of the DPRK Law on Wholly Foreign-Owned Enterprises	March 1994
Regulations on Free Trade Ports	April 28, 1994
Regulations on Foreigners' Stay and Residence in Free Economic and Trade Zones	June 14, 1994
Law on the DPRK's External Civil Relations	September 6, 1994
The Notary Public Law of the DPRK	February 2, 1995
Implementing Regulations for the Law on Equity Joint Ventures	July 13, 1995

Law/Regulation	Date of Promulgation
Regulations on Forwarding Agency in Free Economic and Trade Zones	July 13, 1995
Customs Regulations for Free Economic and Trade Zone	June 28, 1995
Regulations on Transfer and Mortgage of Buildings in Free Economic and Trade Zone	August 30, 1995
Implementing Regulations for the Law on Contractual Joint Ventures	December 4, 1995
Bookkeeping Regulations for Foreign-Invested Enterprises	December 4, 1995
Regulations on Naming of Foreign-Invested Enterprises	February 14, 1996
Regulations on Registration of Foreign-Invested Enterprises	February 14, 1996
Regulations on Processing Trade in Free Economic and Trade Zones	February 14, 1996
Regulations on Engraving and Registration of Logos (Common Seals) for Foreign-Invested Enterprises in Free Economic and Trade Zones	March 28, 1996
Regulations on Development and Management of Industrial Estates in Free Economic and Trade Zones	April 30, 1996
Regulations on Advertising in Free Economic and Trade Zones	April 30, 1996
Regulations on Entrepot Trade in Free Economic and Trade Zones	July 15, 1996
Regulations on Contract Construction in Free Economic and Trade Zones	July 15, 1996
Regulations on Tourism in Free Economic and Trade Zones	July 15, 1996
Regulations on Agents for Foreign Investors in Free Economic and Trade Zones	July 15, 1996
Bookkeeping Regulations for Foreign-Invested Banks	July 15, 1996
Regulations on Certified Public Accounting for Foreign-Invested Enterprises	July 15, 1996
Regulations on Currency Circulation in Free Economic and Trade Zones	July 15, 1996
Regulations on Border Quarantine for Free Economic and Trade Zones	June 18, 1996
Regulations on Traffic Inspection at the Boundary of Free Economic and Trade Zones	July 15, 1996
Regulations on Registration of Vehicles in Free Economic and Trade Zones	July 15, 1996

likely to change the investment environment in the DPRK only slightly. The reservations of foreign investors regarding the North Korean "market" might be reduced, but only in terms of the economic risks involved.

Political uncertainties will not be resolved. Often, they constitute the more important factors affecting foreign investors.[46] North Korea's attempts aimed at adopting an open-door policy by creating free economic zones are no more than emergency measures meant to keep the sinking economy afloat. This limited opening will be of little use since it is meant to seek only the economic fruits of liberalization without initiating adequate political and social reforms.

The almost total lack of market mechanisms in North Korea constitutes yet another obstacle impeding foreign investment. China, for example, created propitious conditions for the formation of indigenous capital, thereby giving rise to nascent free enterprise. This created an initial basis for joint ventures and free economic zones. But, as long as unfavorable macroeconomic conditions persist in the absence of comprehensive reforms, few foreign investors will be attracted to the North Korean free trade zones, let alone its general "market." Thus, the dire shortage of foreign exchange will remain another serious obstacle on the way to realizing the structural adjustment plans of the North Korean authorities.[47]

Because of all of the above-mentioned shortcomings, North Korea appears to be attractive almost exclusively to the Korean residents in Japan, who are sympathetic to Pyongyang politically and possess significant financial resources.

As for light industry, the government has been emphasizing its development as a priority since the 1980s, but to no effect. There are no grounds for believing that Pyongyang will be any more successful in managing the future of its light industry.

In the absence of significant capital investment and new technological inputs, the major potential source of growth for the DPRK's economy will remain the growth of the labor force. But, in the 1990s, the labor force's prior 2 percent growth rate was reversed, and the labor force started to decline, with concomitant effects on GNP growth. In reality, given the operation of other adverse factors, such as the declining productivity of the existing industrial structure, the North Korean GNP will quite probably continue to decline. In the absence of radical reforms aimed at increasing the efficiency of land use and preventing land degradation, unfavorable conditions in the agricultural sector will continue to prevail in the late 1990s. Hence, the agricultural sector's share of GNP can be expected to decline to around 20 percent by 2000.[48]

Because of the ineffective management approach and lack of receptivity to innovation and technology from abroad, industry will continue to deteriorate and decline as well.

Prospects for foreign trade are even more dismal. Increasing exports by means of further reducing domestic consumption is impossible. As a result, the volume of foreign trade will likely continue to decrease substantially into the next century. Only arms exports are likely to continue to grow. Without large-scale financial assistance from abroad, imports will also continue to decrease further.

As for the more distant future, the DPRK's leadership should brace itself for substantial changes in the country. Kim Jong Il certainly does not enjoy his father's prestige. He had nothing to do with the revolution. Despite his assumption of new titles in October 1997 and September 1998, he is not "the father of the nation."

There is the strong possibility that a large proportion of officials may not want to live under the cult of Kim junior. Some believe that Kim Jong Il is not the right man for the job. Others realize that the time has come to do away with the leader's cult and to start reforming the DPRK.

There are certainly no guarantees that an organized opposition will emerge and succeed. Kim junior has a great deal of power. Nonetheless, his position on the top of the power hierarchy will not be an easy one to maintain. Moreover, it appears clear that his regime cannot last long without changes. The primary reason for this assessment is the dismal economic outlook for the DPRK.

Sooner or later, North Korea will have to open itself to the outside world. Pressure for change, which only the unquestionable authority and unlimited power of Kim Il Sung could keep in check in the past, will build up. Today, his heir will have to broaden relations with other countries. The DPRK will have to acquire extensive credit from rich countries. But they will not be generous with North Korea unless Pyongyang institutes fundamental reforms, both political and economic.

In any case, it is obvious that the DPRK will have to undergo a process of reform in order to produce high-quality export items, to be able to import essentials, to maintain at least a minimum rate of growth, and to prevent the collapse of its economy and widespread starvation among its people. As for the consequences of drastic reform for the North Korean regime, it is unlikely to survive this turmoil. Thus, change may take a sudden, uncontrollable character.[49]

8

The Natural Disasters of the Mid-1990s and Their Impact on the Implementation of the Agreed Framework

Alexandre Y. Mansourov

Analysts of international affairs rarely consider the impact of natural phenomena on state behavior. But history provides a number of examples in which nature, not human choice, has determined the course of policy and has decided the fates of empires and nation-states. Desertification stopped the southward expansion of the ancient Egyptian dynasties; the volcanic eruption of Mount Vesuvius buried the city of Pompeii; and just a few successive years of extreme drought, followed by an insect invasion, forced many inhabitants of the Mayan states to abandon their glorious cities. Year after year, hurricanes in the Caribbean wreak havoc on local economies, leaving them in permanent need of foreign assistance for rehabilitation: that is, as long as other states are willing to help them survive. North Korea may be no exception in this regard. This chapter explores the long-term effects of the natural disasters on the Korean Peninsula in the mid-1990s and the impact they have had on the behavior of the North Korean regime domestically and in world affairs.

The 1995 and 1996 Floods and the Agreed Framework

According to North Korean government reports, the floods of 1995 struck 75 percent of the country, including 145 counties and cities in all eight provinces. Pyongyang estimated that 5.2 million people—one quarter of the country's total population—were affected. Of these victims, approximately half a million people, or 100,000 households, lost their homes and were displaced. The floods also caused severe damage to agricultural production: reportedly, 1.9 million tons of the year's harvest was lost. Moreover, one fifth of all arable land was inundated and covered with sand and, hence, lost for the 1996 planting season. Critical facilities and key infrastructure—such as schools, hospitals, health clinics,[1] roads, bridges, irrigation networks, energy providers, the water supply, grain warehouses, and industrial facilities—all suffered considerable damage. On the ground, field representatives of U.N. humanitarian relief agencies observed especially severe damage in North Hwanghae, North Pyongan, and Chagang provinces, including the Amnok River basin along the Chinese border around the city of Sinuiju, the Chongchong River around the cities of Pakch'on and Anju, and the area south of Pyongyang around the cities of Pongaen, Rinsan, and Chonggye-ri.[2] The Democratic Peoples' Republic of Korea (DPRK) government estimated that the total cost of the short-term and long-term damage to the economy was about $15 billion.[3]

Natural disasters of equal magnitude befell North Korea the following summer as well. On July 29, 1996, the Korean Central News Agency (KCNA) reported that the DPRK "had suffered extensive damage from heavy torrential rains, especially in the southern and western parts of the republic in late July."[4] On August 7, 1996, a DPRK Ministry of Foreign Affairs spokesman stated that "as a result of heavy rains in late July, a total of 117 cities and counties in eight provinces had suffered extensive damage."[5] After touring the flood-damaged areas of North Korea, one International Federation of Red Cross and Red Crescent (IFRC) representative to the Republic of Korea (ROK) National Red Cross Society remarked that the floods and torrential rains of 1996 hit hardest in the areas of Chagang-do, Hwanghaenam-do, Hwanghaebuk-do, and Kaesong City. A total of 290,000 hectares of farmland were damaged, and rice production was expected to be 30–40 percent lower than the yearly national average.[6] Overall, the storms killed at least 200 people, left 3.27 million people homeless, and inflicted more than $1.7 billion in damages.[7] Crops failed in 1997, by contrast, due to severe drought, adding to the ongoing food crisis.

In the summer of 1998, the floods returned. On August 25, 1998, the KCNA reported that "winds of 13 to 18 meters per second and heavy rainfall caused much damage to agriculture and infrastructure facilities in Nampo, Pyongyang, P'yongsong, Sinuiju, and other parts of the west coast." It went on to state that "according to conservative estimates as of mid-August, more than 74,000 hectares of arable land were inundated, and some 4,250 houses were destroyed or submerged in Kaesong City, Kangwon, South Hwanghae, South Hamgyong and North P'yongan provinces, and other areas. Schools, hospitals, and other public buildings covering a total of 25,600 square meters were damaged." Aside from the unprecedented rainstorms, unusually cold weather was registered on the eastern coast between mid-July and early August, causing below-average temperatures. The KCNA reported that "this year's cold weather had a fatal influence on the agricultural sector." It added that the rice harvest in southern border areas would "decrease by more than 60 percent," and that "the yields of maize and other dry-field crops would also drop sharply in these areas." The KCNA concluded that "damage in land administration, city management, and other sectors of the national economy was also severe."[8]

The torrential rains and severe floods that struck North Korea's western seaboard in the summers of 1995, 1996, and 1998 and the droughts of 1997 had considerable implications for the nation's economy, domestic politics, and foreign policy. Among other things, they appear to have had a negative impact on Pyongyang's ability and willingness to fulfill its obligations under the October 21, 1994, Agreed Framework and the Kuala Lumpur Joint Declaration of June 12, 1995.

To begin with, North Korea's main nuclear infrastructure is located in Yongbyon and Pakch'on cities, both of which fell in the pathway of torrential rains and rapidly rising waters from the Kuryong River in 1995–96. These storms reportedly inflicted physical damage on a number of nuclear facilities located at the Yongbyon Nuclear Research Complex. Hans Blix, director-general of the International Atomic Energy Agency (IAEA) at the time, expressed "general concern over the situation" in Yongbyon at the IAEA Board of Directors meetings in Vienna in December 1995 and 1996.[9] The exact figures on the total damage to the DPRK's nuclear program from the 1995–96 floods are not available. According to anecdotal accounts from the IAEA inspectors who visited North Korea in fall 1995, however, the basements of the critical assembly and nuclear fuel storage facilities in the Bungangri district of Yongbyon were completely flooded. Their power generators were severely damaged and a number of heat boilers were rendered inoperable. Moreover, the emergency power generators intended for such crisis situations failed to function properly.[10] Whatever power was generated was used primarily to pump the muddy water out of the flooded basements and to power the heavy machinery needed to clear out debris and sand left by the floods. Any leftover power was used to keep the local hills on which some of the nuclear facilities were hidden from eroding and to repair the structural damage inflicted upon various installations in and around the Yongbyon Special Administrative District. The loss of power was so extreme that, from time to time, North Korean nuclear engineers resorted to using flashlights in the flooded basements to examine the damage created by rising water.[11]

Thus, it is evident that nature itself inadvertently commanded North Korea to adopt a nuclear freeze. Simply put, there was probably not enough power to restart the nuclear reactors from September 1995 to February 1996 and from September 1996 to February 1997, even if there had been the political will to do so. In addition, the cold winter that followed created extremely difficult working conditions at the nuclear facilities and literally froze some of the necessary equipment.

Furthermore, the postflood cleanup and repair efforts must have increased the overall maintenance costs at the already nonoperational nuclear facilities intended for a shutdown and future dismantlement. Also, highly skilled technical personnel and considerable material resources had to be diverted periodically from Yongbyon to other damaged areas in its vicinity in order to help the local population and industries cope with the aftermath of the storm.[12]

From May 1994 through December 1996, the IAEA maintained a "continuous inspector presence" in the Yongbyon area. After November 1994, the IAEA performed inspection activities at the following key nuclear facilities subject to the "freeze": the 5 megawatt (MW) reactor, the 50 MW nuclear power plant (under construction), the radiochemical laboratory, the nuclear fuel-rod fabrication plant, and the 200 MW Taechon nuclear power plant (under construction). The IAEA conducted inspections in 1995 and 1996 at nuclear facilities not subject to the freeze. During these inspections, the Pyongyang refused to allow some inspection procedures, such as monitoring nuclear waste at the radiochemical laboratory and measuring the plutonium content of the spent fuel at the 5 MW reactor.[13]

Two key issues hampered the technical discussions (three sets of talks in 1995 and three in 1996) between North Korean nuclear personnel and IAEA inspectors throughout this trying period. The first was the urgent need to preserve the information required for the IAEA to verify the DPRK's initial declaration of its nuclear material subject to safeguards. The importance of this issue was heightened by the fear that the floods might have destroyed or damaged some relevant information or evidence regarding the sites in question. The second issue concerned the installation of the monitoring equipment at nuclear waste tanks in the reprocessing plant. Despite its extremely difficult position, Pyongyang refused to provide the IAEA technical experts with full accounting information and did not allow them to start monitoring the liquid nuclear waste. Consequently, the IAEA director-general time and again stated to the IAEA Board of Directors and IAEA General Conferences in 1995 and 1996 that despite the fact that monitoring of the freeze continued, "the agency is still unable to verify the correctness and completeness of the initial declaration of nuclear material made by the DPRK and, therefore, is unable to conclude that there has been no diversion of nuclear material in the DPRK."[14]

Moreover, on more than one occasion, visiting IAEA inspectors raised concerns regarding physical security at some nuclear installations hit hard by the floods. Some outside observers began to worry about a heightened danger of a nuclear accident at one of the Yongbyon nuclear facilities. Despite North Korea's considerable efforts to maintain tight security in and around the Yongbyon Special Administrative District, there were allegations of unauthorized entry by local inhabitants seeking shelter from the surrounding floods. This, in turn, led the international community to believe that the physical security of the North Korean nuclear installations may have been compromised.[15]

The only substantial progress that the IAEA reported in its technical cooperation with Pyongyang has to do with the canning of irradiated fuel rods extracted from the 5 MW graphite reactor. By the end of 1996, about 50 percent of these rods (out of 8,000) had been put into special containers for long-term storage,[16] and by now all of them have been canned. Ironically, this was the result of the only break that the North Korean nuclear personnel got amid the deteriorating situation on the ground, due to a small infusion of the U.S. government funds and technical expertise called for in enabling protocols to the Agreed Framework.

Specifically, in early 1995 NAC International of Norcross, Georgia, a nuclear clean-up company subcontracted by the U.S. Department of Energy (DOE), sent a team of nuclear engineers to Yongbyon to prevent the corrosion of spent fuel rods by purifying the water in the storage pools in which they were stored. Under its $7 million contract with DOE, NAC International scooped up enough sludge from the bottom of the pool to fill twenty 55-gallon cans and began to clean each of the 8,000 three-foot-long fuel rods and place these fuel rods into 360 canisters. Having filled the cans, NAC International heated them, pumped them dry of all moisture, and injected them with inert argon gas to keep the fuel stable. The cans were then lowered into the concrete pool and loaded into specially designed storage racks. The company, in collaboration with its North Korean counterparts, had already built four canning assembly lines at poolside.[17] Initial doubts and a lack of timely and adequate U.S. Congressional appropriations[18] created delays in the purchase and delivery of necessary equipment and instruments, as well as travel and labor reimbursement. This situation helped fuel ill-founded suspicions on the North Korean side and prompted it to threaten occasionally to restart its reprocessing plant if the canning operation were to stall. In November 1996, primarily because of the lack of promised funding, Pyongyang finally carried out its threat and suspended canning operations. In mid-January 1997, however, following an international uproar and the rapid infusion of a new dose of funds from the United States, it promptly resumed canning.[19] By the end of 1997, almost all 8,000 spent fuel rods had been safely treated and canned.[20] All in all, this part of the implementation program proved to be one of the least controversial and most productive. Nevertheless, in April 1998, North Korea again refused to finalize the canning process, still allegedly some 200 rods shy of completion, due to complaints over KEDO's slow delivery of heavy fuel oil. While this threat made international news, KEDO officials knew that the canning process had actually been completed and that only some sludge remained to be removed from the bottom of one of the cooling ponds.[21]

Moreover, it is evident that, along with the data generated by the monitoring equipment installed by the IAEA inspectors at key facilities in Yongbyon and the continual presence since early 1995 of U.S. nuclear specialists at the Yongbyon Nuclear Research Complex reassured the Western governments, the IAEA, and the international community that North Korea was continuing its nuclear freeze.

Overall, the natural disasters of 1995 and 1996 appear to have had a dual impact on Pyongyang's ability and willingness to implement the nuclear accords. On the one hand, they deprived North Korea of some of the material resources and reserves necessary to resume operation of its nuclear reactor on short notice. In this way, the natural catastrophe bound Pyongyang to its nuclear commitments and crippled its ability to quickly, and without notice, reverse its nuclear freeze.

On the other hand, by damaging some of the nuclear infrastructure in Yongbyon and Pakch'on and mercilessly exposing many of the deficiencies and shortcomings of the North Korean nuclear program, the natural disasters of 1995 and 1996 seem to have changed the agenda of the nuclear policy debate in Pyongyang. In particular, in late 1994 to early 1995, after having successfully negotiated and signed the "landmark" Agreed Framework, nuclear policy makers in Pyongyang, while proud of their own indigenous achievements in the nuclear energy field, eagerly anticipated Western-financed plans aimed at the light-water reactor (LWR) construction and the expansion of the local nuclear industry, underpinned by Western technology and paid for by Western money. Occasionally, when doubtful about Western commitments to comply with the provisions of the Geneva accords, Pyongyang felt confident enough to resort to threats of unilateral abrogation of the nuclear freeze. After the devastating storms of 1995–96, however, North Korean euphoria over the nuclear modernization evaporated and gave way to an overwhelming preoccupation with repairing and maintaining the existing nuclear facilities. This became evident through hints from Pyongyang and its surrogates that the West should strive to assist Pyongyang in keeping its Yongbyon Nuclear Research Complex from falling apart, rather than seeking to fulfill its highly controversial commitment to transfer two 1,000 MW LWRs to the North by the year 2003.

Broader Economic Implications of the Floods

The national calamities of 1995–96, arguably the most devastating in North Korea in the past hundred years, had profound economic, political, military, and social implications. Floods and torrential rains during two successive summers dealt a final blow to the moribund Three-Year (1994–96) Economic Adjustment Plan passed by the Workers' Party of Korea (WPK) Central Committee in December 1993. They exacerbated the already enormous difficulties faced by the North Korean economy both on the supply and demand sides and tightened already smothering macroeconomic bottlenecks, especially in transportation, energy supply, and the labor force.

In order to alleviate popular hardships and regain control over the worsening economic situation, the government established a cabinet-level organization known as the Flood Damage Rehabilitation Committee (FDRC) to serve as an organizational vehicle for implementing Kim Jong Il's directives and policies to cope with the consequences of the 1995–96 natural disasters. As such, its activities were said to be personally supervised by Kim Jong Il.[22]

As its first step, the FDRC ordered all *ri* (town), county, and provincial administrative committees in the affected areas to evaluate local damage and submit damage assessment reports to its headquarters in Pyongyang. Next, the FDRC formed regional and provincial emergency committees responsible for disaster prevention and warning and postdisaster rehabilitation and launched a nationwide campaign to collect donations (clothes, housewares, personal utensils, books, foodstuffs, etc.) from the people living in the unaffected areas for their compatriots who bore the brunt of these natural disasters.[23]

The delivery of donations from the unaffected provinces of eastern North Korea to the flood-stricken west coast provinces proved to be a logistical nightmare. Usually, foodstuffs, especially rice, and textile goods are shipped by rail from the provincial centers in the south to the localities in the north. After the floods and torrential rains wiped out the entire crop in the Yonbaek plains and more than 60 percent of crops in the Jaeryong plains, however, food donations had to be brought to the flood-stricken regions from the east coast from behind the mountains. Here the transportation system was much less developed.[24] Moreover, many railroads and railroad bridges in the southwest and northwest had been washed away,[25] requiring rice shipments and other donations to be delivered by truck. This was a challenge in itself because: (1) hundreds of kilometers of highways and local paved roads were also destroyed; (2) gasoline and wood fuel were scarce, especially in the winter; and (3) the available trucks were being used, primarily to evacuate people from dangerous zones in the eastbound direction and to carry construction materials and army construction personnel in the westbound direction.[26]

Having collected rice, donated clothes, and medicine, the FDRC repeatedly had to confront the Central Transportation Authority and flex its muscles in order to obtain enough trucks to carry donated supplies from their collection points to their destinations.[27] This was difficult in an already highly mobilized and rigidly planned economy such as North Korea's, because most of the 3,000 trucks presumably at the FDRC's disposal were already assigned to transport various goods around the country and could not be reassigned by middle-level bureaucrats without explicit orders from above. Therefore, the FDRC had to use its political influence in Pyongyang to press the Administrative Council into passing emergency transport reallocation measures and reassigning enough vehicles to fulfill its mission.

Furthermore, as a result of growing decentralization, often euphemistically described as "deepening self-reliance," and a worsening food situation on the ground, several officials from the central government, usually the FDRC staff, had to accompany the food convoys through the provinces. In this way, they ensured a "smooth journey" and made sure that trucks carrying rice did not "lose their convoys" and that local officials did not extort "voluntary contributions" to local communities from centralized convoys headed to the flood-stricken areas. This problem tended to be more acute whenever such convoys were known to be carrying foreign donations.[28]

Another problem the FDRC faced was the collapse of the centralized distribution system in the flood-affected areas. This worsened the malnutrition crisis.[29] In the North, peasants usually receive shipments of rice once a year after the harvest is collected, whereas urban families receive their allotted rice twice a month at local rice distribution centers. With the 1995–96 summer storms destroying more than 1.9 million tons of rice, including the rice stored in private rural households and urban warehouses, the population became wholly dependent upon centralized food supplies intermittently delivered from other provincial storage facilities until the next harvest. As a result, food rations were reduced dramatically to the bare minimum necessary for physical survival. Reportedly, the average North Korean could count on receiving only 9 kilograms of 35 percent broken rice a month from the government. This amounted to 9 kilograms per fortnight, or about 300 grams per person per day (or 150 grams per meal, under the current motto "Let's have two meals a day").[30]

According to Trevor Page, the World Food Program representative in North Korea, the food situation in the country was "extremely grave" and had approached that of a famine by the end of summer 1996.[31] Aware that its distribution system had failed to function properly because there was nothing in the pipeline to distribute, the government could not help but turn its back on the various attempts by the local population to procure its own food by any means necessary, as long as these efforts did not lead to social unrest and political dissent. As a result, local authorities often got away with levying their own quasi-taxes on all the traffic passing through their domains. One astute foreign observer described the situation on the ground as generally "stable but not quite."[32]

In order to strengthen its mobilizing and coordinating roles, the FDRC established liaison offices at all the central government ministries and local administrative committees concerned. Pyongyang authorized the FDRC to "borrow" personnel from the Foreign Ministry, the Institute of Peace and Disarmament, and several economic ministries (which may inadvertently reflect the paucity of skilled and reliable personnel at the middle-upper bureaucratic level).[33] In fact, not only was the FDRC apparently empowered to coordinate the relief and rehabilitation activities of other government bodies inside the country, but it was also designated to serve as the primary interface between various North Korean bureaucracies and the international community interested in providing humanitarian assistance to the flood-stricken North. When Pyongyang issued two international appeals for humanitarian assistance, it made the FDRC responsible for receiving and dealing with representatives of international nongovernmental and private relief organizations. Despite these mobilization efforts and enormous emergency spending, some North Korean government officials admitted that it would take a few more years for the country to get back on its feet again and for life to return to normalcy for those who were directly affected by the floods in 1995 and 1996.[34]

Not only did these natural disasters wreck the DPRK's economy, they also appear to have forced its leaders to reconsider some of the long-standing principles on which its political system was based. The ideological dogma of self-reliance and tremendous national pride had to give way to previously unthinkable ideological concessions and political sacrifices as the North Korean leadership reluctantly requested humanitarian aid from abroad. In a way, this policy change reflected a newly formed domestic consensus. Today, everyone in North Korea, from the central government leaders to local party officials to laymen, acknowledges the severity of the food problem, the fact that food rations have been halved or even reduced by two thirds, and that international humanitarian assistance is indeed needed.[35] Pyongyang has admitted that the heavy rains in July–August 1998 destroyed up to 60 percent of crops, with the result that many people are facing the prospect of starvation, creating a desperate need for international food aid.[36]

Nevertheless, these types of admissions have been extremely difficult for the regime. Pleading for donations from former and current enemies is a de facto repudiation of the traditional policy of *juch'e*, which is tantamount to an unforgivable loss of face for North Korean leaders. Kim Jong Il must have realized that this humanitarian decision would provide his domestic opponents with plenty of ammunition. Some opposed international aid on ideological grounds, and others did not want to permit the continual presence and on-site monitoring of aid distribution by representatives of international humanitarian organizations. Moreover, opponents could accuse Kim Jong Il of betraying North Korea and of selling out his country to its enemies. He also must have anticipated that his external opponents would eagerly use it against him in their anti–North Korean propaganda campaigns, which they did. Despite these concerns, in September 1995, he decided for the first time to request international assistance and open up the countryside to international relief workers.[37]

Regretfully but hardly unexpectedly, the initial results of Pyongyang's first appeal to international aid proved disappointing.[38] This brought about a great deal of conservative pressure on Kim Jong Il to halt the appeal work.[39] After considering the "appeal freeze" idea for a few weeks in January 1996, Kim Jong Il apparently decided to stay the course and, furthermore, issue a second request for international assistance in June 1996.[40]

It is noteworthy that the North Korean leadership never attempted to link its willingness to duly implement the Agreed Framework with the Western governments' response to its humanitarian aid appeals. In late 1995, after the West realized the scope of the natural disasters that had struck North Korea, some policy makers in Washington and Seoul expressed concerns that Pyongyang might decide to hold the implementation process hostage to the issue of humanitarian aid. Thus far, however, this worry has proved to be groundless. On the contrary, Pyongyang has made it clear on several occasions that, whatever the outcome of its international humanitarian aid appeals, successful completion of the LWR transfer project will remain its highest priority.

Military and Security Implications

In the aftermath of the national catastrophes of 1995–96, a number of senior military officials in Pyongyang reportedly raised some disturbing questions about the military security of the country.[41] The storms inflicted heavy damage on some of the army and air force infrastructure along the eastern seaboard. Although repair work on these damaged military installations was quickly begun, progress was apparently slow in the making. To some extent, this could be attributed to the fact that from day one, the Korean People's Army (KPA) Supreme Commander in Chief Kim Jong Il ordered many KPA units, which would otherwise be involved in the repair of military facilities, to assist the civilian population in rebuilding homes, highways, railroads, bridges, and certain key industrial facilities.[42] In addition, due to a chronic lack of funding for the repair of damaged airstrips and aircraft maintenance facilities, the KPA General Staff in late fall 1995 redeployed a few dozen fighter aircraft to air force bases located in unaffected areas, which happened to be closer to the demilitarized zone.[43]

In the past few years, there has been growing domestic frustration over the government's inability to provide material resources—including sufficient food and fuel rations, imported spare parts and replacements for obsolete military equipment—to its military. These shortfalls, when added to concerns over the continual massive diversions of military personnel from active duty and field and classroom training to civilian construction and reconstruction projects have induced some hard-liners within the North Korean military establishment to begin to question the KPA's ability to prepare for war.[44] Some went so far as to claim that the DPRK was too weak to defend itself against foreign enemies.[45] Consequently, fearing that hawks within the South Korean government might be tempted to take advantage of a militarily vulnerable North Korea, hard-liners in Pyongyang reportedly urged the government to assume a more belligerent posture vis-à-vis the South and the United States.[46]

Still, it would be presumptuous to argue that these genuine concerns about military preparedness, expressed by some KPA officials, reflect a growing military influence within the DPRK leadership. Even more dubious are the claims of some analysts that, after the death of his father, Kim has become merely a symbolic leader of a regime completely dominated by the military.[47] More likely, Kim Jong Il remains in full control of all power structures in North Korea, including the KPA. At the same time, he remains fully aware of the views and concerns of the military leadership as well. Consequently, in order to appease the military and silence his critics, Kim has continually courted senior military commanders, often allowing more public exposure than they had enjoyed during his father's tenure. Kim Jong Il has been fastidiously grooming his loyalist military cadres for more than a decade. Nevertheless, he does not seem to have yielded any authority to his military advisors in the policy-making arena.

Following the deaths of former DPRK Defense Ministers O Jin U and Choe Gwang in February 1995 and February 1996, respectively, Kim Jong Il received the opportunity to radically reshuffle the senior military command and set up a new balance of power among various military services. These drastic changes were ratified by the tenth Session of the Supreme People's Assembly (SPA) in September 1998.

At the institutional level, the new DPRK Constitution passed by the tenth Session of the SPA made the National Defense Commission the paramount organ of state power in the DPRK, albeit still accountable to the SPA. The National Defense Commission (NDC) consists of the chairman, the first vice-chairman, several vice-chairmen, and members. The chairman of the NDC was empowered to direct and command all the armed forces and to guide defense affairs as a whole. The NDC was given the following duties and authority: (1) to guide the armed forces and defense building of the state as a whole; (2) to set up or abolish national organizations in the defense sector; (3) to appoint or remove major military cadres; (4) to institute military titles and confer promotions at the rank of general or higher; (5) to proclaim a state of war and orders for mobilization; and (6) to issue decisions and orders.[48] Consequently, supervision of the Ministry of Defense was transferred from the cabinet to the NDC.

On the personnel level, NDC Chairman Kim Jong Il is the only civilian among its members, and he is supposed to guarantee civilian control over the KPA. Within the NDC, a careful balance of power has been created between various military branches. A senior air force officer and director of the KPA General Political Department, Vice-Marshal Cho Myong Rok, was appointed the first vice-chairman of the NDC, in essence, the second person in the chain of command after Commander in Chief Kim Jong Il. For the first time in KPA history, a new defense minister, Vice-Marshal Kim Il Chol, was selected not from the army, which constitutes the bulk of the KPA and previously dominated the decision making on military issues, but from the navy. The army was left to head the KPA General Staff, with Vice-Marshal Kim Yong Chun being appointed the chief of the General Staff.[49] These appointments can be interpreted as a carefully crafted power-sharing arrangement whereby each military branch acts as a check on the aspirations and demands of the other. Obviously, this reflects an attempt to curtail political influence of the army. However, it is doubtful that there is any significant new role for the North Korean navy command in the DPRK's military doctrine other than to serve as a check on the ambitions of the army generals. What is more significant is the elevation of the status of the air force in the overall military command structure. Undoubtedly, this must be a reflection of the growing importance of its rocket forces, bomber wings, and the alleged nuclear installations at air force bases as strategic deterrents against the ROK in the DPRK's military doctrine.

All in all, these new institutional and personal arrangements seem to be designed to fulfill two goals. First, they must assure Kim Jong Il that no other civilian leader from the central government might exercise any authority over the military and use it in order to mount a credible challenge against him, backed by the armed forces. Second, they allow Kim Jong Il to play off various military interests one against the other, thereby keeping them divided and preventing army leaders from exercising too much influence in the political process, not to mention orchestrating an army-led mutiny against him.

Since late 1994, Kim Jong Il has intervened more directly in managing the military affairs of the country. He frequently inspects KPA units (altogether exceeding 600[50]) located both along the DMZ and in the rear areas: (1) to promote the combat readiness and boost the morale of his troops during the "cold peace," a time marked by ideological ambiguity and material hardships; (2) to ensure that the military budget will favor the KPA units in greatest need and those that play indispensable roles in national defense by drawing scarce material resources from the central budget to those particular units; (3) to reassure his domestic constituency that he can talk tough to the "Yankee imperialists and their fascist puppets"; and (4) to convey a stern warning to foreign nations that the KPA, despite its current difficulties, remains fully prepared to fulfill any combat orders from the Supreme Commander, whatever they may be.[51] In other words, in the wake of the recent natural disasters, his attitude toward the military may be best described as one of empathy and a pledge of support, mixed with expectations of perseverance and strength on the part of the military.[52] Indeed, some outcomes of domestic policy debates in Pyongyang reveal that, whenever Kim Jong Il decided on a key issue related to the implementation of the Agreed Framework, he consistently opted for the more pragmatic and future-oriented stance advocated by the economic officials and diplomatic personnel, rather than the KPA's more orthodox and conservative positions.

For example, back in early 1995, despite reported pressure from his generals,[53] Kim Jong Il did not allow the military to divert heavy fuel oil from its intended purposes to military use to the chagrin of some KPA hawks and the delight of Western statesmen. Even today, in spite of increasing supplies of heavy fuel oil from the West and against the background of a deteriorating fuel situation in the KPA, this policy still stands. Moreover, despite some Western speculations,

Kim Jong Il did not allow the KPA to divert any rice or medicine from foreign donations to military stockpiles.

Trevor Page from the WFP has stated recently that he believes that "the KPA may have rice stored from the previous Korean harvest, but foreign donations designated for the flood victims are not being diverted." It is noteworthy that, in response to such speculations in the Western media, FDRC spokesman Li Jong Hua stated unequivocally that "these rice donations will not be, cannot be, and are not given to the military." An emotional rebuttal of these allegations came even from a KPA colonel who participated in the Hawaii talks with the United States on the MIA issue in March 1996. He flatly rejected the notion that the military could have received abundant rice supplies at the expense of the civilian population by exclaiming "How could this happen when my mother, father, brothers and sister are all civilians? Do you think we could allow our relatives to starve without food?"[54]

The issue of food diversion arose again in September 1998, when the organization Doctors without Borders suspended its operations in the DPRK, charging that North Korean authorities were diverting relief food and medicine to military units. The FDRC representative told the KCNA that these accusations were groundless and urged other nongovernmental organizations to continue their humanitarian efforts in the North.[55]

Kim Jong Il has also overruled the objections of military opponents adamantly opposed—for ideological and security reasons—to the idea of having ROK-made LWRs installed in Sinp'o. This will inevitably be accompanied by the inflow of thousands of suspicious Korean Peninsula Energy Development Organization (KEDO) employees, mostly of South Korean citizenship, roaming freely around the North Korean countryside and enjoying diplomatic immunity. Still, Kim Jong Il authorized his representatives at the United Nations to sign the LWR Supply Contract on December 15, 1995, and the first implementing Protocol on Consular Protection and Diplomatic Immunities for KEDO-employed personnel on May 22, 1996.

Commenting on Pyongyang's attitude toward the April 1996 DPRK-KEDO protocol on privileges and immunities, which governs KEDO employees and contractors while working in North Korea, then-KEDO Executive Director Stephen Bosworth said, "I don't think very many countries would contemplate willingly giving up the degree of sovereignty that in fact North Korea has given up to KEDO in this implementing protocol." He added that the North Korean leaders accepted KEDO's position that "We could not run a project with several thousands of South Koreans living and working in North Korea without some way to assure ourselves they would not automatically be subject to North Korean laws."[56]

This record indicates that although Kim Jong Il is still likely to play the military card in future nuclear negotiations with the West,[57] he is not going to allow his top military brass to disrupt the ongoing nuclear negotiations on the implementation of the Agreed Framework and jeopardize his first and only diplomatic achievement in the post–Kim Il Sung era, namely, the nuclear package deal.

Impact of the Floods on Political Developments

The natural disasters of the mid-1990s affected impacted succession politics in Pyongyang as well. For instance, the floods seem to have diverted attention away from the second anniversary of Kim Il Sung's death on July 8–12, 1995, and kept Kim Jong Il out of the commemorative events. The entire North Korean leadership was apparently shocked by the magnitude of the natural disaster and quickly became preoccupied with drafting emergency measures to assist the flood victims. Furthermore, torrential rains sweeping the entire country were not very conducive to massive, prolonged outdoor gatherings and rallies. As for Kim Jong Il's failure to appear at a small-scale commemorative rally in Pyongyang on July 9, he was not sick, nor did he feel politically threatened. Rather, he was simply trying to be a responsible leader and direct all of his attention to controlling the flood damage at once.[58]

Once the disastrous consequences of the summer floods became obvious, they presented a major challenge for Kim Jong Il's leadership in the domestic political arena. His father, the first "Great Leader," was no longer around to accept responsibility and pacify the people. Moreover, Kim Jong Il could not blame the destructive schemes of cunning imperialists seeking to stifle North Korean socialism and topple the regime. This time he was on his own.

Moreover, he was faced with a string of bad luck over a three-year period. Within eight months, his mythical father, Kim Il Sung, and his key patron within the military, Defense Minister Marshal O Jin U, had both passed away. In the summers 1995 and 1996, his country was wracked by floods and torrential rains, reportedly, the worst in the last 100 years. In late 1995, a Taiwanese freight vessel carrying rice donations to North Korean flood victims inexplicably sank not far off the coast of North Korea. His ex-wife was rumored to have defected to South Korea in February 1996.[59] A year later, the WPK Central Committee Secretary Hwang Jang Yop, the chief architect of the *juch'e* ideology, fled to South Korea. Not only have Kim Jong Il's leadership skills been challenged, his very character was being tested on an almost daily basis. It appears as if it is a make it or break it situation for Kim Jong Il. Either he will have to emerge out of the current domestic crisis with a new, dynamic personal charisma, or he, along with the debris left by the recent floods, will be swept away by the ongoing crisis.

Paradoxically, the devastating floods in North Korea provided opportunities for rapid promotions for those North Korean middle-level party and state bureaucrats who managed to prove themselves capable of responding promptly and adequately to the emergency situation. Purging political enemies and incompetent workers at a time when their own management failures had become so visible and were so much resented by the previously content general public became harder.

Orthodox approaches could no longer alleviate popular hardships and rehabilitate the economy. New, more pragmatic elites advocating more liberal approaches to the economy and regulation of social life may therefore be in the making.

A final political question one must address is the relationship between natural disaster and state survival in the North Korean context. The floods of the mid-1990s served as an extraordinary external shock to the polity and economy of the North, which had already been under tremendous strain caused by the preceding macroeconomic shocks of the early 1990s, caused by the annulment of Soviet and Chinese preferential treatment and the disintegration of its geopolitical alliance system. Will the current, unreformed North Korean political-economic system survive this horrible onslaught inflicted upon it by acts of nature in its various incarnations? Or will the North Koreans opt to abandon their Korean-style socialist system in the wake of its startling failures to cope promptly and efficiently with the natural disaster? Just as the Chernobyl nuclear accident in 1985 proved to be a harbinger and catalyst of the forthcoming *glasnost* and *perestroika* in the former USSR, the North Korean floods of 1995–96 may already be serving as the initial catalyst for political liberalization and economic reform in the DPRK.

According to recent accounts by foreign diplomats, visiting dignitaries, international relief workers, and merchants, Pyongyang is still struggling to recover from these disasters and the new round of flooding in the summer of 1998, which decimated its rice supplies and destroyed much of its basic infrastructure along the western seaboard.[60] Hence, all other policy issues, including its nuclear policy, tend to be relegated to the back burner of the policy-making agenda.

Notably, in a visit to North Korea in May 1996, then–U.S. House Congressman (and now U.S. Energy Secretary) Bill Richardson came to the conclusion that "this food problem was ... of such overwhelming importance to North Korean officials that I was left with the impression that they would find it difficult, if not impossible, to focus on other issues until they have found a way to deal with this first."[61]

Conclusion

The natural disasters of 1995–98 wreaked havoc on an already struggling North Korean economy. Not only did these natural disasters deplete an already scarce reserve of state resources available for the modernizing North Korea's nuclear industry, but they also forced North Korean leaders to focus their efforts on providing food and medicine for the starving population and repairing and rehabilitating the flood-damaged economic infrastructure. To be sure, in the short term, when nature struck, raising the specter of a North Korean Pompeii, implementing the nuclear agreements became less of a priority for Kim Jong Il and his associates. Nevertheless, as time has passed, the regime has not fully abandoned its efforts to squeeze new concessions from the West when opportunities have presented themselves.

Part III

Political and Military Factors behind the North Korean Nuclear Program

9

Nuclear Blackmail and North Korea's Search for a Place in the Sun

Alexander Platkovskiy

As a result of extensive and ongoing diplomatic efforts by the United States, the acuteness of the North Korean nuclear crisis has been alleviated recently. While some problems still remain, the main concern of outside observers today is the internal political situation in the Democratic Peoples' Republic of Korea (DPRK). Kim Il Sung's death in July 1994 and the drawn-out process of formalizing the authority of his successor, Kim Jong Il, raised questions as to where the regime was moving, what surprises it might present to the outside world, and how long it would survive. Kim Jong Il's consolidation of power in August 1998 seems to have answered some of the these questions, but certainly not all of them. Alongside these political issues are serious fears arising from the fact that Pyongyang has not given up the idea of playing the nuclear card yet one more time. No one dares to rule out that option totally, thus underlining the fact that while the Agreed Framework has dulled the edge of the current nuclear crisis, it could become sharp once again depending on developments in North Korea.

Several questions deserve analysis in this regard: (1) Will we see a new round of nuclear blackmail on the part of the North Korean regime? (2) What might cause such a step on the part of Pyongyang? and (3) What might be the results of renewed crisis?

This chapter attempts to answer these questions by putting them into a broader historical context with regard to North Korean politics. Such an approach provides guidelines to the likely thinking of the North Korean leadership, as well as to factors that might trigger problematic behavior in the future.

The sources used in this analysis include interviews with the North Korean people, its officials, and Russian diplomats in Pyongyang and Moscow.

Nuclear Blackmail as a Means of Adapting to Changing Conditions

Until the mid-1980s the regime of Kim Il Sung functioned in relative tranquility. Internal enemies had long since been done away with, and control over North Korean society had become absolute. The Soviet Union and China, North Korea's staunchest allies, kept external threats at bay. At the same time, Pyongyang was quite skillfully pursuing various intrigues in its relations with Moscow and Beijing, striving to get the most favorable treatment from each of these ideological rivals.

With Mikhail Gorbachev's ascent to power in Moscow, the quiet times for Kim Il Sung and his regime became a thing of the past. After flying to Moscow and having talks with the new leader of Soviet communism, Kim Il Sung—by all appearances—realized that he would have to prepare himself for new times and hard tests. Gorbachev's policy of *glasnost* (openness) in the press, his political reforms, and the improvement of Soviet relations with the United States and China presented the most serious challenge to the North Korean regime in the entire postwar period. The leadership began to feel suddenly that it was losing the ground under its feet. Kim Il Sung certainly understood what dangers were in the wind. As developments unfolded, it became obvious that sooner or later Moscow would cease to refer to Pyongyang as its "bulwark of socialism in the Far East" and then would turn away from it. The regime would eventually find itself being treated as an outcast, rejected for its sins by its erstwhile allies and enemies alike.

At that moment of reckoning, while the socialist camp still existed and the concept of communism had not burned out, Pyongyang undertook a desperate attempt to strengthen its position, save its international prestige, and catch up with the times. Despite the "rampage of revisionism" in the socialist countries and the activation of forces branded in Pyongyang as counterrevolutionary, the regime made a decision to open its doors and host a World Youth and Students Festival in 1989.

The Collapse of Illusions

In 1956, Soviet Premier Nikita Khrushchev had instituted a thaw in the Soviet Union. That year, he also initiated an unprecedented undertaking for the Soviet Union under the name of the Youth and Students Festival, when thoroughly scrutinized representatives of so-called progressive youth came to Moscow to participate in various festivities. Despite the restrictions of the selection process, after the period of Stalinist isolationism, the Soviet people for the first time got an opportunity to make contact with foreigners and even to converse with them, all with the blessing of the government. From that moment on, Moscow lost the ability to close the Iron Curtain again, and ideological doubts began to emerge within Soviet society. This process dragged on for several decades in the Soviet Union, but nevertheless did come to its logical conclusion in the reforms of Mikhail Gorbachev, himself a student of the Khrushchev era.

In August 1989, the World Youth and Students Festival took place in Pyongyang. It was a similarly unprecedented event: for the first time the regime had agreed to admit a large number of foreigners into the country, including some who had been "poisoned by imperialist propaganda." More than 40,000 guests from all over the world stirred up life in Pyongyang for two weeks. While a majority of the city's residents had been temporarily evacuated from the capital, those remaining in the city were undoubtedly shaken, as the foreigners carried with them totally forbidden information about life abroad. Also, North Koreans witnessed for the first time a situation in which their rulers were treated with neither fear nor worship by the young foreign visitors. At Pyongyang Stadium, in the presence of almost 100,000 spectators (including the national leader and his son), foreigners demonstrated with antidictatorial slogans. This produced a major scandal for the regime, even though security forces quickly curbed the dissent at the meeting.

What were the goals of the festival, and who stood behind it? There are no clear answers to these questions, but it seems that there is a substantive relationship between the festival and current events in North Korea. Officially, its aim was declared as raising the banner of *juch'e* even higher in the world arena and demonstrating the successes of North Korean socialism. Notably, Kim Jong Il himself was in charge of carrying out the festival. From conversations with senior staff members of the Working Youth Union during the preparation and conduct of the festival, the author managed to learn that the major reason for staging this politically risky event lay not so much in ideological purposes, but rather in its extreme importance to the succession of Kim Jong Il to the leadership. Kim junior had been successfully persuaded that this great event would bring him the international recognition he needed as the future North Korean leader, which would be of tremendous importance for the fate of the Korean revolution. At the time, Kim Jong Il was a little-known figure even among the socialist countries, and he himself apparently suffered from this lack of prominence. Moreover, party ideologues behind the festival maintained that new approaches were required in the international arena at this time and that, in order to boost the reputation of Korean socialism, the regime needed to demonstrate a certain flexibility and readiness to hold a dialogue even with its enemies.

Notably, in confidential conversations, some youth personnel officials proudly maintained that the festival was just a beginning, and that it would be followed by new North Korean policies of *perestroika* (restructuring) and *glasnost* à la Gorbachev.

It is possible that those figures in the leadership who suggested the idea of the festival and carried it out may have harbored some hidden motives. One should not rule out the possibility that they made their calculations based on the assumption that contacts between the festival's youth personnel and their foreign counterparts might leave some traces, thereby sowing the seeds for future changes. In any case, for many hundreds of young activists and festival officials, the experience was unique. Subsequently, as the majority of those cadres entered the ranks of North Korea's *nomenklatura*, their views and ways of thinking will, no doubt, play a role in the ongoing evolution of the regime.

The irony is that the youth festival became a prologue to the collapse of the socialist regimes in Eastern Europe. The Berlin Wall came down a little more than a month later. This meant that the plans for the consolidation of the international prestige of North Korean socialism had already been ruined. At the same time, Kim Jong Il also lost his chance to break out of his former isolation and non-recognition as heir to the "great leader." But by this time it was already too late. A chain reaction of the collapse of socialism in Eastern Europe and its death throes in the Soviet Union had been set into motion. This was a new challenge to the North Korean regime, which was still harboring delusions as to the temporary nature of the difficulties being suffered by its partners. The massacre of Ceausescu in Bucharest threw Kim Il Sung's residence into a shock state, and for quite some time his son was unable to regain his composure. This account came to the author through Soviet diplomats well versed in the court affairs of Pyongyang.

Events unfolded rapidly, as history made a sharp turn. The North Korean regime felt defenseless in the face of the approaching threat. In fact, it seemed like its very survival was at risk under these new conditions. There was a need to come up with a reliable means of guaranteeing its stability and security.

Nuclear Weapons as a Means of Dealing with the Changing World

The first reports in the Western news media about Pyongyang's conduct of a clandestine nuclear weapons development program appeared in 1987–88. U.S. spy satellites had recorded evidence of suspicious activities in the locality of Yongbyon. There was no confirming data, but suspicions were growing. At that time, numerous military delegations from Iran had begun to visit North Korea. These contacts intensified to such an extent that diplomats from some Arab countries became suspicious that military-technical cooperation was being established between Pyongyang and Teheran. The subject of these negotiations was kept secret, but the same sources in Pyongyang soon reported that exchanges of technologies and specialists in the missile and nuclear fields were taking place, and that Pyongyang was on the threshold of creating its own atomic bomb.[1]

It is difficult to establish when Kim Il Sung got the idea of developing a North Korean nuclear arsenal. Obviously, he authorized extensive efforts in this direction and they produced some concrete results. Though there is no proof of the regime's success in creating the bomb, it is only reasonable to assume that Kim Il Sung instinctively expected bad times to come and sought a means of protecting his power for the relatively long period that might be necessary for transferring state power into his son's hands. Such means could be found only in an absolute weapon, whose possession would guarantee world respect and provide opportunities for bargaining effectively within a hostile international environment. The psychological factor was of no less significance—the regime was in need of such support. But, it seems likely that Kim Il Sung's attempts to obtain a nuclear weapon or, as some argue, only the means to bluff that he had one, were primarily a reaction to the changes in the outside world. The regime found itself facing an unpleasant choice: to agree humbly with the loss of support from its traditional allies or to "lie low" and to wait for the storm to clear. But these scenarios involved enormous risks. Isolated, rejected by both the West and its former allies, and having lost its friends in the Third World as well, the regime could be threatened by collapse from the inside, as those dissatisfied with these changes would inevitably emerge within the ruling elite. On the other hand, there was another option: to struggle toward the goal of a "place in the sun" and to ensure the survival of both the regime and the Kim dynasty. Nuclear blackmail became the means of conducting that struggle.

The First Showdown

In 1990, Kim Il Sung for the first time decided to lift the curtain partially from his nuclear secrets. Moscow was determined to pursue its rapprochement with Seoul. Gorbachev had already met with President Roh Tae Woo in San Francisco and the establishment of diplomatic relations with the North Korean regime's mortal enemies was not far away. In January 1991, Soviet Foreign Minister Eduard Shevardnadze went to Pyongyang to clarify Soviet intentions. As he later reminisced, there he had to endure his hardest and most exhausting negotiations ever. His partner at the negotiations, the former DPRK Foreign Minister Kim Yong Nam, had a task set by the leader: to prevent Moscow's recognition of the South Korean "puppet regime" at all costs. And, just as Shevardnadze was about to get up from the bargaining table, frustrated by the North Koreans' lack of understanding, Kim Yong Nam produced his trump card. He declared that if the Soviet Union decided to proceed in its betrayal of North Korea, the country's leadership would consider itself free of its obligations not to develop weapons of mass destruction, would support Japan in its territorial dispute with the Soviet Union over the Kuril Islands issue, and would open negotiations with the United States on establishing direct diplomatic relations.[2]

It is difficult to judge how effective this blackmail threat was in the North Koreans' attempt to prevent the inevitable. One has to put oneself in their shoes. They felt that they had been not simply betrayed, but cornered as well. They were at a loss as to how to cope with the radically changed world around them in which people who had just recently called themselves their trustworthy friends were now abandoning the regime. Clearly, their blackmail attempt was conducted out of desperation. Of course, this attempt to halt history failed a few months later, as Moscow and Seoul signed an agreement establishing diplomatic relations. From that moment on, the North Korean regime shifted even more fully to engage in a struggle for its own survival, using all possible means.

Pyongyang Triumphant

The darkest times had descended over Pyongyang. The collapse of socialism in Moscow and the disintegration of the Soviet Union had dealt a heavy blow to the North Korean economy, which was already in a desperate situation by that time. The cessation of Russian oil shipments and the shift in trade terms to a hard currency basis presented the toughest challenges. News began to come out of North Korea about factories standing idle for lack of energy and raw materials, about further reductions in already small food rations, and even about food riots. The regime responded to the economic hardships by whipping up its ideological campaign and by suppressing dissent. One can only guess about the prevailing mood inside Kim Il Sung's residence and among those close to him. The elite, Kim's family in particular, was united by their fear of the revenge that awaited them in case of the aggravation of the crisis and the regime's seemingly inevitable collapse. This fear was of a corporate nature, especially under North Korean conditions, where the members of the ruling hierarchy—top party officials and their family members—are equally answerable for the crimes of the regime. Even those who themselves were victims of repression had engaged in enough criminal behavior that they could not expect any mercy.

During that period, it was the loss of authority and control over the economy, the army, and society—as a result of the catastrophic depression and the impossibility of simply providing for the population's existence—that became the main threat to the regime, which found itself on the verge of collapse. But it was not the first time that the country had lived through such conditions. Perhaps because of this fact, it was possible to believe that the regime might be able to keep the half-starved North Korean people from rising up. However, chronic malnutrition and the witnessing of widespread starvation sharply changes people's psyches by liberating them from the repressive ideological yoke and submitting their will and behavior to a self-preservation instinct. Kim Il Sung, who had himself learned in his youth the lessons of the guerrilla movement, was clearly aware of the consequences he and his associates could expect if food riots were to erupt. Suppression of such riots would require the cooperation of the Korean People's Army, which itself was already beginning to drift from reliable state control.

Under such conditions the nuclear card seemed to be the regime's last hope.

With its help, Kim Il Sung figured that he could prevent and lessen the country's isolation, expand its economic contacts with Western countries, and obtain assistance to overcome the food crisis. The United States was seen as North Korea's main counterpart in this game. Pyongyang did not hide the fact that its leaders were striving to normalize relations with their powerful ideological adversary. At the same time, it seems that North Korean strategists had no illusions concerning U.S. philanthropy and expected that Washington would press for reform of the regime. To prepare themselves to deal with the Americans, therefore, Pyongyang decided to use its strategic advantage: that is, the inability of the United States to obtain verifiable information about what was going on at secret military installations in North Korea. It also tried to scare the White House with its unpredictability and readiness to undertake decisions that seemed irrational and even suicidal.

Soon, Pyongyang took the first step. In March 1993, in response to the International Atomic Energy Agency's (IAEA) demand for North Korea to provide inspectors with access to suspicious installations in Yongbyon, the regime declared that it was suspending its membership in the Treaty on the Non-Proliferation of Nuclear Weapons. In response to a concession from the United States and its consent to start direct negotiations with the North, Pyongyang put its decision to quit the treaty on hold. A game of nerves began, from which the North Koreans finally emerged victorious.

The results of this struggle are now well known. In October 1994, the Agreed Framework was signed, followed by the creation of the Korean Peninsula Energy Development Organization (KEDO) consortium now engaged in the project to supply light-water reactors to replace Pyongyang's planned heavy-water reactors, which were obsolete in their safety design and dangerous from the point of view of their military potential. In a front-page article prepared by Kim Jong Il's staff and published by the three central North Korean newspapers in January 1995, the signing of the Agreed Framework with the United States was described as a "breakthrough event" and as North Korea's greatest foreign policy success in 1994.[3]

Settlement Terms and the Road Ahead

Having expertly applied the tactics of nuclear blackmail and having obtained de facto recognition by the United States, the North Korean regime was far from aiming at just the modernization of its nuclear energy industry. First of all, Pyongyang has gotten a free hand to play a complex diplomatic game based on the open and hidden contradictions among the United States, Japan, South Korea, China, and Russia. The above analysis suggests that the U.S. expectation that it will be able to establish useful contacts with technocrats and potential reformers within the regime through the activities of the nuclear construction program is both unrealistic and even presumptuous. Washington is failing to take into account the fact that Kim Jong Il and his associates will never grant

Americans access to is truly secret military areas. (It did provide access to one suspected site in May 1998, where the U.S. found nothing.) All movements of the "imperialists" on North Korean territory will be strictly controlled by the ever-vigilant security service. The reason is that Kim Jong Il perceives the American "Trojan horse" as a major threat to himself and his regime. All "fundamental and visionary" works written by the North Korean leader since his father's death make clear that the regime intends to wage an incessant struggle for its self-preservation and will block any attempts by outsiders to inject an ideological virus that might launch a process of erosion in the existing totalitarian system.

In the coming years, full-scale efforts at exploiting the differences among the major powers active on the Korean Peninsula is the main field where Pyongyang will attempt to reap its harvest. It is hard to predict exactly how successful Kim Jong Il will continue to be in blackmailing the United States and in obtaining concessions. But a likely tilt to the left in Russian politics and the growing anti-American moods in Moscow and Beijing provide considerable hope to Pyongyang. North Korean diplomacy evidently counts on gaining certain advantages from the Russian-U.S. conflict over NATO expansion. Top Russian military officials make no secret that their response to NATO expansion in Europe may be a reinvigorated search for allies in Asia. It is clear that the key state they have in mind is the DPRK. North Korea might use this scenario to attempt to restore its disrupted military partnership with Moscow. In addition, Pyongyang retains the option of pursuing closer ties also with Japan, where quite favorable prospects could emerge in economic as well as other spheres.

Pyongyang can be expected to prepare new surprises aimed at evoking the fear of the West concerning possible unpredictable terrorist tricks by the dictatorial regime. To date, nuclear blackmail has yielded considerable fruit, and it will be difficult for Pyongyang to refrain from exploiting opportunities to use this approach again. It is likely to be careful in choosing the right time for such moves, which will be hard to predict from the outside. Yet, it is also worth noting that besides nuclear blackmail, Pyongyang has a whole array of other means of provoking its enemies and of getting what it wants out of future negotiations.

10

Military-Strategic Aspects of the North Korean Nuclear Program

Evgeniy P. Bazhanov

This chapter examines North Korean efforts to develop nuclear weapons and means for their delivery, as well as the impact of these military programs on the overall strategic situation in Northeast Asia. Since complete data on nuclear developments in North Korea are not available, this analysis relies on several sources: Russian and foreign news reports; eyewitnesses; estimates provided by Russian diplomats, military experts, and technical specialists; Russian archival materials; conversations with North Korean officials; and the author's own analysis of events.

Available information shows that North Korea, for a number of reasons, has made serious efforts to create a missile force and to equip it with nuclear warheads. Notwithstanding the October 1994 Agreed Framework with the United States on the nuclear program, the Democratic Peoples' Republic of Korea (DPRK) "missile development program" has not halted its development, as seen in the dramatic August 31, 1998, test of a three-stage *Taepodong*-1 missile. The first part of this chapter describes and analyzes the history and parameters of the North Korean nuclear program, followed by a similar analysis of Pyongyang's ongoing missile efforts, including its cooperative programs with other states. Given this background, the second section describes the past, current, and future impact of North Korea's nuclear and missile activities on the overall military-strategic situation in Northeast Asia. The evidence provided here shows that while the Agreed Framework's implementation is exerting a positive impact on the region, additional measures in the security field will be needed to ensure that North Korea does not become the source of future nuclear and missile threats to the Asian-Pacific region and beyond.

North Korean Achievements in Building A Nuclear Capability

According to Russian nuclear officials, by 1994, the DPRK had accumulated 7 to 22 kilograms (kg) of plutonium, sufficient for making one to three nuclear bombs.[1] The plutonium was produced and stored at the Nuclear Research Center located in the "special district" of Yongbyon.[2] In order to produce the plutonium, the North used the Soviet-made 2 megawatt (MW) IRT-2000 nuclear research reactor, which it upgraded through indigenous means to a 5 MW nuclear power reactor, as well as 0.1 MW Soviet-supplied critical assembly.

Although North Korea pursued its nuclear weapons program in several directions at once, the main thrust was focused on producing plutonium. The DPRK was also working to develop an independent uranium enrichment capability.[3] According to Russian Ministry of Atomic Energy (Minatom) officials, Pyongyang tried to acquire nuclear fuel and various nuclear technologies abroad—in both the West and the East. Reportedly, North Korea was to accumulate enough enriched uranium to be able to fabricate a number of small, crude uranium bombs, which it believed it might have the industrial and scientific capabilities to develop.

From 1991 to 1994, the North also exerted considerable energies to produce highly sensitive explosive detonation devices for use in nuclear warheads.[4] An explosives test site was built in the vicinity of Yongbyon.[5] Russian Defense Ministry sources claimed that more than seventy tests had been performed by 1994, when the program was shut down according to the terms of the Agreed Framework signed with the United States, which froze the country's nuclear program.

North Korean Missile Programs

The impetus to develop a missile program in North Korea stemmed from a desire to match South Korea's military modernization efforts. Initially, the DPRK attempted to design a missile with the cooperation of Chinese missile designers,[6] who had developed various types of ballistic missiles, including the CSS-4, in the 1960s and 1970s.[7] This program, however, fell short of initial expectations and was aborted in 1978.

Subsequently, in the late 1970s, the DPRK imported a number of Soviet-made *Scud*-B missiles. Using this technology, Pyongyang designed and tested a *Scud* Mod-A missile in 1984.[8] The *Scud* Mod-A had a range of 280–300 kilometers (km) and could carry a 1,000 kg warhead. Despite this achievement, the Korean People's Army (KPA) failed to adopt the *Scud* Mod-A,[9] and the missile was never deployed.

Instead, in 1985, the DPRK set out to develop a more advanced version of the Mod-A, the *Scud* Mod-B missile. Iran agreed to finance the project in exchange for large-scale supplies of the finished product. By 1987, North Korea had

Military-Strategic Aspects of the North Korean Nuclear Program 103

deployed the Mod-B and shipped about 100 of them to Iran.[10] The Mod-B missile had a range of 320–340 km and carried a 1,000 kg payload. In 1989, Pyongyang upgraded the missile once again, constructing one of the first Mod-C missiles (with a range of more than 400 km). In 1991, the new *Scud* Mod-C missile was exported to Iraq and Syria.[11]

At a production rate of approximately 100 *Scud*-C missiles per year, North Korea is reported to have produced approximately 600 missiles so far.[12] Of these, it has shipped about 500 to the Middle East and armed the Korean Peoples' Army (KPA) with the remaining 100 missiles. By some estimates, the KPA has two missile brigades.[13]

In the early 1990s, North Korea began to concentrate its efforts on developing a completely new missile with a much more advanced engine. With foreign financing and scientists and technology from various sources, the DPRK managed to manufacture two such missiles by 1993.[14] These missiles, referred to as the *Nodong*-1 by foreign observers, were first tested over the Sea of Japan on May 29 and 30, 1993. Though the missiles flew only 500 km, their size and other characteristics indicated that the *Nodong*-1 was capable of flying as far as 1,000 km. The *Nodong*-1 could carry a warhead of 1,000 kg, according to analysts. And, most important, it was the first North Korean missile capable of carrying a nuclear or chemical warhead.[15]

According to some Russian experts, however, the *Nodong*-1 was far from being ready for operational deployment, despite these tests. Serious problems with its engine design and performance, its accuracy, and its guidance system remained. If North Korea were to launch a *Nodong*-1 against a target in Japan, for example, there would be no guarantee that the missile would: (a) reach its destination; (b) start descending in the designated area; or (c) reach the surface and explode. One of the missile's major deficiencies was its lack of stability in flight: it had a tendency to roll while airborne.[16] Moreover, it was slow. While its development may have served some deterrent value, it was not appropriate for military use.

Sometime in the mid-1990s, it appears that North Korea accelerated its efforts to develop yet another family of indigenously produced missiles: the *Taepodong*. Given the obvious limitations of the *Nodong*-1, this missile aimed at crossing several technology thresholds in the area of solid-fuel, missile staging, reentry, and guidance systems.[17]

In an attempt to address the deficiencies of its domestic programs to date, North Korea began to explore the world market for missing elements in its technological design.Reports from the time suggested that Iran was providing financing for research on this intermediate-range missile, with some technical assistance coming from China as well. Indeed, experts pointed to the similarity in features between the Chinese CSS-2 missile and the initial versions of *Taepodong*-1.[18]

As evidence of this cooperative project, DPRK Air Force Commander Cho Myong-Rok visited Iran in early spring 1994. He inspected the Iranian missile test site at Shahroud and allegedly discussed future testing of a new generation of missiles.[19]

Moreover, North Korea reportedly began to recruit missile and nuclear experts from the former Soviet republics. For example, in November 1993 North Korean Major General and Counselor at the DPRK Embassy in Moscow Nam Gye-Bok was expelled from the embassy for attempting to recruit Russian scientists to work in North Korea.[20]

The same month, the Russian Pacific Fleet signed a contract with a Japanese trading company for the delivery of *Golf* II-class submarines capable of carrying ballistic missiles, ostensibly for use as scrap metal. The submarines were sent to North Korea,[21] where they were most probably scrapped. However, Western defense experts noted that the submarines' launch tubes could be adapted to fire a modified *Nodong*-1 missile.[22] In March 1994, Western news media reported that Russian counterintelligence had arrested three North Korean diplomats for illegally attempting to purchase new types of weapons. Some Western reports claim that a large number of Russian scientists unable or unwilling to move to the DPRK continued to assist the North Koreans in the implementation of their missile program by sending their research results and other valuable information to the North by electronic mail.[23] Russian Defense and Foreign Ministry officials, however, have dismissed these allegations.[24]

In August 1998, North Korea shocked the world by breaking out of the test moratorium it had observed since 1993 and firing a three-stage *Taepodong*-1 missile across the Sea of Japan and over the Japanese main island of Honshu. The missile traveled 1,646 km and released a small satellite, which failed to achieve orbit.[25] The first two stages of the missile were liquid-fueled, but the third stage was reportedly a solid-fuel booster. Although it appears that the third stage failed, the accomplishment of the first two stages was impressive enough. Western experts have speculated that the missile has a potential range of some 2,000–2,500 km,[26] and could lead to the development of a 4,000-km *Taepodong*-2 missile. Russian military experts are more skeptical of North Korea's capabilities to move quickly beyond the August test to longer-range systems. Senior Glavkosmos expert Gennadiy Khromov argues that North Korea lacks the technical expertise to jump the next set of hurdles for a multiple-stage rocket of intercontinental range.[27] He also denies that Russian experts played any role in the North Korean missile program, but did not rule out Chinese participation.

There are also reports that as part of the ballistic missile program, the DPRK is producing a variety of mobile launchers and transport and support vehicles. Various types of machinery and equipment, including Russian-made MAZ-543 trucks, are being imported from Russia, Japan, Austria, and Italy.[28] Reportedly,

Military-Strategic Aspects of the North Korean Nuclear Program 105

the DPRK is producing its own mobile launchers and transport support vehicles from designs of Fiat trucks with chassis and cranes made by the Austrian company Palmfinger.[29] They also are said to be purchasing Japanese Nissan trucks and redesigning them as *Scud* mobile launchers.[30]

Regarding North Korean missile exports to the Middle East, Russian and foreign experts note that Pyongyang initially delivered completed systems to these clients.[31] Presently, parts are shipped separately and the missiles are assembled by the customers themselves. Because of Western opposition to the missile trade, Pyongyang now delivers its weapons systems by transport aircraft rather than by ship. According to Western experts, in some instances, "This was accomplished with the assistance of the Russian private sector, which calls into question the Russian government's ability and/or willingness to control the DPRK's missile proliferation."[32] However, officials at Rosvooruzhenie, which supervises Russian military exports, categorically deny any role of the Russian private sector in the development of North Korean missiles. Indeed, they insist that Moscow is against providing assistance to Pyongyang in this field and that no organization inside Russia can ignore this regulation.[33] Still, delivery of missiles and missile parts by air is much harder for the international community to detect and control.

Given the available evidence from Russian sources, and taking into account the strengths and weaknesses of North Korea's economy, its geography, and its military potential, it seems safe to predict that the DPRK will require no less than fifteen years to develop an intercontinental ballistic missile (ICBM) capable of carrying a nuclear warhead. Even this prediction assumes that it will find adequate financing for such a program, which is not a foregone conclusion.

Regional Implications of North Korea's Nuclear Developments

The attempts of the early 1990s by Pyongyang to lay the foundations for a nuclear capability were followed carefully by the United States, South Korea, Japan, and by the International Atomic Energy Agency (IAEA). The U.S. government alleged that Pyongyang had been developing a secret reprocessing facility in the vicinity of Yongbyon since 1986. According to the United States, North Korea was trying deliberately to avoid IAEA safeguards.

Pyongyang, however, refuted these accusations, claiming that only civilian nuclear facilities—safeguarded and inspected by the IAEA—were operating in Yongbyon. Washington discounted the North Korean explanations as false and stepped up its demands that Pyongyang admit IAEA inspectors and place all of its facilities in Yongbyon under international safeguards. The Russian government, however, was too preoccupied with its various internal crises to pay much attention to such obscure developments in the DPRK. Thus, while certain quarters in Russia's Minatom and the defense and intelligence communities did report on nuclear developments in North Korea, the Russian government issued no official accusations at the time. It is important to note that many defense and foreign affairs officials did not believe (and do not today) that Pyongyang was actually attempting to develop its own nuclear weapon. Skeptics claimed that the Americans had invented these allegations mainly to exert pressure on an unreformed communist state. The North Koreans, they believed, were simply using the U.S. preoccupation with the nuclear issue in order to extract political and economic concessions from the West.[34]

Despite having joined the Treaty on the Non-Proliferation of Nuclear Weapons (NPT), however, Pyongyang had only signed a safeguards agreement with the IAEA in 1991 and only in 1992 had they finally allowed the initial IAEA inspection teams to enter the country. By February 1993, the IAEA had conducted six rounds of ad hoc inspections. These missions led to the following conclusions:

- A huge radioactive chemical laboratory, which looked like a reprocessing facility of factory size, was found to be under construction at Yongbyon. This was a violation of the joint denuclearization declaration of the two Koreas, agreed to on December 31, 1991.
- Plutonium obviously had been extracted from the nuclear reactors outside of official, IAEA-sanctioned refueling several times.
- The nuclear reactor in Yongbyon was similar to the model used in Chernobyl, and thus deemed unsafe.
- The IAEA inspection teams were being denied access to two unreported nuclear waste sites suspected of being part of the DPRK military nuclear program.

The IAEA concluded that there were some "significant inconsistencies" between the initial report presented by North Korea and its own assessment of the situation. The IAEA demanded an inspection of the above-mentioned undeclared facilities reported to be waste storage sites for nuclear reprocessing.

Pyongyang adamantly refused to grant this permission. In February 1993, under direct pressure for the United States, the IAEA Board of Directors passed a rather tough resolution demanding its right to conduct a special inspection at the suspect sites. The DPRK responded on March 12, 1993, by announcing its plan to withdrawal from the NPT, "for the sake of protecting the national sovereignty of the DPRK in protest against the resumption of the joint Republic of Korea (ROK)–U.S. 'Team Spirit' exercises and imposition of the IAEA special inspections at military facilities."[35]

As a result, even the very limited attempts by North Korea to develop a nuclear capability heightened international tensions in the Far East in 1993. Large-scale U.S.–South Korean Team Spirit maneuvers, intended to rehearse military operations against the North in a simulated nuclear conflict, resumed. In addition, the U.S. Congress and the U.S. public raised the possibility of delivering a preemptive surgical strike against North Korean nuclear facilities. Some members of the U.S. Congress urged the United States to impose tough economic sanctions against the DPRK, including a total embargo on oil supplies, industrial raw materials, and food in order to compel the DPRK to return to the NPT.

Moreover, Pyongyang's actions caused relations between North and South Korea to deteriorate to a dangerous level. Seoul once again felt it necessary to build up its military potential and prepare itself for a worst-case scenario on the peninsula. Meanwhile, Japan aborted its overtures toward the North and refocused its efforts in the defense field. Another consequence of Pyongyang's nuclear activities came with Russia's decision to halt all forms of military and nuclear cooperation with the DPRK. Thus, due to its nuclear ambitions, dark clouds started to gather around Pyongyang.

Nevertheless, in the final analysis Pyongyang's nuclear gamble eventually paid off. In October 1994, the DPRK signed the Agreed Framework with the United States, entitling it to sizable dividends in the military-strategic, political, and economic spheres. This pact obligated the United States: (1) to establish diplomatic ties with North Korea; (2) to renounce the threat of use of nuclear weapons against the North and to extend guarantees of nonaggression; (3) to refrain from undermining the existing regime in the North; and (4) to establish an international consortium to replace North Korea's planned graphite-moderated reactors with light-water reactors, while compensating Pyongyang for its energy losses of the DPRK in the meantime through the yearly provision of 500,000 tons of heavy fuel oil.

For its part, North Korea agreed to remain a party to the NPT and allow implementation of a safeguards agreement to freeze and, eventually, dismantle its graphite-moderated reactors. Pyongyang consented to replacing its planned gas-graphite reactors with light-water reactors. Moreover, North Korea agreed to accept renewed regular and special IAEA inspections of those nuclear facilities that will continue to operate once the shipments of the light-water reactors are completed.

It is important, however, to remember that the DPRK's compliance with its obligations is conditional upon the implementation of the Pyongyang-Washington agreement as a whole, and that the IAEA inspections have been postponed for at least five years. This in itself may be perceived as a continuing violation of North Korea's IAEA safeguards agreement.

Despite this accord, one cannot rule out the possibility that the DPRK will continue its efforts to develop a nuclear weapons capability. Given the closed nature of the North Korean state and society, Pyongyang will be able to continue research, technology production, and importation of materials under the veil of secrecy. The regime is concerned enough about its future that it will try to maintain its nuclear capability as a means of ensuring its survival.

As with other spheres, it is not easy for North Korea to make concessions in the nuclear realm. Such concessions lend themselves to the very real possibility that North Korea might fall into economic and, eventually, political dependence on South Korea and the West and will be compelled to expose its populace to foreign ideological influences. This would call into question the very existence of the communist regime.

Regional Security Implications of North Korea's Missile Programs

Although there has been frequent international criticism, especially from Seoul, Tokyo, and Washington during the past several years, foreign criticism of North Korea's missile programs has reached a new crescendo since the August 1998 *Taepodong*-1 test. Now that Pyongyang has tested an intermediate-range missile, it can now threaten (besides South Korea) all of Japan, U.S. bases there, and even, in theory, Beijing and the Far Eastern region of Russia. Continued North Korean efforts to develop a missile capability may well spark a strategic arms race in Northeast Asia. In response to Pyongyang's actions, the United States and Japan have accelerated their joint research efforts to develop theater missile defenses. Japan has also announced that it is going to launch a fleet of four spy satellites, so that it can better track North Korea's efforts in the missile field independently. This has given rise to fears of Japanese remilitarization throughout Asia. Indeed, both China and Russia have denounced the joint Japanese-U.S. Theater Missile Defense (TMD) program as potentially threatening to regional security and as another rung upward in the regional arms ladder.[36] Meanwhile, the North Korean test may stimulate South Korean or Japanese acquisition of longer-range missiles capable of striking deep into North Korean territory. If the DPRK succeeds in its efforts to acquire a sizable missile potential, South Korea and Japan will likely follow suit.

But North Korea is being strongly pressured by its sole ally, China, not to deploy the *Taepodong*–1 or to proceed with further tests. China is not eager to see the United States' new weapons deployed into East Asia, which may at some point be directed against Beijing. In the early 1990s, China reacted in a similar fashion to the *Nodong*-1 test. One analyst pointed out that the deployment of that missile would "provoke a new cycle of arms racing not only between South and North Koreas, but also among other countries of Northeast Asia."[37]

Thus, despite its eagerness to acquire new missile capabilities, North Korea is going to find itself surrounded by opponents if it proceeds with this test program and deploys large numbers of missiles.

Conclusion: Prospects for the Agreed Framework

As far as positive prospects for the Korean Peninsula are concerned, it should be emphasized that if the U.S.–North Korean agreement is implemented success-

Military-Strategic Aspects of the North Korean Nuclear Program 109

fully, it will open up new opportunities for cooperation between North and South Korea. For example, the installation of ROK-made nuclear reactors in the North will allow Seoul to play an important role in the overall development of the DPRK's civilian nuclear industry. Through this process, the nuclear problem in North Korea could be solved in a more complete fashion and in less time. Additional opportunities to reduce tensions and strengthen peace and stability on the peninsula will likely result. It would be politically prudent for the outside powers to grant Russia a role in the process by allowing it to export equipment for the DPRK's nuclear industry, as Pyongyang requested early on in the negotiations over the Agreed Framework.

Similarly, it is important to put into practice the joint denuclearization declaration agreed to by North and South Korea in December 1991. This declaration stipulates that both Koreas shall: (1) refrain from testing, producing, importing, possessing, storing, deploying, or using nuclear weapons; (2) refrain from building and possessing nuclear fuel reprocessing and uranium enrichment facilities; and (3) provide for the implementation of these agreements by forming and regularly convening a Joint Nuclear Control Commission. It is also necessary to ensure North and South Korean participation in the Chemical Weapons Convention, the Missile Technology Control Regime, and the U.N. Register of Conventional Arms. Through these efforts, both the DPRK and the ROK could contribute significantly to the nonproliferation of weapons of mass destruction, ballistic missiles, and other modern weapons systems and military technology.

11

Leadership Politics in North Korea and the Nuclear Program

Roald V. Savel'yev

By the late 1990s, Russian analysts have seen their predictions of the early 1990s about Kim Jong Il's likely consolidation of power become a reality. Indeed, some argued that the transfer of all powers over the Workers' Party of Korea (WPK), the government, and the military had already largely taken place before the death of Kim Il Sung in July 1994. Formally, however, the rituals of North Korean politics required that an appropriate mourning period must pass before his son could pick up the formal accoutrements of power. To the bewilderment of many Western observers, it took until October 1997 for Kim Jong Il to be appointed as general secretary of the WPK.[1] Still, Western observers wondered when and whether Kim Jong Il would rise to the top position of power. Finally, in September 1998, the Tenth Session of the Supreme People's Assembly named Kim junior to the position of chairman of the National Defense Commission, which was to be treated as North Korea's head of state after the official "retirement" of Kim Il Sung's former position as state president.[2] Thus, after four years, the predictions of Russian analysts regarding the full transfer of power to Kim Jong Il have finally been realized.

By contrast, the death of Kim Il Sung on July 8, 1994, generated numerous predictions in the Western news media about the seemingly inevitable collapse of the North Korean system—its economic, social, and political foundations—after the Great Leader's departure from the political arena. Experts predicted that the new regime would not last for more than one month, or perhaps a year at most. Later, these predictions expanded to three to four years. In the meantime,

U.S.–DPRK negotiations regarding the nuclear problem, a topic of interest to both parties, continued to proceed.

There were other predictions, too, but these were rarely considered seriously. The most radical predictions were made by the South Korean news media and even some acting politicians. The latter have long been eager to see their predictions of North Korea's demise come true. In Seoul, there is a widespread view that the North Korean regime is on the brink of disintegration, which will be followed by mass unrest. South Korean government ministries and agencies, including the Ministry of Defense, are said to deliberate not only about the prospects for unification but also about the measures needed to prevent the "invasion" of the South by the starving populace of the North, which, they believe, is likely to attempt to break through the defensive barriers erected south of the demilitarized zone (DMZ). They are reported to be especially concerned about the fate of Seoul. Due to its proximity to the DMZ, Seoul might become the destination of choice not only for ordinary people from the North but also for North Korean medium- and low-ranking cadres who have been suffering from shortages in the North as badly as the laymen and might want the plentiful South to share all the benefits of its current lifestyle with them. While South Korea's recent economic troubles have reduced some of Seoul's hubris regarding the North and lowered the perceived desirability of a sudden collapse, these views are still prevalent.

While not entirely accurate, such fears as those expressed in Seoul are not completely groundless. Conditions of permanent economic decline have become a normal part of the DPRK's domestic life, leading to the natural assumption that there must be political unrest. Even when Kim Il Sung was alive, North Korean society witnessed the emergence and disappearance of various opposition figures and groups at different stages of his rule. Under current conditions, however, one cannot detect any strong or well-organized opposition group to the ruling regime emerging spontaneously. The general political situation in the country may be judged as stable in the sense that the administrative *apparatchiks* at the central and local levels monitor the situation closely, keep it under tight control, and function in their customary mode of operation. Organizing any dissent at the local level is extremely difficult because of enduring and omnipresent totalitarian controls and the operations of the DPRK's repressive security organs. Also, it is hard to find any force within the North Korean leadership itself that is interested in any unpredictable unfolding of events. None of these people who are part of the ruling elite, are close to it, or are seeking to join its ranks, are interested in any escalation of social and political conflicts, since they all fully understand the inevitability of purges and repression against party cadres should they lose power, similar to those that followed dramatic social and political transformations in the East European countries. The same views are shared by many cadres employed at the middle and lower levels of the administrative and party organs, not only in Pyongyang but also in the provinces.

Given this background, the current chapter provides a detailed analysis of internal political conditions within the DPRK's ruling elite. Drawing on the author's long familiarity with many of the individuals involved, the chapter also provides information on the key interactions among these elites. Finally, it considers how the Agreed Framework and the nuclear issue fit into this political context, examining who are the potential winners and losers if the deal is a success or a failure in the coming years.

The Disposition of Forces with North Korea's Top Leadership Circles

At present, one can state that, contrary to previous predictions, the North Korean regime has survived the worst systemic crisis that the country has faced since the collapse of socialism in Eastern Europe and the Soviet Union from 1989–91. Moreover, would dare to predict with any great deal of confidence exactly when any dramatic transformations are likely to be initiated in the DPRK and whether or not these will come in the widely expected forms. In this connection, one may ask what factors continue to sustain the viability of the North Korean regime under the present difficult conditions and diminish the possibility of popular unrest, public dissent, and social anarchy.

Examining the available evidence, one of the factors that ensures the continuous survival of the regime is the very peculiar and well-built cadre system that Kim Il Sung managed to set up in the country with the assistance of leaders and his comrades in arms, who consented—voluntarily or otherwise—to his permanent leadership of the party and the state.

Over the course of several decades, Kim Il Sung perfected the efficiency of the recruitment, training, and placement of the party and government cadres that were designed to lay the foundations for the preservation of the existing political direction and of continuity within the top leadership. Today, when some time has already passed since the death of the first "Great Leader," one can state that it is this cadre system that has become one of the key factors that continues to guarantee the firmness of the positions of the North Korean elite.

Since July 1994, the situation in the highest echelons of power in the DPRK has remained essentially the same and has been controlled by a group of leaders who agree that during this very difficult transitional period—in accordance with the strategy devised by the deceased Great Leader—Kim Jong Il must be the supreme state authority in the country. This, by itself, will preserve the equilibrium in society and protect it from unnecessary disturbances. The recent decision to end the period of continuous mourning for Kim Il Sung and to preserve his presidency in perpetuity is another means of keeping his legacy—and thereby Kim junior's authority—alive.

It is worth examining the internal politics that led to these decisions, however, because they are not as simple or as well planned as they first appear. Prior to and after the WPK's Sixth Congress, Kim Il Sung worked persistently to prepare the populace, the armed forces, the cadres, his close associates, and relatives to accept without delay and hesitation his judgment as to who should succeed to supreme power in the DPRK, both formally and in substance. Bearing this goal in mind, he made an unprecedented step by bringing back from exile and nominating as one of the vice presidents his younger brother, Kim Yong Ju, who was known to have had conflictual relations with Kim Jong Il and had been previously persecuted. Apparently, all the efforts by Kim Il Sung to ensure a smooth succession while he was alive proved to be insufficient for his son after his death. It was simply not possible, and even unseemly, for his son to take over fully and without any impediments all the supreme posts that had been previously occupied by his father. It appears that the tense economic situation, the necessity of choosing a course of further development, the formulation of new foreign policy priorities, and the desire to find the executors for optimum relations with the South, the United States, and other countries, compelled the ruling elite and Kim Jong Il himself to amend Kim Il Sung's designs and search for forms of governance that would better suit the new domestic challenges and structures of supreme state power in the DPRK and be more palatable to the outside world.

Against the background of the continuing food shortage and declining living standards of practically all strata of the population, the announcement of the election of a new president was unlikely to increase Kim Jong Il's popularity or bring any additional gains to himself or his supporters. Correspondingly, they still had to resort to an old tactic adopted by the ruling elite a long time ago: that is, to continue to prepare the North Korean populace and low-ranking cadres psychologically for future changes.

According to the DPRK's 1972 Constitution (as amended in 1992), the president is elected by the Supreme People's Assembly (SPA) of the DPRK. But the constitution did not specify the procedure in the event of retirement or death of the head of state. When Kim Il Sung was still alive, such a provision could not even be mentioned in the constitution. This trick allowed the endless postponement of the final resolution of this question. It was deliberate that the SPA itself was not convened in the first four years following Kim's death. Formally, legislative functions were performed by the Standing Committee (Presidium) of the SPA; however, in reality, they were performed by the Central People's Committee and Administrative Council (the government) of the DPRK. Finally, new elections were held in July 1998 and the tenth SPA was convened in September. To the surprise of many analysts, the tenth SPA adopted a new constitution, which helped to complete the transition of Kim Jong Il to power and institutionalize his authority.

It is worth noting that the question about another, no less important, post of the WPK's Central Committee general secretary had been resolved in Kim's favor a year earlier, helping to lay the groundwork for the events of September 1998. In October 1997, an all-national party conference convened and elected Kim Jong Il to the new post of general secretary of the whole WPK (not the Central Committee). This left the precedent that Kim junior was leaving the position of general secretary of the Central Committee open, as a means of both broadening his support within the party and of showing his loyalty to the legacy of his father. This move also might have been intended as a signal to potentially assertive local party organizations that Kim Il Sung intended to represent the whole party directly (that is, not through the Central Committee). In this way, Kim junior's decision to put aside the Central Committee position could be seen as a well-calculated preemptive move, while simultaneously diminishing the importance of the Central Committee apparatus itself as a seat of power and potential opposition.

Official Plenums of the WPK Central Committee, notably, have not been held since December 1993. Nonetheless, North Korean propaganda continues to stress that Kim Jong Il pays special attention to the smooth functioning of party organs, including the apparatus of the WPK Central Committee. There was a time when Kim Jong Il was not "the Great Leader" but was rather referred to as the "Center of the Party." This highlighted the necessity to rally all party cells around Kim Jong Il and for the party cadres to fulfill his instructions conscientiously, despite his young age.

Compared with many of the North Korean elite, Kim Jong Il can still be seen as a relatively young statesman (he was born in February 1942). Meanwhile, North Koreans believe that age represents a measurement of mental capacity and largely determines the degree of respect due to any person, especially one who claims the right to be recognized as a civilian or military leader. Indeed, because of his relative youth, Kim Il Sung once experienced difficulties in communicating with the populace, his party associates, and South Korean authorities. Within the WPK, those difficulties often escalated into conflicts that were labeled as factional struggles and often resulted in harsh repressions of those who laid claim to power. Today, elites remember well that past intraparty disputes always led to disturbances, which would be especially dangerous under conditions of instability on the peninsula and risk the possibility of "external forces" interfering with the country's domestic affairs. With this threat in mind, today's North Korean elite is searching for means of ensuring its own survival.

In this situation, North Korean leaders, including Kim Jong Il himself, are not inclined to shake up the senior government and party personnel drastically. They prefer to rely on the skillful and well-tested cadres of the past. Kim Jong Il's credo in this regard is best represented by an article written on his behalf by the WPK's Propaganda and Agitation Department and published in the North Korean news media in December 1995, called "To Respect the Senior Generation of the Revolution Is the Highest Moral Duty of Revolutionaries."[3] The article was aimed at both domestic and foreign audiences. Kim Jong Il's purpose was to attract to his side the statesmen of the elder generation who had been in power for many decades and had many supporters in the Central People's Committee (CPC), Administrative Council, the SPA, and in the "power structures" and party organs. Kim Il Sung's personal experience was steeped in this tradition. Namely, since the late 1970s when Kim Il Sung had already consolidated his leadership positions, he did not block the emergence of different age-based groupings. He controlled them and pitted one against another, thereby enhancing his own personal authority and powers. Only time will tell whether Kim Jong Il will be able to use this tactic as skillfully as his father once did.

Groups within the Top Leadership

Based on this line of analysis, one can distinguish the following groupings within the North Korean leadership. Despite their very old age, within the DPRK's leadership one can still find functioning conservative veterans such as Honorary Vice-Presidents Pak Song Chol, Lee Jong Ok, Kim Byong Sik, Jon Muu Sop, and Kim Yong Ju (although the last is passive and has, in fact "retreated into the shadow").[4] In contrast, the highly influential former chief of Kim Il Sung's personal bodyguards, Lee Ul Sol,[5] the Chairman of the SPA Standing Committee, Yang Hyong Sob,[6] and WPK Politburo member Kye Un T'ye are constantly visible. At the same time, it has become obvious that due to their growing age, these statesmen have gradually relinquished their active duties and often attend to protocol functions only. Still, their opinions bear influence when Kim Jong Il makes important decisions regarding relations with other countries, including the United States, as well as on military and civilian aspects of the nuclear program.

Another group of officials actively influences the decision-making process regarding both domestic and foreign affairs and aspires to become indispensable around the future leader. Correspondingly, they work hard to portray themselves to both the population and to those within the party and government circles as people closely associated with Kim Jong Il. One such person is SPA President and former Foreign Minister, Kim Yong Nam.[7] Presently, complicated conditions demand that a considerable workload be placed upon the "brain center" of the governing team that was formed during Kim Il Sung's days. This extends to the executive apparatus of the WPK Central Committee and its secretaries, Kim Gi Nam, Kim Guk T'e, Choe T'ye Bok, Kim Yong Sun,[8] Han Song Ryong,[9] Chong Byong Ho, Pak Nam Gi,[10] and others. The last three secretaries of the WPK Central Committee are responsible for the development of industry, including the military-industrial complex, as well as for the development of the nuclear energy sector. Dr. Choe T'ye Bok, another WPK secretary, who had played only a secondary role within the WPK Secretariat from December 1993 until June 1994, has become active again. Currently, he is in charge of the advancement of the natural sciences and of academic and educational institutions. Within this team, Kim Guk T'ye and Kim Gi Nam, who have been dealing with matters related to domestic and the foreign policy public relations course chosen by the North Korean elite, are assigned to guide new internal political processes and confine them within acceptable limits. With the obvious support of Kim Jong Il and to the chagrin of other officials of the younger generation, Chang Song T'aek (Kim Jong Il's brother-in-law) visibly strengthened his position. He is currently the head of the organizational department of the WPK Central Committee and, due to his influence there, he has gradually gained control over other departments of the central party apparatus. Previously, he headed the Department of the Three Revolutions, which dealt with organization and propaganda work among the youth. His main objective there was to enhance the position of young Kim Jong Il. According to Japanese media reports, the rise of Chang Song T'aek propelled his elder brother, General Chang Song U, to the post of commander of the army district that constitutes the third line of defense of Pyongyang.

Several high-ranking officials within the North Korean leadership are assigned to handle foreign-policy issues. Among them, specifically among the party-industry *nomenklatura*, a spirit of competition bordering on confrontation often prevails. This does not bode well for their careers. However, even during Kim Il Sung's days, such a prevailing spirit was considered to provide good incentive for them to fulfill their duties. This group is led, without a doubt, by Chairman of the SPA Presidium, Kim Yong Nam. Alternate Member of the WPK Central Committee Politburo Kim Yong-sun, known for his ability to adapt quickly to new political realities, also supports new trends in the DPRK's foreign policy. Rich in experience in dealing with representatives of various countries, Kim Yong-sun is responsible for questions related to unification and inter-Korean dialogue. First Deputy Foreign Minister Kang Sok Ju, known for his activity in the U.S.-DPRK talks, presents considerable competition to both Kims. One may regard him as one of the leaders most inclined to improve North Korean relations with the West, primarily with the United States, with the aim of bringing economic and political gains to the DPRK. However, thus far, he has remained only an active executor, not a decision maker setting the tone in foreign policy making. A rising star within the DPRK's foreign policy establishment is Kim Yong Nam's protégé and newly appointed Foreign Minister, Paek Nam Sun. A professional bureaucrat, Paek is seen as a middle-of-the-road minister who is a reliable executor of the policies of the supreme leadership.

Representatives of other groups within the top North Korean leadership remain rather influential in the post–Kim Il Sung period, as well. The largest of these groups consists of the administrative-industrial *nomenklatura* that helped North Korea build the so-called *juch'e* economic system. This group of officials, considered to be middle aged, is more homogenous. Their intellectual capacities and educational levels are higher than those of the elder generation. Many representatives of this group received their educations in the Soviet Union, China, or East European countries. Despite harsh restrictions, they are familiar with methods of management and administration previously unknown to North Korea. Most of them have considerable experience in leadership positions at the center and at the local level. Among others, they include Yon Hyong Muk,[11] So Yun Sok, Kim Dar Hyon,[12] Ji Gil Song, and Choi Mun Song. Of this group, some were members of the WPK Central Committee Politburo and Secretariat, others used to head the Administrative Council and Combined Industrial Committees, and still others served as ministers and deputy ministers. When working in the periphery, they are responsible for top provincial administrative, economic, and party bodies. Their intellectual potential and influence on the general situation in the country have yet to be exhausted. At a given moment, any one of them can be returned to the highest posts in Pyongyang, following the long-established principle of rotation. Most of the above-mentioned high-ranking officials maintain close relations and kinship ties within the ruling clan and, apparently, are helping one another during the current transition period.

At present, leading members of the former Administrative Council (recently elevated by the Tenth SPA into a cabinet department of the DPRK) include Hong Song Nam and Choe Yong Rim. These men are considered to be the most powerful among the ranks of the administrative-managerial elite. The Tenth SPA named Hong Song Nam premier of the DPRK cabinet and appointed Choe Yong Rim as the DPRK's new prosecutor-general. Their well-recognized leader, former Premier Kang Song San, was considered especially close to Kim Il Sung and may maintain an amicable relationship with Kim Jong Il. He has traveled abroad extensively, including to Moscow, where he had informal meetings with then–Soviet President Mikhail Gorbachev. Recently, however, Kang Song San appears to have retreated from his duties and has ceased to take part in important official functions as the premier. According to the North Korean government, his health condition prevents him from actively attending to his responsibilities. Nevertheless, there are signs that suggest other reasons why he may have been deprived of the possibility to perform the full scope of the duties of the prime minister. The "escape" of his relative to South Korea and Mr. Kang's contacts with the former Soviet leader may be partly responsible for his being currently out of favor with Kim Jong Il. Moreover, he may disagree with other leaders on a number of issues, including the state of the North Korean economy and the ways and means of general cooperation with the United States and South Korea, especially on the nuclear problem. Kang Song San's duties during his absence have been placed upon Hong Song Nam, a very experienced administrator and politician, whose approach to matters of foreign economic cooperation with other countries is pragmatic. The Tenth SPA confirmed Hong Song Nam as DPRK premier.

Many influential officials within the North Korean leadership related to the late president and his son by kinship ties require special attention as well. According to South Korean sources there were, in 1993, about two dozen people in some way related to the two Kims that held significant party and governmental positions. To be sure, these officials are inclined to do their best to preserve supreme powers in the DPRK for themselves. Still, it is unlikely that they all regard themselves simply as obedient servants. This would be even more unlikely in the event of a dramatic transformation in North Korea's domestic situation. It is well known that Kim Jong Il has very tense relations with his stepmother Kim Song Ae and his stepbrother Kim P'yong Il. Their personal differences have crossed over into the political realm. Kim Yong Ju and Kang Song San, also considered to be related to Kim Il Sung by his mother's kin, harbor personal animosities as well.

At present, Kim Jong Il and his supporters are staking their fortunes on the North Korean military and other "power houses." The appointment of Admiral Kim Il Chol as defense minister has not changed the disposition of cadres within the top military brass. It has only confirmed the recent tendency of preserving the levers of command over the armed forces in the hands of the senior generals (this tendency emerged when Kim Il Sung was still preparing Kim Jong Il to manage the party and governmental organs). In the meantime, a process of natural attrition is under way.[13] It is to Kim Jong Il's benefit when key positions within the military are given to new experienced military leaders who have mastered new methods of commanding the troops both in times of peace and in times of war.

In 1996–97, the following generals rose relatively quickly through the ranks of command: Vice-Marshals Cho Myong Nok (chief of the Korean Peoples' Army [KPA] Main Political Administration),[14] Kim Yong Ch'un (chief of the General Staff),[15] Lee Ha Il (head of the Central Committee's Military Department), and Generals Park Chae Gyong, Nam San Nak, Kim Myong Guk, and Kim Ha Gyu. Vice-Marshal Lee Ha Il and the four generals always accompany Kim Jong Il on his visits to military units, which, according to local traditions, symbolizes the new "Great Leader's" personal confidence in them. Some of the above-mentioned military leaders held high-ranking posts even when Kim Il Sung was alive. Immediately following his death, Cho Myong Nok rushed to strengthen his position further by calling upon the military to "cement its loyalty" to Kim Jong Il. Some other generals from this "new" generation acted in the same vein.

As a whole, the top military brass occupy leading positions within the government and constitute the core cadre of such major influential power structures as the Central Committee's Military Committee and the DPRK's new National Defense Commission.[16] They are also represented within the WPK Central Committee Politburo and the DPRK cabinet. In addition to the "old guard," the new military appointees frequently involve themselves in the formulation and evaluation of agreements concluded between the DPRK and the United States over the Korean Peninsula Energy Development Organization (KEDO), and earlier agreements with South Korea (for instance, the agreement on the denuclearization of the Korean Peninsula). Their voice in these and other matters is no less influential than opinions of the leading politicians and economic officials.

The problem of increasing the efficiency of the government has become especially urgent in the post–Kim Il Sung period. North Korea must search for energetic cadres that remain mindful of new realities. Thus far, such transformations have taken place only in country- and city-level administrative structures that remain under central and local control. For example, the top- and middle-ranking personnel in Pyongyang's city administration and party establishments has been fully rejuvenated. Similar changes took place in departments and sectors of the Central Committee apparatus, as well as in the ministries and state committees. Representatives of the middle-aged generation were promoted to the first- and second-tier roles within the central civilian structures.

Among them, children of the senior North Korean party and government officials are promoted especially rapidly. The sons of Kim Ch'aek, So Ch'ol, Lim Ch'un-ch'u, Lee Ul Sol, and Choe Yong Gyong occupy responsible positions. In this way, a transfer of power to Kim Jong Il has been cemented by cadre continuity along this line, too.

According to North Koreans who defected to the South (including an air force pilot who fled in his fighter plane to the South), the privileged position of the graduates of the elite institutions—such as the Mangyongdae School, Kim Il Sung University, and Kim Ch'aek University—generates visible social polarization in the army and society at large. Such assertions are not groundless. Therefore, the *nomenklatura* and propaganda organs treat widespread concerns about social justice and egalitarianism seriously. The official news media often addresses this question as well. Indeed, one popular motto urges cadres of all levels to "Become closer to the masses!" On June 18, 1996, the newspaper *Minju Choson* wrote: "If leaders aspire to achieve a privileged position and fail to live in one breath with the population, then the ordinary people will not trust their leaders."[17] Measures are being undertaken with the aim of fighting the red tape within the party, the government, and the administrative-managerial apparatus.

All indications suggest that the present cadre policy will not be changed in the foreseeable future. Kim Jong Il has already become the fully recognized leader of North Korea. Nevertheless, he remains interested in the support of both the "veterans" and the middle-aged and younger generations of officials. Such a dynamic situation is essential for the ruling elite as well, since this complex entanglement of interests, more or less, preserves their current positions and appears to be instrumental in protecting the country from domestic turmoil and foreign invasion.

Attitudes toward Various Agreements on the Nuclear Problem

In the DPRK, all work and research related to the implementation of the nuclear program of any orientation is conducted under strict government control and with the cooperation of rather influential agencies.

Since the beginning of the 1990s, the main part of the North Korean nuclear program has been implemented by the Ministry of the Atomic Energy Industry (MAEI), a department based on the Committee on Atomic Energy (CAE) and established in December 1986 under the auspices of the DPRK's Administrative Council. The MAEI is headed by WPK Central Committee member Choi Hak Kun, an authoritative researcher and industry organizer. For a long time, Choe's deputy was Professor Pak Kwang O, rector of the Kim Il Sung University and researcher in experimental physics there.

The MAEI guides the work of North Korea's nuclear research centers and its nuclear energy industry, taking into account the country's need for a peaceful nuclear energy program and an adequate defense. Some analysts of North Korea argue that there was a time when Kim Jong Il was responsible for guiding the work of the CAE and later MAEI, as well as the major North Korean nuclear research center in Yongbyon. Using his influence and authority, these organizations were able to intensify their activities, ensure their priority status in material-technical procurement for the nuclear laboratories and design bureaus operating in Yongbyon, and secure the secrecy and security of their work.

Apart from the experts at the MAEI, it seems that in the DPRK there are various establishments and groups of officials concerned with the nuclear problem and interested in prospects for cooperation in the nuclear sphere with other countries and international organizations, including the United States and the KEDO consortium. Among these organizations are the government ministries and agencies in charge of the economy and planning. The North Korean military, including representatives from the air force, navy, and rocket forces, is also interested in the nuclear question. Another group of advocates is composed of researchers working, either directly or indirectly, with the nuclear program, as well as specialists from the electric energy industry. There are also some organizations and officials tasked with providing the nuclear industry with foreign technology and foreign scientific information to accelerate development in this field. Suborganizations include the foreign economic committees, the Academy of Sciences, and intelligence agencies. The management and engineering personnel of the uranium mining and refining enterprises occupy a rather isolated place in this network. Their views, as well as the opinions of the "miners" of foreign nuclear technologies, are taken into account only when technical and industrial production questions are considered and bear little impact on the formulation of major decisions in the nuclear sphere. These decisions are made by the Politburo, the Central Committee's Military Committee, and the new DPRK National Defense Commission.

North Korean economic officials and politicians appear to have already determined their positions on the prospects for further cooperation with KEDO and its member-states. They are aware of the economic and foreign policy gains that the DPRK could derive from improving cooperation with the United States, even if it means that Pyongyang will have to forgo its clandestine nuclear program. The participation of other countries in the KEDO project, however, is less attractive to the DPRK. If the light-water reactor (LWR) project is fully implemented as designed, however, the North Korean economy could boost its energy supply considerably.

Pyongyang fully realizes that the success of the KEDO deal rests upon the ability of South Korea, Japan, and the United States—the major financial contributors—to deliver on their part of the bargain. Having dealt with Seoul extensively, the North Korean leadership first refused to accept LWRs from South Korea, even though the reactors were said to be based on U.S. technology. Pyongyang's negative stance was not based solely on its ambitions. North Korea is conscious of the fact that South Korea could at any time use its participation in KEDO to exert pressure on Pyongyang. Such South Korean actions would likely be supported by the United States. Even if this support is not going to be unequivocal, Pyongyang will likely have to stand alone against its opponents in any negotiation. This fear was confirmed yet again in September 1996, when a North Korean submarine became stranded in South Korean territorial waters near Kangnung, and its crew was tracked down and killed by the South Korean army. This incident marked the beginning of a new cycle of intra-Korean confrontation and immediately delayed the resolution of issues related to KEDO. Moreover, every time a contentious issue arises, the United States tends to suspend contacts in other spheres, as well, including an agreement between Pyongyang and Washington to exchange liaison offices.

It is important to bear in mind that numerous practical issues have been left unresolved since North Korea signed the Kuala Lumpur agreement with the United States in June 1995 and concluded the LWR-supply contract with KEDO in December 1995. According to foreign news reports, as of the end of 1998, only seven of the required ten protocols had been signed, including protocols on privileges and immunities of the KEDO personnel, on transportation, on communications, and other matters. The central question over financing for the main project is still not fully resolved and formally documented. Thus far, heavy fuel oil has been supplied to the DPRK in irregular shipments. The numerous other problems that remain unresolved, however, simply serve the purposes of those within the DPRK who are skeptical of U.S. and South Korean willingness to adhere to the Agreed Framework.

Conclusion

There are several factors that may be of concern for both the opponents and advocates of the DPRK's cooperation with the United States, including in the nuclear sphere. These are as follows:

First, a majority of the North Korean population greatly resents the United States. Many people believe that the United States is the greatest obstacle to Korean unification and a staunch ally of the "puppet authorities of the South." From this point of view, any cooperation with the United States will either be fruitless or insincere. The North Korean government, which echoes the same sentiments, cannot help but take into account this popular attitude. Those favorably disposed to the idea of cooperation with the United States are, in the event of a worsening situation in the DPRK, likely to be blamed and resented by various strata of the North Korean population. The DPRK has rich experience in this respect.

A second concern is that the DPRK has no real guarantees that it will receive modern LWRs that meet the advanced standards of technological safety. North Koreans believe that the United States deliberately deprived the DPRK of the right to select the type of the reactor and its country of origin not only because of financial considerations, but also because of the U.S. desire to prevent Pyongyang from accumulating any weapons-grade plutonium from an old-style reactor. Those who oppose rapprochement with the United States could legitimately argue that the DPRK's national interests were infringed upon not only on political and moral grounds but also because South Korea could easily manipulate North Korea in the future by limiting the supply of nuclear fuel and spare parts for the LWRs. Moreover, North Korea's prestige will suffer greatly in the eyes of those Koreans who believe that Pyongyang should work toward developing the industrial and energy potential of a unified Korea.

Third, the middle-aged generation of North Korean officials thinking about these prospects believes that the real gains from the LWR-transfer project might not arrive until at least 2003. As a result, they regard any interruption in the implementation of the project with a certain degree of reticence, because they consider it almost useless in the short term. At this point, if viewed in isolation and outside the context of broader relations with the United States, the LWR-transfer project causes only worries and inflicts material costs that are by no means covered by the KEDO contributions. Moreover, as time goes by, the North will have to repay the debts that will be incurred in the course of the LWR-transfer project (as well as the cost for the construction of the electricity grid, highways, and other infrastructure facilities). The DPRK has already amassed a huge foreign debt. Therefore, those North Korean officials who are in charge of the country's financial health cannot but worry about the prospect of additional future debts.

Fourth, full implementation of the KEDO project may have negative consequences for the North Korean regime, creating moral and economic dilemmas. On the surface, it appears as if the United States acted autonomously in engaging in direct discussions with Pyongyang, occasionally even diverging from Seoul's opinion. Nonetheless, U.S. government officials do not conceal their objective of transforming the social foundations of the North Korean regime under the guise of resolving the nuclear problem, which undoubtedly corresponds to the interests of South Korea. Since the beginning of the negotiation process with the United States and KEDO, the leaders in Pyongyang have been fully aware of this and intend to localize the negative consequences of future cooperation with the West.

Fifth, the military, primarily the generals and senior officer corps, has its own views on the nuclear problem and its role in U.S.–North Korean relations. Given the obsolescence of most types of armaments used by the KPA and the escalating military build-up in South Korea, the North Korean military elite appears to be in favor of removing the restrictions placed upon the DPRK by its agreements with the United States and of acquiring a nuclear weapon. Any such weapon of mass destruction would serve as a powerful deterrent to or check on the possible use of force by South Korea. At the same time, the North Korean military must take into account the current economic situation in the country and its negative impact on the military-industrial complex. Under present conditions, one can state that a majority of the senior military officials at the Ministry of National Armed Forces and the KPA are compelled to consent, albeit temporarily, to a redistribution of state funds in favor of developing a peaceful energy complex with the assistance of other countries, including the United States, as well as the capabilities of KEDO, if these prove to be real.

Sixth, the KEDO project, while primarily serving the interests of the United States, South Korea, and, to some extent, Japan, also affects the interests of Russia. This project may become disadvantageous to both Pyongyang and Moscow, in light of the 1985 intergovernmental agreement on Soviet assistance in the construction of nuclear power stations in the same Sinp'o area of the DPRK. This agreement has yet to be annulled. Construction of two nuclear power stations (i.e., both Russia's and KEDO's) within the same very limited geographical area is practically unfeasible and technically impossible from the standpoint of nuclear safety. One cannot rule out a rising fear in Pyongyang that KEDO's LWR-transfer project might hamper the stabilization of North Korean–Russian relations, especially when full-scale cooperation resumes in other directions as well.

Moreover, one can envision a set of circumstances under which senior North Korean officials may demand a more reserved and conservative approach to the LWR-transfer project and construction of the nuclear power plant in cooperation with the United States and KEDO.

These points highlight the importance of keeping track of the political groups that are both served and potentially hurt by the KEDO deal. The outside world should not simply assume that it is only dealing with Kim Jong Il, for there are in fact a number of other influential actors (albeit less well known to outsiders) whose views will be taken into account as the implementation of the deal (replete with problems) continues.

Part IV

The International Context of the North Korean Nuclear Program

12

North Korea's Decision to Develop an Independent Nuclear Program

Natalya Bazhanova

Any analysis of events in North Korea shows the importance of external factors in influencing (and changing) Pyongyang's security thinking in the late 1980s and in shaping its decision to launch a secret military nuclear program. This chapter begins with an analysis of the impact on North Korea of the Soviet Union's adoption of *perestroika* (or "new thinking") in foreign policy. It then examines Chinese influence, South Korean behavior, U.S. policy in Korea, and, finally, changes in Japanese attitudes, all of which helped trigger North Korea's interest in acquiring nuclear weapons. This study draws on Russian scholarly and governmental sources, as well as on official statements and policy pronouncements by North Korean leaders appearing in the North Korean press.

North Korea began to develop its nuclear industry with the assistance of the Soviet Union. In 1956, Pyongyang and Moscow concluded two agreements concerning nuclear research, and they signed additional protocols in 1959. All of these agreements limited cooperation between the two countries to the peaceful use of nuclear energy.

From the early 1960s to the early 1980s, Pyongyang's activities in the nuclear sector appeared within the parameters of its bilateral agreements with the Soviet Union and provided little evidence of an attempt to develop military uses for nuclear technology. There was no evidence that the Democratic Peoples' Republic of Korea (DPRK) possessed the scientific or material wherewithal to make such an attempt. Moreover, Pyongyang's security agreements with Moscow and Beijing provided an adequate nuclear umbrella for North Korea and reduced any desire it might have had to move in the direction of military applications. Although many Western (and even some Russian) analysts claim that the nuclear weapons ambitions of Kim Il Sung's regime developed in the late 1970s, neither their arguments nor the data they provide to back these allegations can—in the author's view—be substantiated. Even in 1985, when Moscow had to apply pressure on Pyongyang to make it join the Treaty on the Non-Proliferation of Nuclear Weapons (NPT) and consent to eventual full-scope inspections by the International Atomic Energy Agency (IAEA), the Soviet government did not appear to be aware of any steps by the DPRK to utilize nuclear energy for military purposes.

North Korea's military tendencies in the nuclear sector—according to the evidence presented here—developed sometime later, toward the end of the 1980s. However, no available North Korean document pinpoints Pyongyang's decision to embark upon a military nuclear program or even argues in favor of such a policy. Nevertheless, analysis of the general strategy of the North Korean leadership and of the political situation inside and around the DPRK strongly suggests that Pyongyang began to entertain these ideas in the late 1980s.

Perestroika and "New Thinking" in the Soviet Union

With Gorbachev's ascent to power in Moscow in 1985, Soviet–North Korean cooperation intensified, including in the development of peaceful nuclear energy. From the very outset of his rule, Gorbachev denounced inequality in relations among socialist countries and sympathized with smaller states' intentions to rely on their own resources and to have freedom of choice and action. He expressed readiness to increase economic aid to such partners and to extend his military, political, and moral support to them.

Moscow modified its attitude toward Pyongyang accordingly. Officials in the Kremlin stressed that the DPRK was a strategic ally that was extremely important to the national interests of the Soviet Union and security in the Far East.[1] Development of friendly ties with Pyongyang was declared as one of the priorities of Soviet foreign policy.[2] Moscow began to offer its support to North Korea more actively and openly. Soviet pronouncements, modeled on those issued from Pyongyang, were increasingly critical of U.S. attempts to form a trilateral Washington-Tokyo-Seoul military alliance and to include South Korea in closed economic groupings. Contacts between the Soviet Union and the DPRK in the military, scientific, cultural, and sports fields quickly intensified.[3]

Kim Il Sung reacted with satisfaction to changes in the Kremlin and to the new Soviet policies.[4] Pyongyang expressed its approval of Moscow's domestic and foreign policy initiatives and was very receptive to advice, opinions, and requests made by the Soviet leadership. Pyongyang upheld Gorbachev's initial foreign policy proposals in the Asian-Pacific region and agreed to the expansion of the Soviet military presence in the DPRK. Soviet–North Korean military cooperation reached such a magnitude that it provoked a response from China. In February 1986, the Chinese foreign minister warned that intensification of Soviet–North Korean military interactions could very well become "the fourth obstacle" in Beijing-Moscow relations. Pyongyang's approval of flights by Soviet military aircraft over North Korean territory was interpreted in China as a new step in Moscow's strategy to encircle it. The Chinese stated that Soviet naval visits to North Korean ports contributed to the deterioration of the political atmosphere on the Korean Peninsula and forced the United States to take countermeasures and increase its military presence in the Far East.[5] In their contacts with U.S. officials, Chinese leaders complained that the rapprochement between the Soviet Union and the DPRK would inevitably lead to increased belligerence on the part of Pyongyang and continuous deterioration of the political situation in the Asia-Pacific region.

But after Gorbachev's trip to Beijing in May 1989 and the resumption of normal relations between China and the Soviet Union, China withdrew its reservations regarding Soviet–North Korean military cooperation. The Chinese now approved of cooperation between Moscow and Pyongyang, insisting that the DPRK should rely on the Soviet Union and East European countries as potential sources of aid.[6]

Simultaneously, however, irritating problems developed with Soviet–North Korean relations: These were consequences of growing differences between the Soviet Union and the DPRK in various spheres. The following factors influenced Soviet–North Korean relations the most:

1. *Diverging foreign policy orientations.* The foreign policy positions of the Soviet Union and the DPRK began to head in separate directions on several key issues, including approaches to inter-Korean ties, the geopolitical situation in the Asian-Pacific region, conceptual views on the contemporary world, and the overall state of international relations. Pyongyang finally decided to reject Soviet "new thinking" as heretical and "dangerous for socialism." What troubled Pyongyang most, of course, was Moscow's refusal to support the DPRK on all issues automatically—to see Asia through North Korea's "eyes," so to speak—as it had in the past. In materials meant for internal propaganda and in conversations with foreign representatives, North Korean leaders began to make derogatory remarks about Gorbachev's foreign policy.[7] In turn, Moscow became critical of Pyongyang: it characterized North Korea's behavior in the international arena as "irrational and dogmatic." Experts in the Kremlin stressed that Pyongyang refused to recognize new realities and clung to obsolete notions of class and ideological struggle. Kim Il Sung progressively saw the outside world as a threat to his regime, to be rebuffed by all means possible.[8]

2. *Events in Eastern Europe.* The North Korean leadership was shocked by the events in Eastern Europe. Kim Il Sung and his associates were afraid that these events could be repeated in North Korea. Pyongyang was inclined to blame Gorbachev not only for welcoming East European revolutions but also for instigating them.[9]
3. *The democratization of Soviet society.* Pyongyang disapproved of the democratization of Soviet society, seeing it as a serious ideological challenge. Soviet public opinion progressively rejected the social model of the DPRK as Stalinist and totalitarian. Prior to the late 1980s, North Korea had enjoyed the support and understanding of the Soviet Union as a state professing similar ideological values. By the late 1980s, however, the Soviet Union no longer supported Kim Il Sung; the North Korean regime was considered a doomed anachronism. Had the East European scenarios been repeated in North Korea, they would almost certainly have been approved by the majority of the politically aware Soviet public.[10]
4. *The Soviet–South Korean rapprochement.* North Korea's irritation with Moscow grew as the Soviet–South Korean dialogue intensified. Commenting on Kim Yong-sam's visit to the Soviet Union in April 1990, the North Korean press called it "a hostile, criminal act."[11] Negotiations between Gorbachev and Roh Tae-woo in San Francisco in June 1990 were described as "unforgivable, criminal dealings" in Pyongyang.[12]

Other factors also contributed to the deterioration of Soviet–North Korean bilateral ties. In 1988, the Soviet Union had reduced its military aid to North Korea and afterward it rejected all of Pyongyang's attempts to strengthen cooperation between the two ministries of defense. The Kremlin stated to North Korean officials:

> The new Soviet arms supplies to the DPRK are likely to provoke the United States to extend similar assistance to South Korea. As a result, tensions on the Korean Peninsula are likely to worsen, thereby threatening the DPRK's security and doing harm to the DPRK's national interests.[13]

The volume of trade and economic cooperation between the Soviet Union and North Korea declined as well. In the past, Moscow, driven by strategic and political motives, had often sacrificed its interests to prop up its North Korean ally but, by 1988, the Soviet leadership had no desire or ability to do so. The Kremlin stressed that no matter how hard the Soviet Union tried to help its neighbor, the problems would not be solved unless North Korea reformed its economy, worked to end the confrontation and arms race under way on the Korean Peninsula, and shed its semi-isolationist policy regarding business contacts with developed countries.[14]

North Korea's Decision to Develop an Independent Nuclear Program

As a result of changes within the Soviet Union and the shift in Moscow's foreign policy, particularly its position toward the Korean Peninsula, North Korean authorities concluded that the Soviet Union was no longer an ideological, military, and political ally of the DPRK. Pyongyang could not rely on the Soviet Union as a guarantor of North Korea's security or as a source of the military and economic aid necessary to maintain the balance of forces on the Korean Peninsula. This new understanding became painfully clear by January 1990 after an extremely confrontational set of talks between Soviet Foreign Minister Eduard Shevardnadze and DPRK Foreign Minister Kim Yong Nam in Pyongyang. As a result, it became necessary for North Korea to find a replacement for Soviet security assistance and political support.

Disagreements and Policy Problems between the DPRK and China

Kim Il Sung's worries also deepened at this time because of changes in Chinese society and in Beijing's foreign policy. The Chinese leadership sounded more and more optimistic about the situation on the Korean Peninsula and did not consider it explosive. While the North Korean news media published articles about a growing threat of war and extreme tension produced by the actions of the United States and its allies in the region, Chinese newspapers opined that preconditions for détente existed and that the rival parties did not want conflict in Korea.[15] Beijing strongly recommended that Pyongyang redouble its efforts to find a path to reconciliation with the South and to build mutual trust and peace on the peninsula. The Chinese stressed the necessity for the DPRK to display restraint and patience. Beijing toned down its criticism of Seoul and its ally Washington. China opted for a neutral position on the 1987 incident involving the bombing of a South Korean civilian jumbo jet. The woman who was convicted of this crime admitted that she was a North Korean agent. However, Pyongyang categorically denied its involvement in this terrorist act.

Unofficial contacts between China and South Korea continued to grow by leaps and bounds and became more open than ever. Whereas in 1986 the volume of trade between the two countries was only $1.5 billion, by 1988, it had doubled. A number of South Korean firms were now investing in the Chinese economy. The Chinese media informed readers of a democratization drive underway in the South.[16] At the same time, Beijing and Pyongyang differed in their appraisals of socialist construction and did not feel enthusiastic about the economic practices of one other. They held different views on the United States and Japan, and Pyongyang was apprehensive about China's dealings with these countries.[17]

In the economic field, China remained the DPRK's most important partner after the Soviet Union. China provided for many basic needs of the country, including deliveries of cheap oil, extended financial assistance, and material aid. However, Pyongyang was not satisfied with the scale of cooperation, especially because it sensed Beijing's intention to reduce its aid to a minimum. China felt an aversion to economic ties with the DPRK because of its general economic inefficiency and the failure of the North Koreans to fulfill their contractual obligations.

So, on the whole, the North Korean leadership was losing faith in China's support for Pyongyang's regime and its foreign policy. The role of Beijing in meeting the military-strategic needs of the DPRK was not significant either. Moreover, if, in the past, North Korea could exploit Soviet-Chinese conflicts, now, with the growing rapprochement between the two giants, such possibilities were diminishing. Pyongyang no longer was able to extract concessions from Beijing by making gestures toward Moscow. Not only did the Chinese stop worrying about the prospects the Soviet–North Korean rapprochement, but they also acted as if it were conceding to the Soviet Union the right to be the principal partner of the DPRK, especially in the economic sphere. This situation began to trouble Pyongyang. The North Korean news media provided little coverage of Gorbachev's visit to Beijing in May 1989 and depicted it mainly as an event of solely bilateral significance.[18] Kim Il Sung got worried about a Moscow-Beijing "conspiracy" that involved improved ties with South Korea. Normalization of Chinese–South Korean relations in 1992 confirmed the worst fears of Pyongyang. The DPRK had now almost completely lost its second major ally.

The Position of the Republic of Korea

Throughout the second half of the 1980s, Pyongyang persistently advanced various military initiatives to Seoul. Among them, a special place was occupied by proposals concerning transformation of the Korean Peninsula into a nuclear-weapon-free zone. Before daring to begin development of a nuclear program, the North Korean leadership hoped to pressure their potential adversary, Seoul, to renounce the deployment of U.S. nuclear forces in the South and its plans to create its own atomic weapons.

Back in March 1981, the Socialist Party of Japan and the Worker's Party of Korea had signed a joint declaration calling for a nuclear-weapon-free zone in the area covering the Korean Peninsula and the islands of Japan. Later, in an interview with the Japanese journal *Sekai*, Kim Il Sung reiterated this proposal, calling for a total prohibition of nuclear weapons tests, production, deployment, and use on the Korean Peninsula and in Northeast Asia.[19]

On June 23, 1986, an official statement by the DPRK government confirmed Pyongyang's readiness to create a nuclear-weapon-free zone and thus not to test, produce, store, import, or transfer nuclear weapons on or via its territory. In return, Pyongyang asked the United States to stop shipping new types of nuclear armaments to South Korea, withdraw its nuclear weapons from South Korea, and renounce any plans to use such weapons in Korea.[20] In a statement dated July 14,

1987, the Ministry of Foreign Affairs (MFA) of the DPRK called upon the United States and South Korea to undertake practical measures aimed at removing U.S. nuclear weapons from the South. The MFA statement reiterated that the DPRK had no intention of building its own nuclear weapons or of deploying foreign nuclear weapons on its territory.[21]

Pyongyang stressed that having joined the NPT, it expected the nuclear states to take measures to introduce a nuclear-weapon-free status for the Korean Peninsula.[22] In November 1989, the DPRK proposed trilateral negotiations with the United States and South Korea to achieve an agreement on the withdrawal of U.S. nuclear weapons from the South and, then, through a dialogue between Pyongyang and Seoul, to work out and sign a declaration on turning the Korean Peninsula into a nuclear-weapon-free zone.[23]

However, Seoul and Washington rejected Pyongyang's initiatives on the nuclear issue. South Korea and the United States also did not express any enthusiasm about Pyongyang's other military initiatives, which they considered to be pure propaganda, but which the DPRK advanced quite seriously. Instead, as observed in the North Korean capital, Seoul continued vigorous attempts to build up its military potential. Seoul was launching a new program to substantially modernize its armed forces. In 1990–95, the South Korean government planned to spend more than $24 billion for these upgrades, including $4 billion on the acquisition and joint assembly of 120 modern fighter planes with the United States. Meanwhile, spending on U.S. troops stationed in South Korea rose from $2.2 billion to $2.6 billion.[24]

At this time, Pyongyang received intelligence information that South Korean authorities, behind Washington's back, were laying the scientific and technological foundations for launching production of nuclear weapons in the event of a military conflict.[25] While it is impossible to verify the accuracy of these reports (compared with more reliable information regarding a South Korean bomb program in the 1970s), the mere fact of their existence contributed to an increased sense of vulnerability in Pyongyang.

U.S. Policy in Korea

From the point of view of the North Korean leadership, the American threat to the DPRK had not lessened by the end of the 1980s; on the contrary, it had grown stronger. Pyongyang perceived Washington as getting dizzy with its recent successes in dismantling the socialist system in the Soviet Union and Eastern Europe and suspected that the United States would continue its offensive against socialism in the Eastern direction. The antigovernment uprisings in Beijing and other Chinese cities in the spring of 1989 increased Kim Il Sung's worries. The North Koreans, following Beijing's official line, treated the events in China as the result of premeditated actions by the United States and the West. Pyongyang also blamed Moscow and the Chinese leadership itself for opening China too much to Western influence.[26]

In any case, North Korean leaders were convinced that the United States was in an aggressive mood and would try to "finish off" socialism in the East. They also realized that international support for Pyongyang had slackened.[27] Such conclusions were confirmed by the behavior of Washington in the political and military spheres. Politically, the United States continued to ignore North Korea, refused to engage in a dialogue with it, and did everything possible to maintain the international isolation of the DPRK. Washington also intensified discussions with South Korea, Japan, its other allies, and even with the Soviet Union on how to promote reforms in North Korea. From Pyongyang's perspective, these talks translated into how to undermine and destroy the ruling regime in the DPRK. Various U.S. political leaders, news media correspondents, and academicians continued to attack and personally slander Kim Il Sung, his son, and the entire North Korean leadership.

Militarily, the United States continued to build up its naval and air force presence in the region, upgrade the qualitative level of its armaments, and engage in policy coordination with Japan's armed forces. Specifically, the United States refused to reduce its military personnel in South Korea and, for the first time, agreed to provide South Korea's armed forces with ultramodern technologies and its most sophisticated military systems. Supplies of conventional weapons from the United States to South Korea also increased significantly.[28]

Meanwhile, the U.S.–South Korean "Team Spirit" joint military exercises continued. The partners simulated intrusions into the North and "surgical" operations aimed at destroying the military potential of the DPRK. Other military games like "Milgon," "Focus Lens," "Toksuri," and "Double-headed Eagle" were organized as well. Commenting on the largest of these maneuvers, the annual Team Spirit exercises, the North Korean media stressed the extraordinary scale of these exercises in terms of the number of troops, variety of weapons, and transportation systems involved. North Korea was especially concerned because Team Spirit assumed the possible use of nuclear weapons on the Korean Peninsula.[29]

Pyongyang continued to ground its military planning on the assumption that the Pentagon maintained nuclear weapons in South Korea. The North Korean news media time and again underlined the thesis that "The main firepower of the U.S. Army troops located in the southern part of the country consists of nuclear mines and warehouses with nuclear artillery shells and air-delivered bombs scattered around the country."[30]

To be sure, in the late 1980s, the United States took certain steps to relax tensions vis-à-vis North Korea. In October 1988, the United States lifted some of its sanctions against the DPRK, allowed scholars to visit Pyongyang, food trade between North Korean and private U.S. companies, and official contacts between low-level diplomats of the two countries.[31] Furthermore, U.S. and North Korean diplomatic officials held several meetings in Beijing during which they discussed the problem of the return of the remains of the U.S. soldiers who had died during the Korean War (1950–53). In Panmunjom, the U.S. representative in the Korean Military Armistice Commission proposed to discuss confidence-building measures in the demilitarized zone.[32] Nonetheless, Pyongyang considered these U.S. initiatives as nothing more than maneuvers to cover up the true aims of Washington—the destruction of the DPRK through political, ideological, economic, and military methods. From Pyongyang's standpoint, the U.S. threat kept growing.

Japan's Policy on the Korean Problem

In the late 1980s, the North Korean leadership was becoming increasingly concerned about growing South Korean–Japanese military cooperation. The Japanese Defense Agency had dispatched representatives of its army, navy, and air force to South Korea to inspect defense installations, familiarize themselves with military research done in the South, and observe the U.S.–South Korean Team Spirit joint war games. Japan discussed various possibilities of joint production of weapons with South Korea. Certain types of military supplies began to arrive in South Korea from Japan.[33] They initiated exchanges of navy ship calls and contacts between military academies. The two sides started negotiations on Seoul's participation with the Japanese navy in blockading international straits and in patrolling the sea lanes in the 1,000-mile zone around the islands of Japan.[34]

Pyongyang viewed all these activities as a manifestation of the "dramatic intensification of political, economic, and military rapprochement of South Korea and Japan" and as "the provocative policy of Tokyo, aimed at sharpening confrontation on the Korean Peninsula."[35] North Korea was especially worried about the prospects for the formation of a trilateral alliance with the participation of the United States, Japan, and South Korea. The theoretical journal of the Worker's Party of Korea, *Kulloja*, commenting on Japan's consent to allow the U.S. navy and air force to use Japanese facilities in case of war in Korea, stated: "The Japanese imperialists, induced by their overseas masters ... are getting bold again and are ready to burn and kill Koreans."[36] In September 1986, the North nearly panicked over the U.S.-Japanese maneuvers in the Korea Straits, when the two sides rehearsed operations to block the straits, amphibious landing operations, and strikes at surface targets. Pyongyang responded with belligerence to the trilateral (U.S., Japanese, South Korean) military exercises that took place in 1989 and 1990—"Pasex" and "Rimpac," respectively.

Some of Tokyo's gestures addressed to Pyongyang in the late 1980s only contributed to Kim Il Sung's growing fear that Japan had become actively involved in the U.S. efforts to "stifle socialism on Korean soil."[37]

Conclusion: Pyongyang's Rationale for an Independent Nuclear Capability

As a consequence of external factors in the late 1980s, the North Korean leadership came to the conclusion that it had to reconsider its nuclear-free military strategy. It was a matter of the survival of "the cause of socialism on Korean soil."[38] Pyongyang decided to tighten ideological and political controls over its own population and to make the "iron curtain" surrounding North Korea even more impenetrable. Kim Il Sung chose nuclear weapons as a new element in this strategy of survival.

The development of nuclear weapons was deemed useful for a number of reasons. First, an atomic bomb could serve as the most powerful and the cheapest deterrent against open aggression. In the latter half of the 1980s, Pyongyang considerably increased its conventional military forces—from 785,000 military personnel in 1984 to 1,114 million in 1992; and yet it still lagged behind Seoul in its military strength.[39] The only way to narrow the technological gap between the North and the South was to develop a nuclear capability. No matter how crude and imperfect, such a capability could scare away a potential attacker by threatening possible retaliation. It could also check those who might attempt to interfere in the internal affairs of North Korea in order to undermine North Korean socialism.

The leadership in Pyongyang was not original in espousing this logic. Stalin and his successors in the Kremlin followed the same logic as they spent billions of dollars to create a nuclear deterrent against the United States. British Prime Minister Margaret Thatcher used to say that nuclear weapons had saved Europe from another major war after 1945. French authorities advanced similar arguments in defense of their nuclear policies. The DPRK's leaders thus shared this respect for the atomic bomb.

Second, in Korea, the bomb could be useful for domestic consumption as well. Kim Jong Il, the son of the aging leader Kim Il Sung, allegedly took charge of the nuclear weapons program as a way of increasing his personal prestige in the armed forces and within the North Korean leadership as a whole. A nuclear weapon might help the entire ruling class in the North—by increasing tensions in the Far East and by distracting the population's attention from daily grievances. A bomb promised to make people even more scared of the North Korean authorities and consequently to make them more submissive.

Again, the two Kims were not the first to try to use the bomb for domestic purposes. Stalin and Mao Zedong successfully applied similar tactics to suppress dissent at home. Noncommunist dictators, like Mu'ammar Gadhafi of Libya or

Saddam Hussein of Iraq, did and continue to do the same. Even some U.S. presidents have manipulated the issue of nuclear bomb production to enhance their image as strong leaders at home.

A third factor pushing Pyongyang to embark on an independent nuclear weapons program was less rational and more emotional and psychological. The two Kims were furious with the entire international community: the Americans, South Koreans, Russians, Japanese, and Chinese. Thus, Pyongyang wanted to show everybody that it would not bend under pressure, and it would defy any forms of control.

This kind of behavior is akin to French foreign policy in the 1960s, which was aimed at defying U.S. supremacy in European affairs. The difference is that under a totalitarian regime such outbursts of "manhood" and "ambition" can be triggered much more easily and are harder to stop because dictators are not used to opposition inside their countries and are less vulnerable to critics in the international arena. Thus, when dictators do decide to make dangerous moves in world affairs, there is no one around domestically to stop them.

Fourth, a nuclear weapon could serve as a symbol of North Korean achievements and a means of raising the international status of the DPRK. Vanity has been one of the prominent features of the two Kims' regime all along. They have consistently tried to prove to everybody inside and outside the country that their government is the greatest in the world. The Kims perceived nuclear weapons as an important aspect of greatness. Such feelings were reinforced by the examples of other vain dictatorships: in Libya, Iraq, and Iran.

However, the fifth and probably most important catalyst for North Korea's nuclear plans was the desire to use the bomb as a diplomatic card to bring Washington to the negotiation table and to gain substantive concessions from it. Time and again, officials in Pyongyang called upon Washington to negotiate a compromise on the basis of mutual concessions, equality, and reciprocity. But a nuclear weapon would provide the ultimate bargaining chip to bring concessions from Washington. Therefore, nuclear blackmail was an attractive option, even if the military program fell short of actually developing a usable weapon.

There is a school of thought among experts that Pyongyang intended primarily to bluff the West, threatening to acquire a military nuclear potential, rather than actually developing it. The analysis presented here does not subscribe to this view. The intentions of the Kims were serious indeed, even if the two Kims were unsure of whether North Korea would eventually succeed. The very real efforts of the DPRK to create a nuclear arsenal testify to this conclusion.

Today, it remains uncertain whether North Korea has ceased these efforts altogether, despite its pledges under the Agreed Framework. Given North Korea's continuing security concerns, it seems unlikely that we have seen the last of the North Korean nuclear weapons program.

13

North Korea and the Nuclear Nonproliferation Regime

Vladimir F. Li

In the early 1990s, the North Korean nuclear problem became an acute international concern, quickly thrusting it into first place among the other geopolitical controversies emanating from the Korean Peninsula. Successful settlement of this problem will alleviate some of the worst tensions plaguing the vast region of Northeast Asia and create favorable conditions for a fruitful dialogue between the two Koreas, thereby assisting in their gradual reconciliation and reunification on a nonviolent basis.[1] Without a constructive settlement of the nuclear problem, however, it will not be possible to construct a meaningful peace settlement for the Korean Peninsula.

This chapter analyzes the question of what prospects exist for successfully drawing North Korea into full compliance with the international nonproliferation regime. The discussion first focuses on Pyongyang's stormy relationship with the International Atomic Energy Agency (IAEA), whose inspections in 1992 uncovered reasons to question North Korea's compliance with the Treaty on the Non-Proliferation of Nuclear Weapons (NPT). It then examines North Korea's security relations with the major powers of the region, particularly China, attempting to explain Pyongyang's rationale in moving to confrontation with the IAEA. Finally, it examines North Korea's signing of the Agreed Framework with the United States, which has largely replaced the IAEA as the major North Korean partner in deciding the fate of its relations with the outside world. Despite the agreement, however, the analysis presented here concludes that North Korea will continue to cause problems in the nuclear sector for many years to come, largely because it believes that it must use all available tools to its fullest advantage in what it perceives as a life-and-death struggle with surrounding hostile powers.

North Korea's Decision to Join the NPT

At the outset, it is important to note that the NPT of 1970 is one of the major accomplishments of world diplomacy in the second half of the twentieth century. At present, 187 states are members of the treaty, making the NPT the most significant achievement along the protracted and exhausting path toward the limitation of nuclear weapons. It was on the basis of the general principles of the NPT and of the START I and START II treaties of the early 1990s that Russia and the United States committed themselves to remove and dismantle more than 17,000 nuclear warheads from silos, submarines, and bombers. But the international barriers to enlargement of the nuclear club have proven even more effective. When the NPT was signed, it was expected that by the end of this century there would be between twenty and thirty nuclear states. But thanks to the NPT, the total number of the states possessing nuclear weapons has remained limited during the past several decades, which in turn creates the necessary preconditions for the full implementation of the treaty's principles of eventual nuclear disarmament.[2]

However, a small group of states (including the Democratic Peoples' Republic of Korea [DPRK]), initially refused to recognize the NPT regime, as if to leave their hands free for uncontrolled activities for the pursuit of nuclear weapons. The DPRK was particularly stubborn in this regard. For fifteen years (until the mid-1980s), North Korea refused even to hear about joining the NPT, since it believed that U.S. troops in South Korea were equipped with such weapons. Only on December 12, 1985, after increasing pressure from the world community, did Pyongyang agree to join the regime. On that date, a North Korean governmental delegation visited the Soviet Ministry of Foreign Affairs in Moscow to deposit its official instrument of ratification on joining the NPT.[3]

However, this event was not the end of North Korea's nuclear bargaining. With the collapse of the bipolar world system, the demise of the Cold War, the gradual rapprochement between Moscow and Washington, and, finally, the famous Mikhail Gorbachev–Roh Tae-woo summit in San Francisco in 1990, Pyongyang adopted a course aimed not at peaceful nuclear cooperation but at the accelerated development of a costly nuclear weapons program. It has become clear, in retrospect, that North Korea's accession to the NPT was intended mainly to provide a facade of legitimacy, without which the DPRK could not expect to acquire nuclear technology on world markets, including from the Soviet Union.

After its widely publicized act of joining the NPT in 1985, Pyongyang undertook a number of international steps of a propagandistic nature. Most noticeable among them was a detailed statement by the North Korean government on June 23, 1986, on the urgent necessity of transforming the entire Korean Peninsula into a nuclear-weapons-free zone.[4] Its diplomatic edge was pointed mainly at the U.S. tactical nuclear weapons deployed in South Korea. (It is noteworthy that the United States neither confirmed nor denied the presence of these weapons during the Cold War.) Following a series of complex intergovernmental negotiations at the prime ministerial level in the early 1990s, a historic agreement between North and South Korea on reunification, nonaggression, exchanges, and cooperation was signed in Seoul on October 13, 1991. On December 31, 1991, this pact was supplemented by a joint DPRK-Republic of Korea (ROK) antinuclear declaration on converting the Korean Peninsula into a nuclear-weapon-free zone.[5]

On April 9, 1992 (seven years after joining the NPT), North Korea's Supreme People's Assembly finally ratified an agreement on IAEA safeguards, which came into force the next day. The road toward making the Korean Peninsula a nuclear-weapon-free zone seemed to be open. But, on March 12, 1993 (less than a year after the North Korean ratification of this safeguards agreement), Pyongyang demonstratively and arrogantly stated that it was suspending its membership in the NPT. North Korean authorities explained this impudent political and diplomatic démarche by referring to the "extraordinary situation" created by a recurrent series of U.S.-ROK joint "Team Spirit" military exercises and by the "unjust" demands of the IAEA, which insisted upon international inspection of strategic sites "having no relation whatsoever to nuclear activities."[6]

It should be mentioned that the initial reaction of the great powers (except for China) was close to panic. No one among the knowledgeable experts in the United States or in Russia had paid any attention to the possibility of such a North Korean démarche. Usually, the IAEA managed ultimately to reconcile demands of this nature by means of negotiations. Nobody among analysts of international relations seemed to be prepared even to think that the North Koreans might start playing an intricate nuclear bluff. The strategy was difficult to discern because of the especially tight veil of secrecy surrounding the country due to its practice of the *juch'e* model of socialism.

What were the geopolitical factors that led Pyongyang to act so impudently and defiantly, challenging the world community and blatantly ignoring the national interests of its neighbors in the Asian-Pacific region?

The Chinese Factor

The carefully thought out diplomatic maneuvering that resulted in the DPRK's unilateral withdrawal from the NPT was, first of all, closely connected to the nuclear strategy of its ally, the People's Republic of China (PRC), which had managed by this time to find an optimal symbiosis of a totalitarian political system with the benefits of a market economy. For Pyongyang, China represented a potential "nuclear shield," given its obligations on mutual assistance in case of the "aggression of a third party," according to their Treaty on Friendship, Cooperation, and Mutual Assistance. These obligations had been later confirmed in the statement of Chinese leader Ziang Zemin, who said that whatever else may happen in the world, the ties between the PRC and the DPRK would remain as tight as between "lips and teeth."[7]

What was it about the Chinese nuclear potential that caused so much admiration in Pyongyang at that time? China's intensive nuclear development had been initiated in the mid-1950s when the PRC administration allocated special funds for the Scientific and Technological Commission of the PRC, the Second Ministry of Machine-Building, and the Ninth Academy in the province of Qinghai. According to an agreement signed in October 1957, the Soviet Union had also undertaken large-scale obligations to provide China with equipment and to educate nuclear experts. Beginning in the late 1950s, some 950 Chinese physicists and several dozen North Koreans received academic and practical training in nuclear research at the United Institute of Nuclear Research in Dubna. In accordance with this agreement, the Soviet side undertook obligations "to transfer to China not only a sample nuclear bomb, but also the uranium for further enrichment at the Lanchou plant."[8] Thus, China initially developed its nuclear program without building facilities for uranium mining and processing: it expected to rely mainly upon Soviet supplies. Only after the sharp deterioration of Sino-Soviet relations was Beijing finally compelled to pay priority attention to this problem, which was later fully taken into account by Pyongyang when the latter was designing its own national nuclear program.[9]

According to some Russian sources, since the first Chinese atomic bomb test at Lop Nor in 1964, the total amount China has spent on producing atomic weapons is approximately $4 billion, which roughly corresponds to what other states have spent on similar programs. On June 17, 1967, China successfully exploded a three-megaton hydrogen bomb. In subsequent years, in spite of the fact that it formally joined the NPT in March 1992, the PRC continued to conduct nuclear tests, claiming that it was still "far behind" the other nuclear powers in terms of the total number of nuclear explosions.[10]

While all the countries in the Asian-Pacific region—particularly Japan, India, the Philippines, Bangladesh, Indonesia, Malaysia, Thailand, and even South Korea—were deeply alarmed by and actively protested against these accelerated Chinese nuclear preparations, North Koreans turned out to be, in fact, the only state in East Asia that praised the nuclear ambitions of Beijing. This fact reflected Pyongyang's far-reaching geostrategic calculations.

During the Sino-Soviet confrontation, Pyongyang gave its clear preference to Beijing, whose nuclear strategy was called upon to ultimately "convince the world community of China's right to be called a great power . . . to consolidate its influence in the countries of the Communist camp, and to have additional capabilities to support the wars of national liberation and to strengthen China's hegemony in Asia."[11]

However, the North Koreans undoubtedly overestimated the potential readiness of China to encourage the emergence of a mininuclear state along its borders, capable of causing difficult regional problems that could inhibit the development of long-term relations between China and South Korea. In reply to the alarming reports in the world press about secret supplies of Chinese nuclear technology to Algeria, Iran, and other countries, Chinese officials felt compelled to state that "China had no cooperation with the Democratic People's Republic of Korea in the sphere of developing the latter's nuclear program."[12] Moreover, Beijing refused even to express its moral support for North Korea's nuclear ambitions.

Nonetheless, from the very outset Beijing clearly indicated that it was opposed to any sanctions or boycott against the DPRK in connection with the nuclear crisis. China consistently avoided direct participation in international negotiations on this problem, claiming that it should be settled by patient and flexible discussions between the DPRK, the ROK, the United States, and the IAEA. This position proceeded from the PRC's traditional preference for bilateral solutions and its desire to avoid any additional burden of financial expenditures related to the maintenance of regional nuclear security. Only in October 1994, at the Thirty-Eighth Conference of the International Atomic Energy Agency in Vienna, did the PRC announce that it was joining the International Convention on Nuclear Safety and that it was going to contribute an additional $1 million to the IAEA fund for financing nonmilitary projects in developing countries, including North Korea.[13] At the same time, the Chinese side made the qualification that prevention of the proliferation of nuclear weapons and promotion of the peaceful use of atomic energy—albeit mutually complementary—were very different matters that should not be confused.

The above analysis leaves no doubts that under the present balance of geopolitical forces in the Asian-Pacific region, China will not support North Korea's military nuclear program. Moreover, Beijing is likely to demand more persistently than earlier that the DPRK fully adhere to the NPT regime, because any other position might inflict irreparable damage to China's international prestige, which is already at a critically low level (due to its nuclear and missile exports and violations of human rights).

The fact that China is not interested in expanding the membership of the "nuclear powers club" even by offering a seat to its closest ally, opens realistic possibilities to prevent North Korea from becoming a new nuclear threat in East Asia. Although Chinese authorities have repeatedly stated that their ability to influence the DPRK "was extremely limited," in fact, Pyongyang took into consideration Beijing's signals that there would be negative implications from any state's withdrawal from the NPT.

It is common knowledge that one of the basic conditions of the NPT is the strict obligation of the states possessing nuclear weapons never to transfer nuclear weapons or nuclear explosive devices, or control over them (direct or indirect), to anyone and in no way to assist, promote, or impel any state not possessing nuclear weapons to produce or acquire (by whatever means) nuclear weapons or other nuclear explosive devices.[14]

The danger of violating this key provision of the NPT by potential mini-nuclear states is sharply increased in light of the fact that some of these states, in particular the DPRK, have made considerable progress in research and design efforts aimed at creating missile weapons of both intermediate- and long-range that—with some effort—could be equipped with nuclear warheads.

By the end of 1994, North Korean experts, as reported by the South Korean Ministry of National Reunification, were close to completing the development of missiles of 500-kilometer (km) range, of 1,000-km range (*Nodong*-1) and of 2,000-km range (*Nodong*-2). The military-industrial complex of the DPRK already possesses the capability to produce about 100–150 short-range missiles per year. From 1980 to 1993, exports of the North Korean weapons, including missile complexes, particularly to "hot spots" like Syria, Iran, Iraq, and other countries, accounted for approximately $3.25 billion, or nearly 25 percent of all export revenues.[15] Who could guarantee that modernized missiles with nuclear warheads would not become one of the leading items of North Korean export in the future? This question remains to be answered, and China, one of the principal members of the nuclear club, is likely to play a significant role in interpreting the NPT's international guarantees.

Although it may sound paradoxical, the "Chinese card" played an important role in the emergence and escalation of the North Korean nuclear crisis. Without being able to fall back on China and its guarantees of support, Pyongyang would not have dared to challenge the international nonproliferation regime so flagrantly. To some extent, more fuel was added to the fire by some inaccurate Soviet and U.S. estimates of the DPRK's nuclear capabilities.

North Korea's Alleged Capabilities—Russian Estimates

In August 1994, at the request of a committee of the Russian State Duma, a group of Russian experts on international affairs prepared an analytical report on this problem. It revealed that the official Soviet and later Russian estimates of the military nuclear potential of North Korea, based on a secret report by former KGB Chief Victor Kryuchkov to the Kremlin leadership dated February 22,

1990, appear to have been exaggerated because these estimates did not take into consideration all the complexities of the industrial production of nuclear weapons under the extremely difficult economic and political situation in the DPRK.[16]

In the secret report, KGB experts had provided the following assessment to their Kremlin superiors:

> From a reliable source, the KGB has received information that scientific and experimental design work on the development of atomic weapons is actively continuing in the DPRK.
>
> According to this data, the development of the first atomic explosive device has been completed at the Institute of Nuclear Research of the DPRK, located in Yongbyon.... For the time being, tests are not planned in the interests of hiding the fact of the DPRK's production of an atomic weapon from the world community and from international organizations responsible for nuclear controls.
>
> The Committee of the State Security is undertaking additional measures to verify these data.
>
> V. Kryuchkov
Chairman of the KGB of the USSR[17]

The above-mentioned document, written by highly qualified professionals, contains no references to the fundamental differences between such phases of producing nuclear weapons as laboratory research, field tests, and serial manufacture of weapons. This is not accidental. By combining different stages of developing nuclear weapons, despite the fact that each of them requies a different means and time, the KGB was trying to use every opportunity to remind Soviet leaders that it was the most zealous guardian of the national interests of the Soviet Union. Meanwhile, any unbiased observer could easily see that for a long time—especially before North Korea's ratification of the IAEA safeguards agreement in April 1992—the North Koreans had an opportunity to undertake (if they had actually been ready) at least one underground test explosion in one of their underground shelters. Such an action would have brought them much greater geopolitical dividends than the tactics of indecisiveness and delay.

These data suggest that the KGB may have fallen victim to an intricately calculated leak of disinformation from skilled North Korean sources. Despite its acclaimed professionalism, Soviet political intelligence was not immune to serious miscalculations or even obvious misperceptions. A former KGB resident in Japan, S. Levchenko, who defected to the West, makes an interesting admission in this connection in his memoirs: "As some experts have admitted, Soviet and later Russian disinformation sometimes appeared on the table of the U.S. Presidents." In turn, "similar disinformation was sent to Moscow by Soviet intelligence agents themselves. This practice flourished especially in the Brezhnev era.

North Korea and the Nuclear Nonproliferation Regime 145

The then-KGB chief Andropov and other Politburo members often were fed with data that was to their advantage and caressed their ear. The truth was told only by the few who dared to speak up. To a large extent, it did not reach 'the top.'"[18]

Returning to the main theme of this chapter, one cannot help concluding that not only was the Soviet Union deceived, but the United States was also misled by the North Korean nuclear bluff of the late 1980s and early 1990s. The latter point is illustrated by the exaggerated estimates and contradictory reports concerning North Korea's military nuclear capabilities made by some "sufficiently reliable" sources in the United States.

North Korea's Alleged Capabilities—U.S. Estimates

In October 1993, citing sources inside the U.S. Congress and U.S. military intelligence, NBC news insisted that Pyongyang had already accumulated a sufficient quantity of weapons-grade plutonium to produce at least five nuclear weapons. But, nearly one month later, then-CIA Director R. James Woolsey said that the North Koreans possessed enough plutonium to manufacture only one to two nuclear weapons. This meant that there continued to be varying estimates of the North Korean nuclear potential by the Pentagon, the intelligence community, the State Department, and the White House.[19] About half a year later, U.S. analysts Brent Scowcroft and Arnold Kantor insisted that the DPRK was in a position to bring its nuclear arsenal up to six to eight nuclear weapons.[20] To neutralize and extinguish this North Korean nuclear threat, they proposed—proceeding from past experience—to deliver a limited preventive strike against North Korean facilities processing nuclear fuel. This would have reduced the "danger" of radioactive pollution, as compared with the bombing of nuclear reactors. If the North Koreans retaliated by attacking South Korea, thereby unleashing a second Korean War, the United States and its allies would intervene directly into the conflict. Scowcroft and Kantor went on to say that "strategically, we should clearly demonstrate to Pyongyang that we are not to be scared by threats." They stressed that the United States would not be paralyzed by the prospect of war and that, if military engagement was unavoidable, the United States "had to start fighting as soon as possible," before North Korea obtained a more sizable nuclear arsenal.

It is worth mentioning that extremely nervous statements of this kind were by no means exceptional cases. Reflecting the views of some influential business circles in the United States, Vice-President of the Dow Jones Corporation K. House stated in an address to U.S. Congressmen that if the United States were to engage in a direct confrontation with Pyongyang, then the prospective outcome would be war. But if the U.S. were to capitulate and let Pyongyang preserve its nuclear program intact, this would bring about some reduction in the U.S. military presence in South Korea and soon in other parts of Asia, and "would also mean [the] unavoidable failure of our global efforts to stop the proliferation of nuclear weapons and thus would encourage the emergence of new nuclear powers from Tokyo to Teheran."[21]

From May to June 1994, citing influential sources within the Pentagon, the U.S. media carried a series of articles on the active U.S. preparations for possible escalation of military conflict on the Korean Peninsula. According to a *Washington Times* report, Pentagon officials were seriously contemplating the possibility of a new war in North Korea. In this context, the urgent build-up of South Korean troops located directly near the demilitarized zone (DMZ) was considered of paramount importance. U.S. military planners were reported to be considering some operational scenarios involving the dispatch of additional U.S. troops, arms, combat aircraft, and various military equipment to South Korea. Reportedly, they issued orders to check the availability of sufficient stockpiles of so-called bunker bombs, designed to destroy the heavily fortified underground facilities constructed beneath the nuclear complex at Yongbyon.[22] U.S. experts still seem to believe that the current balance of forces on the Korean Peninsula is not to Seoul's advantage: only a few of South Korea's twenty-one divisions are located near the DMZ, while two thirds of the DPRK's military forces (totaling 1.2 million men) are deployed within 60 km of the DMZ.

According to the Pentagon's operational estimates, U.S. troops in South Korea (numbering 37,000) could be quickly reinforced by bringing in the 26,000 Marines deployed on Okinawa. Also, there were two combat aircraft carriers—the *Independence* and *Constellation*—cruising relatively close by in the Pacific. U.S. intelligence services also transported to South Korea an additional number of their personnel to analyze and evaluate electronic and photo-reconnaissance data obtained by U.S. spy satellites, spy planes, and listening devices at sea and on the ground.[23]

Apparently, newspaper publications of this sort and quite similar statements by U.S. officials were aimed primarily at psychological warfare goals rather than those of a hot war. Nevertheless, it is dangerous to play with fire. In this context, the opinion expressed by a group of Russian experts on international affairs appears to be reasonable and well grounded. They pointed out that: "It is well known that North and South Korea still remain the most militarized parts of the Asia-Pacific region, where military conflict might be sparked even accidentally as a result of a mistake or an accident.... The tragic implications of such miscalculations are widely known."[24]

The International Politics of the Nuclear Crisis

In late June 1994, the United States sponsored a U.N. draft resolution on international sanctions against the DPRK to be imposed if the latter refused to return to the NPT regime. The draft resolution stipulated that thirty days after the adoption of the resolution, international economic sanctions, no less severe than those against Libya and Iraq, should be applied to North Korea.

The proposed measures included a ban on takeoff, landing, or entry into the member states' national airspace of any aircraft originating in or headed for North Korean territory, except for those cases when a specific flight was agreed upon by the United Nations beforehand and pursued for humanitarian purposes.

The United Nations, and its member states and affiliated agencies, would also be obliged to suspend all ongoing assistance and to ban any additional economic assistance to the DPRK, including supplies of any goods and services pertinent to such aid. All the states that have effective bilateral treaties with the DPRK on mutual assistance in the sphere of security would have had to suspend the operation of these treaties as well.

Further, according to the draft resolution, all states would have had to curtail considerably the number and level of North Korean personnel in diplomatic and consular missions on their territory and at North Korean missions to various international organizations and would have had to limit and control the movement of the remaining North Korean personnel on their territory. Concerning the DPRK missions to international organizations, the host states were recommended to enter into consultations, if needed, with such organizations on possible measures to implement the provisions of the sanctions.

All U.N. member states would have been required to ban any imports from or exports to the DPRK of all types of nuclear, chemical, biological, strategic, and conventional arms, missiles, and missile-delivery systems, including dual-use items, as well as any other weapons and ammunition, spare parts, equipment, subsystems, components, materials, and transportation equipment for these items, after the adoption of the Security Council's resolution on sanctions .

To supplement these all-embracing economic sanctions, the draft resolution stipulated the suspension of all forms of cultural, technological, scientific, commercial, and educational exchanges, and all visits with the participation of individuals or groups officially appointed by the DPRK or representing the country.[25]

It is interesting to note that Pyongyang, as a whole, reacted to the heavy psychological pressure from the United States, Japan, Russia, and other states quite calmly, though the prospect of sharing the same "sinking boat" with Libya, Iraq, and other states subject to international boycott did not appear very tempting. Having China's nuclear, missile, and diplomatic shield as a backstop, North Korea chose the course of action of quite an experienced boxer, skillfully using counter-attacking blows and timely retreats.

Literally just one day prior to the expiration of the three-month waiting period before its ultimate departure from the NPT, June 11, 1993, North Korea officially stated that it was temporarily suspending its withdrawal from the international nonproliferation regime. This news was received with relief in Russia, the United States, and other key members of the nuclear club, who took notice that Pyongyang had reiterated its readiness to continue its cooperation with the IAEA.

As far as the pressure and threats of the nuclear powers on the DPRK were concerned, North Korean authorities qualified them as an encroachment upon basic NPT principles. Instead of focusing on their own duties to comply with IAEA safeguards, Pyongyang shifted attention to the NPT's general political declarations on security guarantees to countries not possessing nuclear arsenals, as well as its requirement that the nuclear powers engage in persistent negotiations on the cessation of the nuclear arms race and on general disarmament under effective international control.[26]

From October 1993 to April 1994, the international nuclear crisis on the Korean Peninsula reached its culmination. On October 19, 1993, the Director General of the IAEA Hans Blix stated that the work of the IAEA nuclear inspectors in the DPRK had been impeded in many respects and that the IAEA had reliable grounds for believing that the North Koreans were "using their nuclear facilities for production of plutonium."[27] This alarming statement prompted the U.N. General Assembly to adopt a special resolution on November 4, 1993, calling for the DPRK to arrange "close cooperation" with the IAEA to implement fully its international safeguards against nuclear weapons nonproliferation.

Although the DPRK expressed its disagreement with the U.N. resolution, alleging that it was groundless, Pyongyang, on January 4, 1994, declared its readiness to assist the work of international inspectors at its seven main nuclear facilities.[28] On March 3, 1994, the IAEA—overcoming significant obstacles—resumed its inspections in the DPRK.

Soon, however, on March 16, 1994, the IAEA declared that the DPRK had refused to allow unrestricted access to the agency's inspectors in Yongbyon.[29] This compelled the IAEA Board of Directors to state officially that the North Koreans were trying to hide something from international inspectors, possibly important nuclear installations that may be involved in military nuclear research or the separation of plutonium.[30]

Following the earlier rounds of intensive diplomatic wrangling, the West now had virtually no doubt that North Korea either had nuclear weapons in its possession already or had created the scientific experimental and industrial basis for producing such weapons. But at the culmination stage of these polemics on April 16, 1994, North Korean President Kim Il Sung suddenly intervened in the diplomatic confrontation. He officially declared that Pyongyang "had no atomic weapons with which to claim the status of a nuclear superpower."[31] But he rejected IAEA demands to inspect two undeclared nuclear installations at Yongbyon, saying that all countries have the right to have "military secrets." Kim Il Sung stressed that the DPRK had no interest in unleashing a new Korean War and was instead trying to settle the nuclear crisis on the Korean Peninsula "through diplomatic channels."[32] This highly authoritative statement, albeit short of clearing up all of the questions and doubts concerning the North Korean nuclear program, opened new opportunities for a settlement. It became evident that Pyongyang had reiterated its commitment to a nonnuclear status and was ready to start negotiations aimed at solving the nuclear crisis on the Korean Peninsula without any preconditions.

The above-mentioned statement of the president of the DPRK was followed by a well-calculated diplomatic move: on April 23, 1994, North Korea decided to open access for the IAEA inspectors to previously closed declared nuclear research facilities in Yongbyon, including the 5 megawatt research reactor.[33] This step considerably decreased tensions between Pyongyang and the nuclear weapon states.

It is worth mentioning that by this time, the geopolitical course of North Korea had been determined considerably by its policy of distancing itself diplomatically from Moscow. To a certain extent, the Russian side had provoked such developments by unilaterally suspending various business contacts with the DPRK and seeking by means of direct pressure to collect large debts (exceeding 3.6 billion gold rubles) from Pyongyang.[34] The North Korean leadership practically ignored a constructive initiative by Russia, proposed on March 24, 1994, to hold an international conference on the problems of the Korean Peninsula with the participation of Russia, the United States, China, Japan, and North and South Korea, as well as representatives of the United Nations and the IAEA. Pyongyang noticed that this well-meaning initiative of the Russian Foreign Ministry was not backed by the rest of the Russian government. For example, there were no indications that Minatom was prepared to resume its exploratory, design, and other preparatory works on the site slated for the construction of the nuclear power stations agreed to in the 1985 Soviet–North Korean deal.[35]

In the deteriorating atmosphere of Russian–North Korean relations, the United States quite firmly and decisively spoke in favor of adopting broad international economic sanctions against the DPRK on June 2, 1994.[36] This proposal was immediately upheld and approved by Japan, South Korea, and France. Russia supported the general idea of sanctions with the reservation that the problem should be finally settled after adopting a decision to convene a multilateral international conference on the North Korean nuclear problem. In an exclusive interview with *Izvestiya*, the head of the Russian Foreign Ministry Andrei Kozyrev stated: "If the DPRK remains stubborn and inflexible, refuses to cooperate with the IAEA, and withdraws from the Non-Proliferation Treaty, then sanctions will be unavoidable. We are in favor of their gradual introduction. One should start by imposing a series of 'soft' sanctions. Then, if there is no result, one should switch to tougher measures. One needs to work out a list of sanctions and simultaneously proceed with preparation of the [Russian-proposed] international conference on the problems of the Korean Peninsula."[37]

However, notwithstanding such heavy diplomatic pressure exerted primarily from Washington and Moscow, Pyongyang was firmly convinced that any form of international boycott of North Korea would be at best of a symbolic nature, as long as China did not participate. By the middle of 1994, China's share accounted for 70 percent of North Korean imports of oil and oil products and nearly 60 percent of imported food commodities.[38] This is why in its reply to severe warnings about sanctions and a boycott, Pyongyang undertook another diplomatic counter-démarche: on June 13, 1994, the North Korean Foreign Ministry published a political declaration about the DPRK's "forced withdrawal" from the IAEA structure, reaffirming its previous decision to suspend the country's membership in the NPT.[39] The principal members of the global nuclear club, with the possible exception of China, now had no doubts that North Korea was acting in the shadow of the Chinese nuclear shield to free its hands in order to begin military industrial production of weapons-grade plutonium and to enter directly the exclusive club of nuclear powers by exploiting the disarray in the world geopolitical situation.

However, it soon became evident that behind this emotional démarche there were purely pragmatic reasons to obtain much-needed hard currency and high technology for reconstructing the country's nuclear energy sector from the leading members of the global nuclear club especially the United States. These intentions were revealed on June 27, 1994, when North Korean newspapers published a special statement by Choi Jung Sun, a representative of the DPRK Committee on the Atomic Energy, approved by the top political leadership.[40]

According to this statement, North Korean authorities had previously agreed to allow the IAEA inspections because they believed that these would be "honest inspections" of North Korean nuclear installations. However, the IAEA began to exert "pressure upon the DPRK," which forced the latter to withdraw from its cooperation with the inspection regime. "The firm position of the North Korean government," Choi's statement continued, "is to develop relations with those international organizations that respect the sovereignty of the republic and maintain relations based on the principle of justice." He pointed out that during the visit of IAEA Director General Blix to the DPRK in May 1992, he was provided with detailed information about the nuclear activity of the DPRK, though later this guest took an "unjust position."

Choi went on to say that the IAEA had deliberately tried to open up non-nuclear military installations, acting upon U.S. directives and using their intelligence reports. Moreover, in September 1992, a high-ranking IAEA delegation had already visited the so-called suspicious nuclear sites upon the request of the agency's Board of Directors. But instead of just surveying the installations, its members tried to undertake a technical inspection, which they had not been allowed to do. The DPRK considered the resulting demands for access to the installations and the reports in numerous foreign newspapers that these facilities were connected with the nuclear program as "unjustified pressure."

Choi pointed out that despite the suspension of its membership in the IAEA, the control equipment installed by the IAEA at the North Korean nuclear facilities in Yongbyon continued to function. In the event that something needed to be changed or replaced in the course of its exploitation, the DPRK intended to hold consultations with the IAEA. Two IAEA inspectors still remained on North Korean territory. Though they were expected to leave the DPRK soon, two new experts were coming to replace them and to continue the inspection activity. North Korean representatives intended to stay in Vienna and maintain working contacts with IAEA headquarters.

Furthermore, Choi's statement contained a fundamentally new idea, which was expressed in a seemingly causal manner. Later, this proposal laid the foundation for the settlement of the nuclear crisis on the Korean Peninsula in the Agreed Framework.

What actually took place? As a high-ranking representative of the North Korean Committee on Atomic Energy put it, the basic economic role of the Yongbyon reactor, built with Soviet assistance, was the development of the nuclear energy sector in order to concurrently develop the national economy. The type of graphite reactor used at Yongbyon was quite appropriate for the DPRK because the country possessed the necessary resources of uranium and graphite.

However, the United States, Japan, and other countries suspected that North Korea needed this type of nuclear reactor to "produce nuclear weapons." In order to alleviate these suspicions, the DPRK advanced the idea of replacing the graphite-moderated reactor with a light-water reactor. According to Choi Jung Sun, this idea was brought to the attention of the U.S. side. He went on to say that the DPRK was not yet in a position to build a light-water reactor because it lacked the "appropriate technology." Though the former Soviet Union had assisted them in building the graphite reactor, its successor, the Russian Federation, was faced with a grave situation itself. As a result, North Korea "had almost no cooperation with Russia in the nuclear sphere" and "did not harbor any hopes about Russia in this respect."

Generally speaking, the DPRK felt no specific technological necessity to replace the reactor and considered such proposals only in the context of its efforts aimed at "removing suspicions about its secret production of nuclear weapons." Although the North Koreans had no great hopes in connection with other countries, they had some general suppositions concerning possible implementation of the concept. Choi hinted that they might be discussed at the then-forthcoming U.S.-DPRK high-level negotiations in July in Geneva, which, in fact, isolated Russia from the modernization of the North Korean nuclear sector.

In conclusion, the representative of the North Korean Committee on Atomic Energy stated that the DPRK had plutonium for experimental use at the radiochemical laboratory in Yongbyon, but he did not specify its quantity. He responded to the critical remarks made by the IAEA and other countries regarding the ongoing replacements of the nuclear fuel rods by explaining that they "were going according to plan and were conditioned by the existing technology" and by the DPRK's need to test operational equipment. Of course, this refueling was being conducted outside of IAEA inspections and left the agency with no chance of making the measurements needed to provide evidence of North Korea's allegedly peaceful intentions.

The statement by the North Korean Committee on Atomic Energy on June 27, 1994, became a turning point in the settlement of the North Korean nuclear crisis. Its ideas were laid into the foundation of the final U.S.-DPRK nuclear negotiations and eventually the Agreed Framework signed in Geneva in October 1994. According to the agreement, the U.S., Japan, and South Korea took upon themselves the obligation to provide approximately $5 billion through an international consortium for replacing the existing and planned heavy-water graphite reactors in Yongbyon with light-water reactors in return for North Korean agreement to remain as a member of the NPT and to allow full IAEA inspections. As some Russian experts summed it up, this was one of the most significant victories of North Korean diplomacy in the entire postwar period.

Conclusion: Explaining the Nuclear Deal and Predicting Its Future Prospects

How can one explain why the nuclear powers, as well as Japan and South Korea, agreed to make major concessions to North Korea in settling the nuclear crisis? It seems likely that, by this time, the principal nuclear powers had come to a more realistic and unbiased evaluation of the DPRK's nuclear potential. Earlier, as noted above, the U.S. Central Intelligence Agency and the Pentagon had provided quite different assessments of North Korea's nuclear capabilities. No less controversial were the reports of the former KGB and of the Military Intelligence Department of the Russian General Staff. While the Soviet KGB had reported in February 1990 that North Korea admitted "the completion of the development of its first atomic explosive device,"[41] in January 1994, Chief of the Russian General Staff General M. Kolesnikov—citing the conclusions of his Main Intelligence Department—issued an absolutely different assessment. General Kolesnikov noted: "According to the information at our disposal, the DPRK today has no nuclear weapons. Nor does it have hundreds of medium-range missiles. The creation of nuclear weapons in the DPRK in the near future is unlikely."[42]

It is noteworthy that the conclusion of Russian Military Intelligence was also rather different from the assessment made by a group of international relations experts at the Russian Academy of Sciences, who estimated "the probability of the existence of nuclear weapons in the DPRK at 50 percent."[43]

General Kolesnikov's statement provided the nuclear powers with the most authoritative piece of information, which deserved more attention than the estimates from other sources. The initial reaction in the diplomatic circles of the nuclear capitals regarding the findings of the Russian General Staff was close to shock. Only some time later did it become evident that this development opened fundamentally new perspectives for diplomatic maneuvering. If the North Koreans had no nuclear weapons (and if they were not going to produce any), they could be talked to in normal diplomatic language, and one could negotiate serious agreements with them. It was the United States that grasped this fact first. While working out a draft of the Agreed Framework, U.S. negotiators took everything into account, including the apparent distrust of Pyongyang toward the then-Russian leadership, which (as opposed to China's leadership) had unilaterally resolved to denounce the July 6, 1961, Treaty on Friendship, Cooperation, and Mutual Assistance between the Soviet Union and the DPRK.[44]

During the decade of heightened tensions surrounding the nuclear crisis on the Korean Peninsula (1985–95), Pyongyang managed to demonstrate flexibility and agility in its foreign policy strategy, characteristics not often seen in the policies of totalitarian regimes. Its position in the region was substantially strengthened. The role of the United States and Japan increased too. The position of China remained unchanged but sufficiently stable. Only the position of Russia was undermined and weakened due to a number of subjective factors.

The October 1994 Agreed Framework is a major achievement of recent international nonproliferation diplomacy. Nonetheless, it would be unrealistic to claim that it represents a comprehensive solution to the nuclear crisis on the Korean Peninsula. In fact, the foreseeable future may well bring new diplomatic démarches from North Korean diplomacy, possibly including new threats to withdraw again from the NPT and the IAEA.

The nearest and most plausible pretext for action of this nature might be the DPRK's insistence on some new additional conditions—technological and financial—needed to implement the Agreed Framework. The rationale for these demands may be that the DPRK's nuclear energy sector is likely to suffer major losses as a result of the mandatory changes in its nuclear plans. The DPRK may believe that the Korean Peninsula Energy Development Organization (KEDO) participants, having taken upon themselves the major expenses involved in modernizing North Korea's nuclear industry could be ultimately forced to bear additional expenditures to keep the basic agreement alive. Thus, in October–November 1995, the DPRK demanded that KEDO agree to supply additional equipment needed for the distribution of nuclear energy to the rest of the North Korean power grid, as well as other technologies needed for meaningful use of the two light-water reactors. Simultaneously, the North Korean side pointed out that the DPRK should have "strict guarantees" in relation to shutting down its graphite-moderated reactor. Moreover, Pyongyang insisted on receiving additional "compensation for losses" in the amount of $1 billion for freezing its nuclear activities.

Another pretext for a possible withdrawal or temporary suspension of the DPRK's membership in the NPT might be the activization of military preparations by the South Korean army or the redeployment of U.S. nuclear weapons on South Korean territory. Complete military denuclearization of the peninsula is one of the key demands of the North Korean side. During the drafting and signing of the Agreed Framework, Pyongyang deliberately shut its eyes to this problem, but the official North Korean media is likely to continue to raise it. Meanwhile, North Korea continues to focus on the fact that it is the presence of U.S. troops—equipped with the most sophisticated weapons and interacting with the formations of the U.S. Seventh Fleet in Pacific—that creates a dangerous disequilibrium of forces on the Korean Peninsula. This is why North Korea feels compelled to undertake special measures to ensure its national security.

Though the DPRK deploys nearly 70 percent of its armed forces south of Pyongyang, including seven mechanized mobile-attack combat corps along a defensive perimeter within 20 km of the DMZ, North Korea reacted with considerable irritation to the U.S. Navy's military exercise "Fall Eagle," which took place close to its borders in October–November 1995. North Korean demands for the dismantlement of the concrete wall along the thirty-eighth parallel and for the repeal of the National Security Law in South Korea became increasingly insistent. The purchase of Russian military technology by South Korea was labeled in Pyongyang as a "flagrant antinational crime hostile to the cause of reunification," as a step toward "unleashing a new war," and as a dangerous act that might undermine the U.S.-DPRK Agreed Framework.[45] In this context, Pyongyang is actively trying to replace the interim armistice agreement between North Korea, China, the United Nations, and the United States with a bilateral U.S.-DPRK peace treaty bypassing South Korea.

Quoting the Seoul Institute of Defense Research, the North Korean press repeatedly stresses that South Korea, as early as the mid-1980s, had already completed technological preparations for the production of an atomic bomb, and that only the intervention of the great powers forced Seoul to suspend those efforts.

A serious impediment to the implementation of the Agreed Framework and another pretext for the DPRK's withdrawal from the NPT might evolve from problems related to the storage of South Korean nuclear fuel on Kurapato Island in the Yellow Sea. As reported by the North Korean Committee on the Environment, South Korea's nuclear industry has already disposed of some 11,000 trillion bukkerels of nuclear waste in the waters around the Korean Peninsula, which has caused considerable damage to the hydrosphere and biosphere of the region. A new nuclear waste storage site on Kurapato Island is being

North Korea and the Nuclear Nonproliferation Regime 155

constructed in a zone of intensive seismic activity and in the pathway of disastrous typhoons and underwater currents. More than 25 million people, not only Korean residents, but citizens of neighboring states as well, live within 100 km of Kurapato. The plan is to store from 3,000 to 15,000 tons of high-level nuclear waste there. Any leakage from this waste that might result from a natural disaster reportedly could be 1,000 times more dangerous than the consequences of the Chernobyl nuclear catastrophe.[46] The North Korean official press insists on linking Pyongyang's obligations under the NPT with the freezing of construction on the South Korean "nuclear graves" on the Kurapato Island.

Finally, another serious pretext for a future North Korean attempt to suspend or freeze its membership in the NPT might be active military-technological (including nuclear) research in Japan. The tragic days of colonial subjugation and humiliation by the Japanese are deeply rooted in the consciousness of every Korean. The North Korean news media constantly makes note of the fact that the current Japanese military budget, although within 1 percent of its GNP, is actually considerably higher than the total military spending of a major power like Russia. The recent statement of then-Japanese Premier Tsutomi Hata that "Japan had the capability to develop nuclear weapons" provoked a stormy reaction in Pyongyang. Following that statement, North Korean authorities demanded that Japan "admit having nuclear weapons." Citing materials by Greenpeace, the Pyongyang news media has claimed that since 1987 the United States has repeatedly transferred dual-use nuclear technologies that could also be used for designing nuclear weapons to Japan, in violation of the NPT.[47] North Korean sources have also sharply criticized Japanese statements in the wake of the August 1998 *Taepodong-1* test confirming their joint research program with the United States on missile defenses, as well as Tokyo's new plan to launch four spy satellites to improve its intelligence-gathering capabilities regarding North Korea.[48]

The above analysis shows that subsurface reefs and rocks still endanger a genuine settlement of the nuclear crisis on the Korean Peninsula. The world community must therefore continue its search for constructive means of achieving a consensus in order to overcome this ongoing threat to peace in Northeast Asia.

14

North Korea's Negotiations with the Korean Peninsula Energy Development Organization (KEDO)

Alexandre Y. Mansourov

This chapter investigates the record of the Democratic Peoples' Republic of Korea's (DPRK) negotiations and other interactions with KEDO in implementing the Agreed Framework from 1994 to 1998. The analysis presented here discusses some of their negotiating tactics and assesses the results that they have been able to achieve thus far in their quest for survival. The chapter also attempts to determine how Pyongyang's strategic calculations and tactics may affect the future implementation of the Agreed Framework.

"Implementing while Negotiating"

Since the conclusion of the Geneva Agreed Framework in October 1994, the DPRK's approach to its implementation has been rather consistent and overall a positive one. It is not a stop-and-go policy dependent on prevailing political winds, which seems to have characterized the Republic of Korea (ROK) policy during the Kim Yong-sam's administration. Nor is it a centrally guided determined massive application of labor, material, and financial resources aimed at rapid completion of a project, resembling a traditional North Korean *soktochon* (speed battle), the analogue of yet another grand communist construction effort. Neither does the implementation resemble an anemic process occasionally reinvigorated by injections of KEDO-arranged financing (a "go as they pay" attitude) or by what the North deems the unreasonable escalation of Western unfair demands (a "when dogs bark, cats meow" approach). In a fashion similar to their negotiating strategy during the Korean war (i.e., "negotiating while fighting"), the DPRK leaders have adopted an approach best described as "implementing while negotiating." They hope to improve the terms of the nuclear package deal, even if only on the margins, while abiding by their commitments and keeping a watchful eye on U.S. actions in this regard. Hence, vigorous bargaining continues up to the present time, but it is accompanied by good faith efforts in Pyongyang aimed at successful implementation of the nuclear package deal in the end.

When in early 1995 senior Japanese diplomat Itaru Umezu was packing his bags in Tokyo to accept his new assignment as one of KEDO's deputy directors in New York, his colleagues at the Japanese Ministry of Foreign Affairs joked that he had the best of both worlds waiting for him: that is, "to be in New York and to have no work to do." This is indicative of the generally low original expectations regarding KEDO's tasks, workloads, and likely success. However, to the surprise of many, what was accomplished in the course of intensive KEDO-DPRK bargaining in 1996 and 1997 "far exceeded the most optimistic expectations of the KEDO officials."[1]

Despite ups and downs and various twists and turns, day-to-day negotiations proceeded in a rather businesslike and friendly atmosphere. The first round of the working delegation meetings between eleven DPRK representatives headed by Ambassador Ho Jong and fourteen KEDO officials headed by KEDO Executive Director Ambassador Steven Bosworth, took place in Kuala Lumpur in mid-September 1995. This was the first formal negotiation between the two sides, and the first time that DPRK and ROK delegates had sat together at one table under the KEDO framework. KEDO delegates from South Korea included a number of officials from the ROK's interagency Office of Planning for the Light-Water Reactor (LWR) Project and the ROK's KEDO Deputy Executive Director Choi Young-jin. This exchange provided one of the first occasions for DPRK and ROK officials to sit together at one table within the KEDO framework and negotiate face to face. It was a remarkable departure from the past practice when Seoul used to express its positions to Pyongyang indirectly through the good offices of Washington since all previous discussions on the nuclear question had taken place at the bilateral U.S.-DPRK talks.

On September 12, 1995, at a high-level meeting attended by the KEDO executive director, his two deputies, and the head of the DPRK delegation, Ambassador Ho Jong, reiterated the official DPRK positions on the scope of the supply contract and terms of repayment and suggested that the supply contract could and should be concluded by October 21, 1995 (i.e., the first anniversary of the Geneva Agreed Framework). KEDO's senior officials rejected such an artificial deadline, as well as Ambassador Ho Jong's "unreasonable" demands to expand the scope of the supply contract and ease the repayment terms. Ambassador Bosworth urged Ambassador Ho Jong to accept the principle of burden sharing and to commit his country without reservation to the principle of full and timely repayment of the loan.[2]

Following these formal high-level meetings, from September 13 to 15, 1995, delegates from the DPRK and KEDO entered expert-level talks at which they compared their respective versions of a draft agreement and went over their positions item by item.[3] But the Kuala Lumpur negotiations stalled over a number of issues of both a substantive and symbolic nature. Regarding the scope of financial obligations under the supply contract, the North Korean side continued to insist that KEDO must bear all the costs involved in the LWR transfer project, including the transportation and assembly of the LWRs, construction of infrastructure for the LWR plant, and training of the DPRK personnel who would work there. It also demanded that KEDO provide due compensation for everything Pyongyang had invested in the development of an independent nuclear power industry over the past few decades. In total, Pyongyang presented KEDO with a bill exceeding $10 billion. The DPRK spokesman stated bluntly that "We have already made it clear to the U.S. side that we will not invest a penny more in this project because we are not certain whether or not the LWRs will be built so well that we will be able to operate them for commercial purposes."[4]

In addition, Pyongyang voiced strong opposition to KEDO's intentions to use the ROK-originated LWR model, to bend the principle of simultaneous action in favor of the international community, and to retain the right to certify the safety standards on the newly built LWR plant. But the KEDO delegation did not budge and reasserted its demand that the scope of the LWR supply project be limited strictly to the KEDO-sponsored assembly of LWRs and to the compensation for minor additional expenses for the site surveys and site preparation. Also, KEDO reiterated its right to choose the LWR model and primary contractor by itself and to examine the safety of the proposed LWRs without North Korean participation.[5]

In anger, the DPRK delegates responded by resorting to some vaguely disguised threat, stating that "to reach a solution, we can take another road, which our dialogue partner is well aware of": that is, the unilateral abrogation of the Geneva accords.[6] At this point, the Kuala Lumpur expert-level talks were temporarily suspended. The two sides agreed only "to make joint efforts to conclude an agreement on the LWR project at the earliest possible date ... as part of the political process to provide light-water reactors to the DPRK under the U.S. DPRK framework agreement of October 21, 1994, and the U.S. DPRK joint press statement in Kuala Lumpur dated June 13, 1995."[7]

On September 27–28, 1995, KEDO held a round of working-level meetings of experts from the United States, Japan, and the ROK, including representatives of the ROK Office of Planning for the LWR Project, (the ROK government-owned electric utility), the ROK Atomic Energy Institute, and the ROK Atomic Energy Safety Institute, at which they finalized Western positions regarding the overall and item-by-item contents of a draft supply contract.[8]

KEDO DPRK expert-level contacts resumed in New York on September 30, 1995. They paved the way for the second round of formal negotiations between the DPRK Ministry of Foreign Affairs (MFA) team, headed by Ambassador Ho Jong, and KEDO delegation, led by Ambassador Bosworth, which took place in New York from October 16 to December 15, 1995. During these talks, both delegations were able to narrow their differences regarding some technical and legal aspects of a proposed draft supply contract. In a joint statement following their conclusion, both sides declared that "progress has been made on some key issues" and reiterated their willingness "to redouble their efforts to conclude a LWR supply contract at an early date."[9]

However, it took two more months for both sides to reach the final resolution of all the outstanding issues, including "technical matters and wording" of the final document.

Once their attempt to impose the October 21 deadline was thwarted by KEDO officials, and feeling no time pressure any longer, DPRK officials, fully aware that it could be their last chance to get what they wanted at this final negotiating stretch, attempted to squeeze as many new concessions out of their Western counterparts as they could, even at the price of delaying the conclusion of the final contract. ROK representatives became rather worried at what they perceived as the North's "salami tactics": that is, the DPRK's continuous demands of additional concessions, backed by its stubborn refusal to sign the final agreement and threats to break off talks with KEDO in favor of their bilateral crisis-driven negotiations with the United States. As a result, as U.S. diplomats had done a year before in Geneva, KEDO representatives adopted the principle of package resolution, i.e., the concept that nothing could be agreed to with the North unless every dispute was settled first. Thus, Pyongyang's salami tactics, when bordered on brinkmanship, occasionally had to confront KEDO's high-pressured "all or nothing" strategy.

Eventually, both sides decided to compromise and settle. Specifically, in exchange for the DPRK's acquiescence to almost halve its initial bill and to drop its demands about technical standards, delivery schedules, and passage routes, KEDO agreed to extend its financial obligations to cover the costs of building auxiliary port facilities, an access road to the LWR plant to be constructed at Sinp'o, as well as local electricity transfer and distribution facilities. By and large, both sides settled the issue of their mutual financial obligations regarding construction expenses that were still to be specified in the final version of the supply contract. Moreover, the parties made another step toward a compromise on the type of LWR model, although not the designation of the reactor. They agreed that it should be selected by KEDO, but with the DPRK's acceptance.[10] However, the issue of the repayment terms and schedule was still not resolved.[11]

Finally, on December 15, 1995, "as a result of multilateral cooperation based on good will, a pragmatic, cautious approach, and hard work in a businesslike atmosphere,"[12] the head of the DPRK MFA delegation Ambassador-at-Large Ho Jong and KEDO Executive Director Bosworth successfully concluded the third round of formal talks in New York and signed the Supply Contract elaborating the provisions of the October 21, 1994, Agreed Framework and the June 13, 1995, Kuala Lumpur joint declaration on the terms and modalities of the supply of two LWRs by the international consortium to the DPRK government.[13] Both sides claimed victory. Interestingly, they both stressed that the significance of the supply agreement was twofold. First, it converted political promises stated in the Geneva accord and Kuala Lumpur declaration into legal obligations binding on both sides under international law. Second, it opened possibilities for expanded exchanges and improving the atmosphere of relations between the North and the South, as personnel, technology, and equipment would flow between the two Koreas throughout the duration of the contract.

The first compromise struck by the DPRK and KEDO negotiators combined KEDO's agreement to select "an advanced version of a U.S.-originated LWR technology currently under production based on reference plants" and the DPRK's reluctant consent to accept a prime contractor to be selected by KEDO, which was understood to be a South Korean company, as long as the overall supervision of the LWR transfer project would be conducted by a U.S. company.[14] In other words, a U.S. supervisor was to oversee how a South Korean prime contractor would build the ROK-made Model 3 and Model 4 of the Uljim-type reactor design at Sinp'o. Subsequently, with Pyongyang's grudging acquiescence, KEDO selected KEPCO as its prime contractor. On July 4, 1997, the KEDO Secretariat and KEPCO reached an agreement on the terms and conditions of the Preliminary Works Contract (PWC),[15] and it was signed on August 15, 1998.[16] Also, without many objections from Pyongyang, KEDO selected a U.S. architecture-engineering firm, Duke Engineering & Services (DE&S), as its technical program coordinator to assist KEDO's Project Operation Division in supervising overall implementation of the LWR project.[17]

The second compromise preserved the principle of simultaneous action and specified that the nuclear cores should be loaded into the LWRs only after the DPRK has fully and beyond any doubt clarified all the questions raised by the International Atomic Energy Agency (IAEA) regarding its compliance record. This would be achieved by allowing the IAEA to conduct special inspections aimed at verifying the amount of processed plutonium initially declared and ostensibly being hidden at two suspect nuclear waste sites at Yongbyon.[18]

In addition, according to the Supply Agreement, both sides agreed that

KEDO would sponsor the purchase, transportation, and assembly of two 1,000 megawatt (MW) LWRs at Sinp'o at anoverall cost to be jointly determined at a later date.[19] Moreover, KEDO acquiesced to assume additional responsibility for building roads at the nuclear reactor site, facilities for unloading construction equipment and materials at a port adjacent to the area, as well as water supply and drainage facilities at the site.[20] The DPRK reluctantly agreed to the KEDO-proposed repayment terms: it is to repay KEDO in cash, cash equivalents, or through the transfer of goods for each LWR plant in equal semiannual installments, free of interest, over a twenty-year term after completion of each LWR (including a three-year grace period beginning upon completion of each plant).[21] Pyongyang also committed itself to sign at least ten follow-up subject-by-subject enabling protocols and additional agreements with KEDO in order to facilitate smooth and timely implementation of the provisions of the Supply Contract and its four annexes.[22]

Subsequently, KEDO has been rather successful in negotiating follow-up agreements with Pyongyang in "lengthy and arduous, but very businesslike talks."[23] On April 8, 1996, in New York, KEDO officials from the United States, Japan, and ROK, headed by Lucy Reed, KEDO's legal advisor, entered into negotiations on a Protocol on the Juridical Status, Privileges, Immunities, and Consular Protection with a DPRK negotiating team composed of five MFA representatives, led by Choe Byong Gwan, the DPRK MFA deputy director general. On April 16, 1996, Dr. Mitchell B. Reiss, senior policy advisor to KEDO's executive director, led a KEDO delegation composed of the ROK representatives from KEPCO and the ROK Office of Planning for the LWR Project, as well as U.S. and Japanese officials, to start the talks on a Protocol on Transportation and a Protocol on Communications with a seven-member DPRK delegation headed by Ri Myong Sik, counselor at the General Department of Atomic Energy. Following twelve weeks of arduous but constructive talks between the representatives of the DPRK MFA, Ministry of Transportation, and Ministry of Posts and Communications on the one hand and KEDO officials on the other, they initialed the Protocol on Privileges and Immunities, on May 22, 1996,[24] and the Transportation and Communications Protocols, on June 14, 1996.[25] On July 11, 1996, KEDO Executive Director Bosworth and the head of the North Korean delegation, Ambassador Ho Jong, formally signed all three protocols in New York, making them an integral part of the Supply Agreement and legally binding on all parties concerned.[26] The Korean Central News Agency (KCNA) reported that, as a result of the conclusion of these first three enabling protocols, "the agreement on the provision of LWRs signed on December 15, 1995, finally entered the stage of implementation."[27]

What is striking about the privileges and immunities protocol is that the DPRK agreed to grant the virtual right of exterritoriality to KEDO, its personnel and members of its delegations, its contractors and subcontractors, to its activity areas,

KEDO-designated transportation routes, its property and assets, and offices and private residences. Having acquired a sovereign juridical status in the DPRK, KEDO is allowed to conclude contracts, purchase and dispose of movable and immovable property, institute its own legal proceedings, and enter into negotiations with relevant authorities of the North. Moreover, it is immune from all forms of legal and administrative action by local authorities against it; it is immune from search, requisition, confiscation, expropriation, and any other form of interference; it is immune from local financial controls, regulations, and moratoria; it is exempt from local taxes and duties, charges and fees, and any restrictions on imports and exports of materials related to the LWR project. It is free to hold, use, remit, and convert currency, and to use codes and secure means of communications. It is not subject to local censorship and is allowed to maintain order in its activity areas.[28] In other words, this protocol created a totally independent quasi-state within the state in the DPRK, a "KEDO-land," with its own KEDO-controlled activity areas where the sovereign rights of the DPRK government are severely restricted, and where KEDO operates its own legal administrative and enforcement systems, its own banking, postal, and communications systems, its localized labor and traffic regulations, and where KEDO-issued identity cards are tantamount to internationally recognized travel documents.

Although such extremely generous treatment offered to foreigners in the land of *juch'e* might seem rather unusual at first sight, one can find at least a couple of precedents that may help us put these current developments into a broader historical perspective. The Korean-Japanese Kanghwa Treaty of 1876 that opened the "hermit kingdom" to the West was the first international treaty signed by the traditional Korean rulers under considerable pressure from the Japanese. It opened Korean ports, granted immunities and various exemptions from taxes and customs duties, and offered exterritoriality rights to the newly built settlements of the "barbarian Nipponese." Also, in the mid-1950s, when the North Korean economy was being rebuilt by hundreds of thousands of Chinese People's Volunteers following the devastating effects of the Korean War, the latter's settlements enjoyed virtual exterritoriality rights in the North until the Chinese withdrew completely in 1958. Thus, from an historical perspective, it is less surprising that the DPRK government acquiesced to KEDO's omnipotent juridical status in the LWR project areas now that the North Korean economy lies in shambles and its leaders are confronted with considerable pressure from the West to open up and modernize their country. So far, Kim Jong Il has not done anything yet that had not been done before by his traditional predecessors, including his father, the Great Leader. When desperate, in order to survive, despite their occasional hot rhetoric, Korean leaders are used to caving in and showing an unlimited degree of pragmatism rather than suicidal inclinations.

What is interesting about the transportation protocol is the fact that it highlights North Korea's so-called China dependency, whereby the air transportation route to be used by KEDO, its contractors and subcontractors for the delivery of LWR project-related supplies and equipment, and transportation of its personnel would have to go to the nearest DPRK airport of Sondok via Beijing.[29] This arrangement reflects growing North Korean dependence on China as a primary route for its outgoing and incoming foreign trade and travel following the collapse of Soviet-DPRK economic ties. Also, it gives the Chinese government a powerful tool to control KEDO-related traffic between the DPRK and the West, thereby indirectly affecting the pace and direction of the evolution of the LWR project as a whole. However, in this regard, it is clear that increasing reliance on Beijing as the key transportation route for the LWR project was the price that Pyongyang chose to pay for its refusal to open a direct transportation link between the North and South across the demilitarized zone (DMZ), as Seoul had initially demanded.

In addition, the DPRK government agreed to open up a number of domestic sea routes for KEDO transportation needs, and agreed to allow visa-free entry and exit for KEDO personnel, its contractors, and subcontractors at the mutually designated points of entry and exit. In return, it was able to secure KEDO's commitments to upgrade the Sondok airport and auxiliary port facilities to bring its own transportation vehicles to the North. Moreover, Pyongyang negotiators were able to get KEDO to commit itself not to allow its personnel, contractors, and subcontractors to smuggle in weapons, explosives, and military supplies (that could be clandestinely channeled to potential local opponents of the regime), binoculars and telescopes (of more than ten magnifications), cameras and camcorders (with more than 24X power zoom that could be easily used for spying), narcotics, and various publications and video-audio tapes (that might be used for propaganda purposes).[30]

What is noteworthy about the communications protocol is the North Korean government's agreement to allow KEDO to set up its own independent satellite and wireless communications systems and exempt KEDO's communications and correspondence from the duties, fees, interference, and censorship of local authorities.[31] In exchange for this concession, North Korea might get its first global satellite communications system and a regional wireless network, along with relevant technologies, equipment, and human expertise, which could be later applied for commercial purposes in other parts of the country.

Following the successful conclusion of the three above-mentioned protocols, both parties entered into the first round of negotiations on a protocol on project site and a protocol on the use of labor, goods, and services, which were held in the Mount Myohyang Resort area from July 20 to 30, 1996. The KEDO delegation consisted of government officials from the ROK, Japan, and the United States, and was led by Dr. Reiss. The North Korean delegation was headed by Choi In Ha, counselor at the General Department of Atomic Energy.

By all accounts, it was a remarkable personal experience for all its participants. Torrential rains that drowned the country in late July and brought the North Korean government machine to a partial halt presented both delegations with a very unpleasant choice of either negotiating under extremely adverse natural conditions or cutting the talks short and parting without any meaningful results. The DPRK delegation asked the head of the KEDO delegation to stay, displayed considerable good will to all KEDO delegation members, and adopted an uncharacteristically pragmatic stance in its negotiating posture. To KEDO's surprise, the DPRK's principal negotiator, Choe In Ha, made a number of concessions on major technical issues (as it turned out later, to the profound displeasure of the DPRK's Ministry of Foreign Affairs), which allowed both sides to narrow down significantly the number of differences in their basic technical positions. Thus, much progress was achieved, despite the limited amount of time at the parties' disposal and significant difficulties that the KEDO side experienced in maintaining communications with third countries. Moreover, at this round of talks the DPRK and ROK technical experts for the first time began direct face-to-face negotiations in various subcommittees within the KEDO framework, occasionally without any formal participation by U.S. representatives.[32]

The second round of talks on these two protocols was held in New York from August 22 to September 26, 1996. Negotiations proceeded in a businesslike manner, without much contention. The North Korean delegation showed a keen interest in acquiring new expertise in the area of atomic energy development and therefore was willing to accept most of KEDO's technical demands related to the procedures and modalities involving the site takeover, site access, and the use of site.[33] In return, in a concession to Pyongyang, KEDO agreed that its contractors will use the monopoly services of a DPRK government-run company for the procurement of local labor, goods, facilities, and other services needed for the implementation of the LWR project.[34] It was obvious that Pyongyang wanted the groundbreaking for the construction of the LWRs to start as early as possible and authorized its delegation, on September 26, 1996, to initial both draft protocols, subject to the final approval of the governments concerned.[35] At this point, both sides were full of confidence that by the end of 1996 they would be able to have finalized all outstanding matters so that the reactor construction process could begin in earnest before the beginning of 1997.[36]

However, on September 18, 1996, the notorious "submarine incident" erupted. A lot has been written about the incident and its implications.[37] The best available evidence suggests that the submarine intrusion may have been orchestrated by those within the North Korean government who did not like the LWR project as a whole and wanted it to fail.[38] Predictably, they got what they hoped for: the South refused to continue any meaningful talks on the LWR project until its demand for an apology was satisfied. Consequently, KEDO interrupted all external activities related to the LWR transfer project with the DPRK. The ROK's espionage accusations and demands were met initially by official silence in Pyongyang and, then, by a series of denials and propaganda-style condemnations of Seoul. Only after the United States stepped into the dispute displaying a marvel of mediation skills and exerting considerable pressure on both sides to compromise[39] was the DPRK government compelled, after arduous negotiations, to take the responsibility for the actions of its submarine and its crew and to issue a vaguely worded official apology with a pledge to refrain from similar provocative activities in future.[40]

There appeared to be intensive debate within the leadership in Pyongyang as to how to deal with the fallout from the submarine incident and the suspension of the LWR transfer project. In hindsight, it is clear that the forces advocating continuous political engagement with the West and sincere cooperation in the implementation of the nuclear deal prevailed in the end. Subsequently, they led the North Korean government to soften its stance vis-à-vis the South, apologize for the unwarranted provocation, and call upon Washington to resume KEDO activities in New York and in the DPRK.

Once the submarine incident was resolved, KEDO and DPRK representatives in New York moved quickly to sign the Protocol on Site Take-over, Site Access, and Use of the Site and the Protocol on Labor, Goods, Facilities, and Other Services for the Implementation of the Light-Water Reactor Project, respectively, on January 8 and February 8, 1997.[41]

Furthermore, following intensive negotiations in the spring and early summer of 1997, KEDO and DPRK concluded a Protocol on Actions in the Event of Nonpayment, an extremely important issue to the Japanese government. Also, on July 2, 1997, in New York, they agreed on a number of complex and highly detailed agreements and memoranda of understanding on various matters furthering the implementation of the LWR project. The agreements on procedures addressed the issues of entry and exit of KEDO persons to and from the DPRK; customs clearance and inspection; quarantine; sea transportation; the use of means of land transportation; banking services from the DPRK Foreign Trade Bank; the establishment of a non-DPRK bank's branch office and joint venture bank; and insurance services for the DPRK Foreign Insurance Company. The eight memoranda of understanding covered general principles on air transport; a more economic and efficient sea transportation route to the project site; telecommunications; postal services; general principles and guidelines for the conclusion of contracts on the provision of labor, goods, facilities, and other services by the DPRK; medical services and medical evacuation routes; an amendment to the protocol on site takeover, site access, and use of the site for the implementation of the LWR project; and determination of the details of the protocol on the site takeover, site access and use of the site.[42] In sum, on the eve of the third anniversary of

Kim Il Sung's death, the DPRK and KEDO made a quantum leap forward in their cooperation in the implementation of the Agreed Framework. Was it Kim Jong Il's way of paying his last respects to his deceased father by finally fulfilling the latter's last foreign policy wish, or was it a reflection of the official end of the three-year mourning period in the North?

Another area where the North Korean government showed keen interest and a spirit of cooperation was in regards to the KEDO-led LWR site surveys. From 1995 to 1996, six teams of nuclear experts, contracted by KEDO,[43] traveled to Sinp'o in order to conduct site surveys and collect relevant geological, climatic, and general data. During the first two site surveys, KEDO relied on an engineering firm, Burns & Roe, to provide technical experts to assist in gathering information and evaluating the suitability of the site for the project. Afterward, following the signing of a Pre-Project Services contract in January 1996, KEPCO provided technical experts for subsequent site surveys and assisted KEDO in preparing the Site Survey Report.[44]

The first site survey took place from August 15 to 22, 1995. Its participants included fifteen people: four South Koreans, seven Americans, and four Japanese experts. The survey team received and reviewed a summary of the results of the site survey conducted earlier by Russian specialists. (It is noteworthy that no compensation was ever paid by the DPRK to the Russian side for its labor and intellectual property rights to this study.) Also, they conducted a sight survey of the seismic, terrain, water, weather, and environmental conditions of the area in order to determine its overall suitability for the project.[45] But, apparently, their stay in the North was cut short because the entire country and its leadership were overwhelmed by the task of coping with the devastating consequences of the torrential rains and floods that had inundated the western seaboard of North Korea in early July and mid-August.[46]

The second site-survey mission, composed of nineteen people (seven South Koreans, nine Americans, and three Japanese), visited Sinp'o from October 24, 1995, to November 4, 1995.[47] This group of experts again reviewed the initial Russian site-survey report.

They also evaluated the results of the site safety survey prepared by the first mission with regard to soil and seismic conditions and the possibility of natural calamities. They also conducted a preliminary feasibility study of the site area with regard to water intake and displacement, electricity, communication, water, roads, construction funds, and other issues. It was upon the recommendation of this second mission that KEDO concluded that Sinp'o was a suitable site to consider for building the LWR plant.[48]

The third site-survey team, composed of fifteen people (eleven South Koreans—one official from KEPCO, one official from the Korean Power Engineering Co., or KOPECO, two KEDO officials, and seven specialists from the Samkyong Technical Services Corporation—two Americans, and two Japanese), went to Sinp'o on December 16, 1995, the day after the supply contract had been signed in New York. It stayed until January 16, 1996. Its members measured the site area and prepared a preliminary 1/1,000 map of the area. They installed seismic gauges in the vicinity of the site. Also, they had preliminary discussions with their North Korean counterparts from the General Bureau of Atomic Energy, the leading North Korean interface with KEDO on technical matters, about the preparatory work for soil and water examination in the area.[49] From the third site survey on, KEPCO, which was chosen by KEDO as the prime contractor to undertake the LWR project on a turnkey basis,[50] assumed the leading role in the site examination and preparation part of the LWR project.

The fourth site-survey mission lasted from January 16, 1996, to February 24, 1996. The twenty-three-member team (nineteen South Koreans—one KEPCO official, two KOPECO experts, two KEDO officials, and fourteen specialists from the Tong-A Consultants Company—two Americans; and two Japanese) performed preliminary boring procedures and seismic tests in order to determine the condition of the base rock, subterranean water, and soils, and measured the depth of nearby coastal waters. It also conducted an oceanic survey.[51] The significance of this mission was that for the first time, all KEDO directors, including its Executive Director Bosworth[52] and his Deputy Directors Choi Young-jin[53] and Itaru Umezu,[54] visited the DPRK all at once and held substantive talks with senior officials at the DPRK Ministry of Foreign Affairs and Ministry of Atomic Energy Industry in Pyongyang and Sinp'o. Moreover, on the eve of their mission, Ambassador Bosworth stated in an interview to the Associated Press that KEDO hoped to start building two LWRs in Sinp'o in the fall of 1996, which created a positive atmosphere for the upcoming talks in the North.[55]

The fifth site survey team, composed of twelve persons (eight South Koreans—five from KEPCO, two from KOPECO, and one from KEDO—two Americans, and two Japanese), visited North Korea from April 25 to May 7, 1996. They collected materials regarding the existing infrastructure facilities in areas near the project site, including the Yanghwa harbor, discussed what additional infrastructure facilities might be needed in the future, and reviewed the economic and social conditions for initial project programs.[56] It was on the eve of this mission that tensions inadvertently increased and hot rhetoric escalated again on the Korean Peninsula.[57] However, "to the great surprise" of the KEDO delegation, "a remarkable thing happened: this latest exchange of barbs and threats by the military on both sides of the 38th Parallel did not affect the course of nuclear negotiations and subsequent site survey missions at all."[58] Obviously, Pyongyang opted to stay the course in implementing dutifully its part of the nuclear bargain.

The sixth site-survey mission went to Sinp'o from July 6 to July 30, 1996.[59] It included nineteen persons (sixteen South Koreans—a KEPCO official, seven KOPECO experts, three professors from Seoul National University, one professor from Hanyang University, and four KEDO officials—one American, and two

Japanese). The North Korean side and KEDO formed five joint teams of experts who conducted a detailed observation of the site with regard to soil, water, environment, weather, and general affairs. Also, following on-site examination, the North Korean and KEDO specialists jointly reevaluated the site-survey materials prepared earlier by Russian experts and currently held by the DPRK.[60] The team composed a preliminary safety analysis report (PSAR) and an environmental report (ER), both required by the supply contract as mandatory steps toward the takeover of the project site from DPRK authorities and their issuance of permission for the LWR construction. In addition, on July 11, 1996, as part of the compensation to the local resident population stipulated in the supply contract, in a solemn ceremony KEDO, led by Park Song-hoon, deputy director of the ROK Office of Planning for the LWR Project, distributed $75,000 worth of humanitarian-aid items to the residents of the project area site who were to move away as a result of its redevelopment.[61]

On the basis of results of the above-mentioned six survey missions, in early 1997, KEPCO prepared a Site Survey Report analyzing the suitability of the site for the LWR project. Overall, the report concluded that the proposed site matched well with the requirements of the reference plant and the site could be fully demonstrated as suitable for constructing the LWR units at the PSAR stage. In March 1997, KEDO delivered an "advance" copy of this report containing the results of the first six site surveys and preliminary site boundary coordinates to the DPRK for its review in March 1997. At the same time, the KEDO Secretariat conducted its own independent review and documented its results in the Site Survey Report, which was then submitted to KEDO executive board members for approval in early July 1997. The final copy of the Site Survey Report was submitted to the DPRK in mid-July 1997, after the site boundary was agreed upon through further negotiations between KEDO and Pyongyang and after the site boundary coordinates were surveyed during the seventh site survey. Subsequently, on July 21, 1997, the North Korean government issued to KEDO the Site Take-over Certificate that allowed KEDO to open the KEDO Office at Sinp'o on July 28, 1997, and to begin work at the site.[62]

Throughout late 1997 and through the spring and summer of 1998, North Korean technical experts continued to work with KEPCO specialists at Sinp'o, gathering the additional meteorological, oceanographic, noise/vibration, air/water quality, and ecological data required to complete Chapter Two of the PSAR and ER. These materials, along with other reports produced by KEPCO, entitled "Establishment of the Site Design Parameters," "Analysis of Design Wave Height," "Analysis of Probabilistic Seismic Hazard," and "Tentative Floor Response Spectra," were included into the PSAR before its submission to the DPRK Nuclear Regulatory Authority (NRA). Now the DPRK NRA has to make the decision on the issuance of the Construction Permit.[63]

In the meantime, negotiations on enabling protocols continued in a businesslike fashion. In November 1997, KEDO and North Korean officials met in Hyangsan, to begin talks on a quality assurance and warranties protocol. A second round of these discussions was held in New York in December 1997, and the third round took place in New York in August 1998. At these meetings devoted to the issues of nuclear safety, DPRK nuclear safety experts shared information about their nuclear regulatory infrastructure and nuclear safety regulations. In turn, KEDO officials provided information concerning codes and standards, a model, and the reference plant's PSAR and ER, and also informed the DPRK side about the details of te KEDO's Nuclear Safety Confirmation System approved by the KEDO Executive Board on February 6, 1998, in its resolution No. 1998-02. Although negotiations on the quality assurance and warranties protocol are still in progress, these meetings have already gone a long way to alleviate the nuclear safety concerns of various KEDO members and the DPRK's concerns about quality control and warranties.[64]

Finally, with further evidence that the implementation of the Agreed Framework appeared to be on track, the DPRK agreed to sign two Records of Discussion, following the conclusion of its technical meetings with KEDO at Hyangsan in January and May 1998. In January 1998, the two sides finalized details involving the most efficient transportation to the site, the location of a new pilot station, the improvement of telecommunications, the cultivation of vegetables at the site, off-site recreational travel by KEDO personnel, as well as the construction of a switchyard and off-site power system for the LWR plants. In May 1998, the two parties agreed on certain technical modifications to the existing protocols, further improving KEDO's telecommunications and broadening off-site recreational travel rights. They also agreed to designate additional DPRK suppliers of labor, goods, and services for the LWR project, and to reach an accord on how to compute wages and salaries for DPRK workers. Finally, they also came to a compromise on the question of who should bear responsibility for building and paying for new road construction near the site.[65]

In general, from 1994 to 1998, the North Korean record of implementation of the new policies designed in the Agreed Framework was quite good. Among its achievements, one could cite the uninterrupted freeze of its nuclear program, which the DPRK has been grudgingly maintaining up to the present; continuous access to and unimpeded monitoring of the Yongbyon nuclear facilities, which Pyongyang has granted to the IAEA inspectors, some officials from the U.S. Department of Energy, and U.S. private contractors; substantial North Korean cooperation with the international nuclear community in resolving technical issues at hand, including safe storage of spent fuel rods and repair of the flood-damaged nuclear facilities; and, in general, good faith displayed by the North Korean officials in negotiating the enabling agreements envisioned in the Agreed Framework.

Among some major problems facing the implementation of the Agreed Framework, one could mention periodically voiced North Korean demands for new add-ons to the already agreed upon $5 billion technical assistance package, including demands for the construction of an electric power grid, port facilities, and a training center for nuclear specialists; occasional flashes of counterproductive hot rhetoric; difficulties in raising the promised financing for payment for heavy fuel deliveries, not to mention the payment for the LWRs themselves; deteriorating general economic environment in the North and in the South; compartmentalization of policy implementation on all sides; and other problems.

Conclusion

To sum up, Pyongyang has adopted an "implementing while negotiating" strategy, following the historical example of the "negotiating while fighting" strategy so skillfully employed by Kim Il Sung during the armistice talks throughout the Korean War in 1951–53. Specifically, today it displays a considerable amount of good faith and patience during trying and uncertain periods of the nuclear negotiations and by and large fulfills meticulously its part of the nuclear bargain, despite a constant deficit of resources and protectionist demands by domestic hard-liners.

At the same time, its negotiators continue their hard-nosed bargaining with KEDO on the nitty-gritty details of the implementation process in hopes of gaining better terms in the nuclear deal for the DPRK, even if only on the margins. A good sign is that no matter how tough North Korean demands occasionally appear to be and how hot their rhetoric often gets, Pyongyang has not crossed the line yet, as it did in March 1993 when it decided to withdraw from the nuclear nonproliferation treaty. Since the negotiations began, all the time it remained at the bargaining table, always willing to come back with more reasonable offers in order to strike a deal.

At present, it is obvious that the DPRK will not have received two LWRs from KEDO by the year 2003 when all the elements outlined in the Geneva agreements are supposed to be put in place. But delays in the implementation of the original package deal seem to have already been discounted by political actors concerned in Pyongyang, Seoul, Tokyo, and Washington. Therefore, as long as the basic original strategic and political-economic calculations on both sides remain unchanged and in force, despite mounting challenges, the entire Agreed Framework is unlikely to collapse due to the occasional inability of the parties to meet certain negotiable deadlines and conditions. As long as policy makers on both sides are preoccupied with the day-to-day problems and conflict resolution and refuse even to speculate about the prospects for DPRK-KEDO cooperation, there are reasons to remain cautiously optimistic about the future of peaceful denuclearization, growing détente, and greater economic and humanitarian cooperation on the Korean Peninsula in the years to come.

15

China and the Korean Peninsula

Managing an Unstable Triangle

Evgeniy P. Bazhanov and James Clay Moltz

Among states active and interested in the Korean Peninsula, China has recently enjoyed the closest relations with both North and South. China has sought to retain the political benefits it sees in continuing close party-to-party relations with Pyongyang, while exploiting new economic opportunities with Seoul through the rapid development of bilateral trade and investment.

Of the two Korean sides, Pyongyang has expressed the greater irritation with this arrangement, since it has had to acquiesce to Beijing's recent decision to engage its archrival in official relations. China's sheer size and importance as North Korea's last major ally, however, have allowed it to weather Pyongyang's criticisms without a break in official relations. As a result, China remains the only major power (besides Russia) that maintains a diplomatic presence in both capitals, and, for economic reasons, Beijing's presence has recently been a far more active one. Indeed, China's unique status puts it in a potentially important position in regards to any settlement on the Korean Peninsula.

The Chinese government's pursuit of a balanced policy between the two sides may have already influenced the Agreed Framework and the current pattern of seeking a negotiated settlement to the nuclear crisis with North Korea. As U.S. analyst Larry Nitsch suggested in mid-1994 (before the signing of the Agreed Framework), China's two-track policy might have served as an example for the United States regarding the necessity of engaging North Korea in order to solve the nuclear crisis.[1] China's Foreign Minister Qian Qichen explained Beijing's prescription to the United States in a speech in mid-1993:

We support dialogue between North Korea and the United States, North Korea and the International Atomic Energy Agency, and inter-Korean dialogue. We hope there is much progress in these talks. We oppose the use of pressure.[2]

The eventual U.S.–DPRK Agreed Framework, signed in October 1994, incorporated many of these principles.

Yet, while China may have pointed to fruitful new avenues for compromise, it has not itself played an especially active role in supporting the actual compromises reached by other powers, at least in a material sense. Instead, it has sought to maintain a detached position regarding the Agreed Framework, while still remaining engaged in trade and political relations with the peninsula. This line has allowed it to maneuver strategically between the two Koreas, providing it with maximum flexibility and minimal responsibility. Beijing's two leading principles seem to be preventing North Korean collapse, while at the same time not pushing too fast for eventual reunification. In this way, Beijing has sought to preserve the status quo on the peninsula,[3] which it sees as offering China the best of both worlds. As Chinese Foreign Ministry spokesman Zhu Bangzao stated in December 1998: "China will continue to play a constructive role in safeguarding peace and stability on the Korean Peninsula *in its own way* [italics added]."[4]

This chapter first analyzes China's official policy on the Korean Peninsula. In doing so, it examines recent high-level relations and Chinese policy pronouncements especially regarding North Korea, which continue to emphasize its balanced approach. The second part of the chapter, however, examines recent undercurrents of unofficial disagreement with the North, which is now causing new strains in the relationship. In its conclusion, the chapter questions whether China's efforts to patch over these problems will remain tenable in a future crisis or whether China instead will be forced to choose sides. How and whether China is able to manage this triangular relationship will undoubtedly influence the future outcome on the Korean Peninsula.

China's Official Policy toward the Two Koreas

Whereas Chinese foreign policy under Mao Zedong was often characterized by aggressive internationalism in the promotion of communist ideals, Chinese foreign policy today places its highest priority on serving the country's national interests, particularly within the East Asian geopolitical environment. In this regard, Beijing remains highly concerned with the maintenance of stability along its borders, even to the exclusion of what would previously have been called its "international communist" duties. As Thomas J. Christensen has noted, "China may well be the high church of realpolitik in the post–Cold War world."[5]

But China's growing awareness of its role in the emerging multipolar world and its growing economic needs have caused an increase in Chinese diplomatic activity. China has normalized relations with South Korea and has broadened its ties with a variety of former adversaries in Asia. At the same time, however, Beijing still maintains an alliance with North Korea according to the 1961 DPRK-Chinese Treaty on Friendship, Cooperation, and Mutual Assistance. This treaty includes a military component pledging Beijing's aid to Pyongyang in case of foreign attack. (A similar treaty with the former Soviet Union expired in 1996.) Although there may be questions of China's eagerness to act on these pledges, Beijing has been an active supporter of the DPRK in the United Nations, blocking several attempts by the West to punish North Korea for particular behavior and making the imposition of meaningful sanctions against North Korea in 1993 impossible, despite its threat to withdraw from the Treaty on the Non-Proliferation of Nuclear Weapons.[6]

China's official policy pronouncements toward the Korean Peninsula stress the need to prevent conflict, keep nuclear weapons off the Korean Peninsula, promote peace talks between the North and the South, and participate in the Four-Party Talks and other multinational efforts. As Chinese Foreign Ministry spokeswoman Zhang Qiyue commented at a press briefing in January 1999: "We hope that the parties concerned can move the four-party talks forward ... on the basis of safeguarding the peace and stability of the Korean Peninsula."[7]

According to Russian expert Leonid Moiseyev, Peoples' Republic of China (PRC) Chairman Jiang Zemin laid out the core guidelines of this policy in a speech during his November 1995 visit to Seoul.[8] Jiang mentioned the following points, noting China's desire to: (1) pursue peace and stability in this area of the world as its main objective; (2) support the independent and peaceful unification of the peninsula; (3) view favorably rational proposals and constructive efforts by the parties concerned in order to relax tensions on the peninsula and improve mutual relations; (4) reject any selfish goals in connection with the Korean problem; and (5) seek to implement the five principles of peaceful coexistence in order to promote the development of good-neighborly and friendly relations with the North and the South for the sake of the preservation of peace and stability on the Korean Peninsula.[9]

In conversations with Russian colleagues, Chinese officials and scholars have emphasized the importance of creating a subregional security mechanism for the Korean Peninsula. But they complain that hostile feelings between the DPRK and the ROK have grown even more intense. Chinese diplomacy has sought consistently to reduce these tensions and also to minimize the threat that the DPRK's current economic crisis might spill over into a regional conflict. Part of this effort has involved trying to reassure Pyongyang regarding China's support for the regime.

At the thirty-seventh anniversary celebrations of the DPRK-PRC Treaty that took place in Pyongyang in July 1998, for example, Chinese Ambassador Wan Yongxiang praised the recent elections of deputies to the tenth DPRK Supreme People's Assembly and expressed his belief that Sino–North Korean relations would remain strong no matter what changes might occur in the international situation.[10] China has also periodically sent its naval vessels to visit the DPRK as a signal to other states that China remains tied to North Korea and will not allow other states to bully it with military threats.[11]

In terms of assistance, China continues to send periodic shipments of food to North Korea, at least in part due to its desire to stave off even further out-migration of North Koreans across the Chinese border into Manchuria than is already occurring. Indeed, Chinese officials have deported considerable numbers of refugees back into North Korea due to its fears about their rising numbers and complaints about their tendency to become involved in criminal activity in order to support themselves.[12] China donated 30 million yuan in food assistance after the summer floods devastated North Korea's west coast in 1995 and in 1996 provided 120,000 tons of additional food aid.[13] In July 1997, China donated 80,000 tons of food to the North to mark the third anniversary of Kim Il Sung's death.[14] Overall, in accordance with the bilateral Agreement on Economic and Technical Cooperation for 1996–2000, the PRC is to provide the DPRK with long-term economic aid worth 500 million yuan.[15] But China's aid is not entirely without strings. For example, China has agreed to continue its shipments of oil and coking coal, but only under the condition that within three years the North Korean party will pay back the debt it has incurred as a result of the prior Chinese deliveries.

According to statistics from the International Monetary Fund, trade turnover between China and North Korea totaled only $809 million in 1997. This consisted of $690 million in Chinese exports to the DPRK and a meager $119 million in North Korean exports.[16] Figures for 1998 indicated an additional 37 percent decline.[17] Bilateral trade involves North Korean sales of steel, copper, zinc, timber, and silicon in return for purchases of Chinese food products, shoes, coal, soap, sewing machines, and televisions.[18] Despite the low volume, however, China constitutes North Korea's top trading partner, giving Beijing potential leverage over its neighbor.

Beijing's main economic interests on the peninsula, now lie clearly in its much larger and more profitable relations with Seoul. In 1997, for example, China's trade turnover with South Korea totaled $24 billion. Bilateral trade figures grew at an average annual rate of 40 percent per year through the late 1990s (with the bulk of this trade consisting of Chinese exports to the South).[19] China now represents Seoul's third-largest trading partner, while the Republic of Korea (ROK) has become the fifth-largest trading partner for Beijing.

However, analysts of the triangular relationship should not overestimate the role of economics. According to Moiseyev, possible U.S. attempts to consolidate or expand its military presence in Northeast Asia, together with a growth in the independent military role of Tokyo and Seoul, for example, could stimulate China to move closer to North Korea once again.[20] Some movement in this direction has been seen in the context of the increasing assertiveness of U.S.-Japanese theater missile defense programs since fall 1998. This military cooperation remains a serious irritant for China, as well as North Korea. For this reason, the close bilateral relationship between Beijing and Pyongyang is one that will continue to merit monitoring and attention.

While China's official policy regarding the Korean Peninsula is to behave as a status quo power, this policy is beginning to show signs of tension below the surface. In fact, as informal discussions with Chinese officials make clear, China's relations with North Korea over the past several years have been anything but smooth. The PRC has felt a need to distance itself from Pyongyang's military ventures (such as the 1996 submarine incident and the August 1998 missile test), while also seeking to maintain its independence in sensitive political negotiations involving bilateral relations (such as the highly publicized defection of Hwang Yong Yap to the South Korean embassy in Beijing in 1997).

In their official communiqués and speeches Chinese and North Korean leaders continue to insist that Beijing and Pyongyang are "as close as teeth and lips" and that their friendship is "eternal and unbreakable."[21] However, there are real tensions that seem to be increasing, not decreasing, with the passage of time.

Unofficial Chinese Views of the Sino-North Korean Relationship

Evidence of difficulties in Sino-North relations cannot be found in official statements, but they are clearly simmering inside of China. Anecdotal evidence of such tensions was revealed during a recent trip taken by Russian experts through China. There they had informal conversations about North Korea with numerous middle-level officials, scholars, journalists, businessmen, and common people. On the basis of these conversations, the following conclusions can be drawn.

At the unofficial level, the Chinese have a very low regard for the North Korean regime. At a dinner in Zhejiang province, the Russians asked a local government functionary if the province had "sister city" relations with foreign provinces and municipalities. The functionary named a whole list of such partners: in Europe, Asia, and America. The guests ventured to ask the Zhejiang official: "What about North Korea?" The question drew loud laughter from this official and all other Chinese participating in the dinner. Explaining the reasons for this laughter, the Chinese called North Korea "a strange country," "an absolute feudal state," and "a land of bizarre events."

On other occasions, Chinese scholars in Fujian province and Beijing described the North Korean economy as "deformed, twisted, and lacking sense and vitality." A journalist from the province of Liaoning questioned Kim Jong Il's legitimacy as the national leader. Many students of Beijing University, when asked if they were interested in visiting North Korea, stressed that they "were ready to go anywhere" but not to the DPRK. Just one fisherman from Hainan island praised North

Korea as "the only true socialist state remaining in Asia." The reforms in China obviously had not brought any positive results to this man.

The experts' second conclusion on the Chinese attitude toward Pyongyang is that many Chinese don't believe that Kim Jong Il's regime has much of a future. A prominent journalist said: "Unless Pyongyang undertakes reforms, the ruling circles will be swept away. The economy of the DPRK does not work any longer." Chinese scholars stress that unless farmers are given land for their personal use, North Korea will continue to suffer from shortages of rice and vegetables. Conditions will become more and more severe. Eventually, riots will take place and the leadership's unity will be undermined. Kim Jong Il will be openly challenged as an inept leader, and the ensuing infighting will bring about the disintegration of its authority. Chinese observers also stress that a modern society cannot function on the basis of one-man, absolutist rule: decisions are bound to be biased, primitive, and unachievable. As for the masses, one Chinese journalist said that they could not display social and economic creativity under such a totalitarian regime. The common view expressed in China is that the days of the present regime in Pyongyang are numbered; in three to five years it will have to give way to a more sensible government and policies.

A third conclusion drawn from conversations with various unofficial Chinese is the most striking. Time and again the Russians asked their Chinese friends what Beijing's reaction would be if, unfortunately, a war broke out in Korea. The answer was unanimous: "We'll stay away from that war." The guests tried to rephrase the question: "Suppose the Americans participate in the conflict and overrun North Korean positions and advance to the border with the PRC?" The typical answer was stunning: "So what? The United States would never dare to attack China anyway. We are too big and powerful for America. Besides, in any case, why would the Americans want to attack us?" The Russian visitors asked time and again: "But don't you care about the survival of China's ally, a socialist state, and an old friend?" The Chinese responded, "North Korea is not our real ally." One Shanghai-based expert explained, "Look, South Korea and the United States proposed recently to hold four-party talks on the Korean issue. Pyongyang refrains from giving a positive answer. Why? Because they don't want China to participate! Is this the behavior of an ally?"

The guests from Russia reminded their Chinese acquaintances that China had sacrificed hundreds of thousands of soldiers in the Korean conflict of 1950–53. Had they forgotten this fact, and were they therefore indifferent to the fate of the DPRK? The Chinese explained: "We have not forgotten. However, Pyongyang pretends that it won the war on its own, as if it was not our Chinese volunteers who saved the North Korean regime from annihilation."

The Russian side responded, "But what about top Chinese leaders? Some of them personally fought in that war. Do they have a special attachment to Pyongyang?" The Chinese responded, "No such leaders are left in the government." One of them added: "Maybe you are right. Maybe our leadership will get involved in the next Korean conflict. However, the entire Chinese people would be against it. This kind of involvement would ruin China's reforms and future." This last phrase summarizes the key point: China's and North Korea's societies have gone so much in different directions that their interests—strategic, political, and economic—are becoming incompatible. As a result, such indifferent (and even hostile) attitudes—as those cited above—exist in China regarding the DPRK and its eventual fate.

According to the Chinese view, North Korea deserves more of the blame for current DPRK-ROK distrust since it is the one dispatching spy submarines, lashing out at President Kim Dae-jung in its propaganda organs, and constantly warning about a coming war. In addition, Chinese analysts worry that the North Korean economy is in such a sorry state that the DPRK may become destabilized.

A signal event in official relations that exemplified the growing rift was the defection of high-ranking North Korean official Hwang Yong Yap in early 1997. The outcome of the Hwang affair, in which Beijing allowed his defection to the South Korean embassy and his safe passage from the PRC seriously undermined North Korea's faith in China as an ally. Despite North Korea's calls for his repatriation to the DPRK, the Beijing government sought to avoid a conflict with Seoul. This became its top policy priority over the wishes of Pyongyang.

More recently, Beijing was quite angry with the August 1998 North Korean missile test. Chinese officials complained that the DPRK did not provide any advance notice to Beijing. When the Chinese side asked for clarifications, Pyongyang gave "a rather rude" answer by saying that each country had a right to develop its space program without outside interference.

The Chinese have strongly urged the North Koreans to "exercise restraint and refrain from doing anything that may cause tensions in the region and spark a new arms race." The Chinese government's position is that unless the DPRK acts prudently, Tokyo and Seoul "may be pushed to a more militaristic stance—intensified cooperation with the USA [and] procurement of more sophisticated weapons." It has cautioned Pyongyang against new tests of its missiles in order "to avoid international complications and to deny American hawks of additional arguments against the DPRK."

A final irritant in Sino–North Korean relations is Beijing's attempt to encourage reforms in the DPRK, offering its own experience in this field. In Kim Jong Il's eyes, the social and economic reforms going on now in China pose a threat to North Korea. In its internal propaganda, Pyongyang brands Chinese reforms as a "dangerous policy not only undermining socialism in the PRC, but doomed to deny the DPRK of its socialist ally and having a [negative] impact on the DPRK." This rejection is so strong that a number of North Korean officials were dismissed and arrested recently for praising Chinese economic policies.

All of these points do not rule out the possibility that Kim Jong Il may someday decide to visit China and that he will be welcomed there (although rumors have circulated that the North Korean dictator tried to tour China recently, but did not receive the requested invitation). Clearly, there are fundamental tensions in the relationship that will be difficult to overcome. On key issues of national security, it is now obvious that Beijing and Pyongyang no longer see eye to eye, and that China will see to avoid a regional conflict rather than side with its erstwhile ally. A few years ago, North Korea allegedly asked China for rights to test a nuclear weapon on the territory of the PRC, but received a negative response.

Overall, Pyongyang is particularly nervous about the apparent renewal of Sino-U.S. relations after President Bill Clinton's trip to China in spring 1998. The North Koreans fear the possible development of a cooperative strategy by Beijing, Washington, and Seoul against the DPRK. Such concerns can be found (albeit in a veiled form) in statements by North Korean officials and in press reports. Unofficially, North Korean officials make the argument to Russian analysts that China has turned into a "junior partner" of the United States and is "on the verge of internal social upheavals."

To punish the Chinese for their "misdeeds," Pyongyang is making moves to improve its ties with Taiwan, including offering to bury Taiwanese nuclear waste in the DPRK. According to Chinese sources, North Korean–Taiwanese relations are gradually acquiring a quasi-official status. In early 1999, unconfirmed press reports suggested that Taiwan may be opening an intelligence office in North Korea, in order to gain information on Chinese missile technologies.[22] Other reports noted the likely opening of direct air links between Pyongyang and Taipei.[23] So far, Beijing has been silent on these matters, but an official protest may follow in the near future.

Conclusions

As shown above, China is no longer the close ally it once was for North Korea during the Korean War. Despite its official pronouncements, the PRC today represents only a "conditional" friend of North Korea. There is increasing evidence that Beijing is fully prepared to draw clear lines when Pyongyang threatens to destabilize China's security or draw it into conflicts with other, higher-placed partners of China.

At present, official Sino–North Korean relations mask a series of problems that exist below the surface, and even some fundamental contradictions in regards to Beijing's simultaneous and flourishing relations with Seoul. Thus, while recent conflicts do not predict a breach in bilateral ties, they leave no doubts about China's increasing wariness concerning its relations with its last Stalinist neighbor.

16

The Korean Peninsula and the Security of Russia's Primorskiy Kray (Maritime Province)

Larisa V. Zabrovskaya

Russia's Far Eastern border regions have a long history of contact with the Korean people. Although this point is often forgotten, Koreans have lived in Primorskiy Kray (Maritime Province) off and on for more than 150 years, sometimes in considerable numbers. The Russian side has adopted various policies over time regarding Korean citizens, sometimes accepting their residence in the region, sometimes seeking to remove them. Nevertheless, the Russian and Korean states have never been at war, and Russians maintain a friendly attitude toward Koreans. In considering any settlement on the Korean Peninsula, whether directly related to the capping of the North Korean nuclear program or not, it is important to take Russian-Korean relations at the local level into consideration. Given their proximity, the two peoples on either side of this international border are bound to influence one another, and the stability of this relationship is bound to affect broader security on the peninsula, in either a positive or negative way.

This chapter begins with a brief history of Korean relations with Russia and proceeds to analyze in detail recent developments at the regional level, particularly with neighboring North Korea. It shows how the security of Russia's Primorskiy Kray is inevitably influenced by the status of its relations with the Korean Peninsula. For this reason, it is important for Russia to be involved in future negotiations toward a peace settlement in this region.

Early History

The Russian State established a common border with the Korean kingdom after the signing of the Treaty of Peking with the Qing Empire in 1860. The treaty recognized the lands "to the east of the Ussuri River as the possessions of Russia."[1] The lower reaches of the Tumen River delineated the shared Russian-Korean border, a length of 17 kilometers (km).

Koreans in Primorskiy Kray

Despite the delineation of the border and strict Korean laws prohibiting Koreans from leaving the domain of their kingdom—under threat of capital punishment—the inhabitants of the northern Korean provinces were not deterred from seeking permanent residence in the adjacent newly Russian lands. Mass migrations of Koreans into Primorskiy Kray were recorded especially during poor harvest years. The tsarist government displayed a positive attitude toward the resettlement of Korean peasants (many of them wheat farmers), since it hoped that they would assist in providing food for the Russian army. Hence, whenever Koreans converted to Christianity, the tsarist government endowed them with free plots of land.

The number of Koreans arriving in Primorskiy Kray rose after 1905, when Japan established its protectorate over Korea and then annexed it in 1910. From that time on, Korean immigration acquired a clearly demonstrated political character. The number of Koreans immigrating to Russia grew during World War I. By early 1915, the Korean population of Primorskiy Kray exceeded 73,000 people.[2] Primorskiy Kray became one of the key foreign-based centers of the anti-Japanese Korean liberation movement. Such chieftains of Korean guerrilla bands as Hong Bom-do and Lee Bo-myun set up permanent bases in an ethnic Korean village located on the outskirts of Vladivostok.[3] In October 1909, one of the members of those guerrilla bands, An Jun-gun (who had resided in Vladivostok for many years), entered China and assassinated the first Japanese Resident General in Korea, Count Ito Hirobumi, during his visit to Harbin.[4] Following that incident, Japanese diplomats in Russia often asked Russian authorities to deport these Koreans as "Japanese citizens" suspected of anti-Japanese activities.

In the 1920s, after the consolidation of Soviet power in the Far East, the border regime on the Soviet Korean frontier was strengthened, and no unauthorized border violations from either side were reported. In the 1930s, slightly more than 100,000 Koreans resided on the territory of Primorskiy Kray with Soviet citizenship. The Korean population constituted a majority in some districts. For example, 95 percent of the population in the Posyet district was made up of Koreans, and more than half of the residents of the Budyennovsk (the present Partizanskiy) district were Koreans.

But in the fall of 1937, the Soviet government began a campaign of mass resettlement of Koreans to Central Asia, officially in order "to thwart Japanese provocations against the Korean population."

Despite this mass deportation, the number of Koreans arriving in Primorskiy Kray rose again in 1947–56, that is, during the formative years of the Democratic Peoples' Republic of Korea (DPRK). But this migration of North Korean youth for employment at Far Eastern industrial enterprises and fisheries was jointly planned and authorized by the Soviet government. By 1961, most of the 25,000 temporary Korean residents had returned to their homeland.[5] From 1954 onward, some of the Koreans who had been previously resettled to Central Asia began to return to Primorskiy Kray, finding work in the agricultural sector of the economy.

In the 1950s–80s, the Communist Party administration and management of Primorskiy Kray lacked wide-ranging autonomous ties with the DPRK. All contacts had to be pursued through the Soviet government and the Central Committee of the Communist Party. In this connection, the DPRK's leadership sought to take advantage of the visits of Soviet leaders to Primorskiy Kray to solve these problems. For instance, when, in May 1966, Soviet General Secretary Leonid I. Brezhnev arrived in Vladivostok to award the Order of Lenin to Primorskiy Kray,[6] Kim Il Sung requested a summit that took place at a Vladivostok resort (19 km from the city).[7] The main subject under discussion was allegedly Kim Il Sung's request that the Soviet Union render military assistance to the North Korean government for pursuing the unification of Korea. This request was allegedly turned down.[8] The Soviet leadership preferred to keep Kim's visit secret and its main purpose secret. Nonetheless, Moscow agreed to provide the DPRK with more economic assistance. Besides shipments of industrial equipment, food, and strategic raw materials, the two sides signed the first long-term Soviet–North Korean agreement on joint lumber exploitation in the Khabarovskiy Kray in March 1967.[9] The DPRK was again allowed to export its labor force to the Soviet Far East.

Although Brezhnev and Kim Il Sung never met again, a new channel of unofficial nongovernmental communication was opened. That is, the North Hamgyong Province bordering the Soviet Union was long headed by one of Kim Il Sung's relatives, whereas during these years (from 1969 to 1987) Primorskiy Kray was headed by V. P. Lomakin, a Moldovan and longtime close friend of General Secretary Brezhnev. Every year, the leadership of Primorskiy Kray exchanged visits with the Communist Party and administrative heads of the North Hamgyong Province. The two provinces developed close ties. Every year, the Primorskiy Kray sponsored a month of solidarity with the struggle of the Korean people for unification of their country. They exchanged visits of workers' delegations.

Primorskiy specialists worked at various enterprises in North Hamgyong Province, sharing their experience with the North Koreans. From the 1970s to 1980s, Primorskiy residents worked at the ports of Rajin and Sonbong, which now constitute the center of the North Korean free economic zone. It was at this time that a wide-gauge railroad was extended from the Soviet-DPRK border to these two ports for use in unloading Soviet vessels.

The intermediary activities of First Secretary of the Soviet Embassy V. V. Mikheyev, who frequented Primorskiy Kray in the 1980s in search of partners for North Korean enterprises and with the purpose of explaining the political and economic situation in the DPRK to the leadership of the Primorskiy Kray, contributed a great deal to the broadening of economic ties between Primorskiy Kray and the DPRK.

From the 1970s to 1980s, military cooperation between the DPRK and Primorskiy Kray intensified. Besides arms shipments, the Soviet Pacific Naval Command established friendly ties with the North Korean navy and exchanged port calls by their respective battleships. The last Soviet visit was to the port of Wonsan in 1986, while North Korean navy battleships called on the port of Vladivostok last in 1987.

After Mikhail Gorbachev assumed power in Moscow in 1985, relations between Primorskiy Kray and the DPRK began to show signs of decline, as a result of the general deterioration of relations between the Soviet Union and the DPRK. The North Korean leadership criticized Gorbachev's policies of *perestroika* (economic restructuring) and *glasnost* (openness in the media) and objected to their importation to North Korean soil.

The Soviet government's decision to abolish barter trade in 1990 played an extremely negative role in trade and economic relations between Primorskiy Kray and the DPRK. Changes in the administration of Primorskiy Kray also had a considerable negative impact. V. P. Lomakin was transferred to another post and the provincial leadership changed quite often, which did not facilitate the formation of trust and long-term ties. These factors, among other pressures, further complicated the situation with the debt the DPRK owed to the Soviet Union as a whole and to Primorskiy Kray in particular.

In the 1990s, having detected a relaxation of controls over foreigners in Primorskiy Kray, the DPRK attempted to set up a transit and sale network for smuggling heroin to and through the region; but Russian law-enforcement authorities thwarted this effort.[10] Nonetheless, North Korean lumberjacks, traveling by train to Khabarovsk are still frequently apprehended carrying drugs. In 1994, there were 16,000 North Koreans employed at the timber-cutting sites. (Some 200 have fled the camps and have attempted to immigrate to South Korea.[11])

From the 1980s to 1990s, the Korean population continued to grow in Primorskiy Kray. In the late 1980s, there were only 8,300 Koreans (0.4 percent of the total number of residents) in Primorskiy Kray; but, by early 1996, there were already more than 18,000 ethnic Koreans (0.8 percent of the total). The Association of Koreans of Russia hopes to resettle more than 150,000 Koreans in Primorskiy Kray by 1998.[12] This plan provoked concern from Primorskiy Kray's Chief of the Police Major-General Alexander Vasileyev. In early 1996, he sent a memorandum to the Primorskiy Kray Governor Evgeniy Nazdratenko, to the provincial legislature, and to the personal representative of the Russian President in the region, introducing a proposal to reduce sharply the return of deported Soviet Koreans from Central Asia and to put more restrictions on the stay of foreign citizens on the territory of the *kray*. The general expressed his apprehension regarding the proposed establishment of a Korean national homeland in the region, which might be later annexed by Korea. In his memorandum, he stated that "South Korean authorities are pursuing a well-coordinated, long-term strategy aimed at increasing the population of Koreans in Primorskiy Kray with the goal of setting up an autonomous Korean autonomous zone. Later on, they intend to have this Korean autonomous zone advance a demand for its unification with the land of their ancestors in one form or another."[13]

According to data from the region's Department of Internal Affairs, in order to implement this strategy in reality, South Korean authorities intend to use the Association of Koreans of Russia to resettle gradually all ethnic Koreans residing in various parts of the former Soviet Union to Primorskiy Kray. According to the general's information, South Korean businessmen, scholars, missionaries, and public figures traveling to the Russian Far East are expected to play an organizing role in consolidating the Korean communities in Russia and strengthening their presence in the regional economy. Judging by the data of the Russian Ministry of Internal Affairs, various religious organizations headed by South Korean missionaries have intensified their activities in the region. As one internal document states: "Some religious communities headed by South Korean pastors and relying on the material assistance of South Korean businessmen, are trying to rent vacant fertile lands in Primorskiy Kray to establish religious labor colonies there."[14]

But the DPRK is also paying attention to its Primorskiy compatriots. General Vasileyev argues that "Primorskiy Kray has become a battleground for social-economic and spiritual influence between the two rival Korean states, the DPRK and the ROK [Republic of Korea], regarding ethnic Koreans." In this context, North Korean authorities are seeking to export into the region the maximum number of the laborers allowed by long-term intergovernmental contracts and to set up an independent Korean workers' community in the southern Primorskiy region, following the model of the North Korean lumberjack communities already existing in Khabarovskiy Kray. Primorskiy provincial police officials believe that the Federal Immigration Service is failing to take account of the difficulties in the socioeconomic situation in Primorskiy Kray and, inadvertently, is promoting competition with the local Russian labor force by freely issuing permits for the use of foreign workers.

Initially, General Vasileyev's Koreaphobic proposals were received with skepticism by the kray administration, for fear that the passage of regulations restricting the entry of foreigners into Primorskiy Kray would cause foreign investment to drop. This, in turn, might lead to the reduction of revenues for budgets at all levels. But subsequent events surrounding the September 1996 murder of an ROK diplomat in Vladivostok have demonstrated that General Vasileyev had legitimate grounds for the above-mentioned demands.

Military Cooperation and Confrontation

The leadership of the DPRK has not remained indifferent to the expansion of Russian–South Korean military cooperation in the 1990s. It is necessary to stress that both Moscow and Seoul fully realized that the character of their military ties and the shipments of Russian armaments to the ROK would provoke the DPRK to undertake unreasonable and dangerous actions, such as declaring the suspension of its membership in the Treaty on the Non-Proliferation of Nuclear Weapons (NPT). Besides prohibiting the International Atomic Energy Agency (IAEA) from inspecting its nuclear facilities in the middle of 1993, the DPRK began to test its *Nodong*-1 medium-range missile. This missile, an advanced version of the Soviet *Scud* missile, has a range of 1,000 km and is capable of reaching any point within the ROK, as well as the territories of Primorskiy Kray, Hokkaido, and Honshu. Foreign news media have also reported after the August 1998 *Taepodong-1* test that the DPRK has continued its work on the development of the *Taepodong*-2 intercontinental ballistic missile.[15] Reports have also suggested that the DPRK continues to engage in activities aimed at developing its own nuclear bomb.

The unpredictability of the DPRK's policy has become a threat to the security of the entire Asian-Pacific region. Retaliatory measures have followed without delay. For example, Japan has begun to work jointly with the United States on developing a theater missile defense system. Some people fear that such research and development efforts might provide an impetus for a new cycle of arms racing in the Asian-Pacific region.

The overly rapid intensification of Russian-ROK military contacts has upset the fragile balance of military and political forces that had formed on the Korean Peninsula during the past several decades. Further deepening of Russian–South Korean military ties could affect the evolution of a North-South dialogue on peaceful unification of Korea in a negative manner. Such contacts indirectly strengthen the position of the ROK and, therefore, put pressure on the DPRK, which might decide to launch a civil war in retaliation. Such a development would be extremely dangerous for Russia, since any civil war on the Korean Peninsula would affect the Primorskiy Kray's population directly.

The joint U.S.-Russian naval exercises conducted in southern Primorskiy Kray since June 1994 near the North Korean border deepened the mistrust of the DPRK's leadership toward Russia. One such exercise was staged in the vicinity of the village of Slavyanskaya in August 1996. As part of the exercise, a large amphibious vessel of the U.S. Navy (*Belle Wood*) called on Primorskiy ports. The naval exercises were code-named "Cooperation at Sea, 1996" and were designed to practice the organization of joint missions with combined Russian and U.S. Navy forces to alleviate regional disasters, such as the use of marines to provide assistance in the aftermath of earthquakes.[16]

But, after the completion of these exercises, the *Belle Wood* went to South Korea in order to participate in a two-week-long joint U.S.-ROK military exercise in the Yellow Sea (near North Korea's western coast) code-named "Ulji Focus Lens."[17] About 620 combat aircraft practiced mock bombing raids as part in these joint exercises.[18] From a North Korean perspective, it would be easy to draw the conclusion that the United States, by staging military exercises both with Russia and the ROK, might want to reanimate General Douglas MacArthur's plan for taking over the North.

Judging from reports published in North Korean newspapers during the military exercises, the situation on the demilitarized zone (DMZ) did not remain calm either. Pyongyang viewed such military activities as "a pre-planned provocation by the Seoul regime aimed at setting on fire a new military conflict on the peninsula with the goal of eliminating the DPRK."[19] The subsequent intrusion by the North Korean submarine into South Korean territorial waters near the town of Kangnung on September 18, 1996, and the immediate declaration by the United States and the ROK of their intention to resume their large-scale "Team Spirit" joint war games testify to the fragility of the political and military balance of power on the Korean Peninsula and to the necessity of caution in joint consultations about the goals and scale of joint military exercises with other countries in the region. The DPRK, now without reliable military allies, cannot observe calmly major naval maneuvers staged by hostile powers near its territorial waters. Russia too should not allow other nations to draw it into war games with the aim of contradicting Russian national interests.

The Fate of the USSR–DPRK Treaty of 1961

The establishment of official relations between the Soviet Union and the ROK in 1990 distanced the North Korean leadership from that in Moscow. Pyongyang interrupted all official contacts with the Kremlin for a considerable period.

However, this long alienation did not serve the best interests of either country. Without Russian raw materials and technical aid, scores of North Korean enterprises built with Soviet assistance have remained idle. Their joint exploitation could bring benefits to the Russian economy. Moreover, these enterprises were built on a compensatory basis: that is, they were expected to export to the Soviet

Union from 60 to 80 percent of their output as repayment for Soviet construction costs. Therefore, it is not surprising that in the mid-1990s, both sides began to look for new political and economic contacts with each other, without giving much publicity to their intentions. For instance, a delegation of the Russian Liberal-Democratic Party (LDPR) headed by Deputy Chairman of the Russian State Duma A. Vengerovsky visited Pyongyang in April 1994.[20] DPRK Deputy Foreign Minister Lee In Gyu visited Moscow in May 1994. He was the first North Korean diplomat of any significant rank to pay an official visit to Russia in three years.[21]

Russian Deputy Foreign Minister Alexander Panov visited Pyongyang in September 1994, as the special envoy of the Russian president, which raised the significance of his visit in the eyes of the DPRK's leadership. During his talks with Deputy Foreign Minister Lee In Gyu, Panov reiterated the readiness of Russia to develop good-neighborly relations with the DPRK and to facilitate the resolution of the nuclear problem by providing Russian-made light-water reactors (LWRs).[22] The North Koreans, although expressing some interest in the Russian LWRs, nonetheless, stated that they would make the final decision after the conclusion of the U.S.-DPRK talks in Geneva.[23]

Delegations of the Khabarovskiy and Primorskiy Krays also visited Pyongyang in March 1995. They aimed at revitalizing regional economic cooperation between the DPRK and these two most economically developed regions of the Russian Far East. Deputy Head of Administration O. Kuzenkov headed the delegation of the Khabarovskiy Kray. During the talks, Kuzenkov signed a protocol on the establishment of a joint working commission to study the possibilities for broadening cooperation. Besides the deepening of trade ties, participants considered a number of opportunities for the implementation of joint projects. In particular, they discussed the prospects for the participation of Khabarovsk regional enterprises in the development of the infrastructure in the Rajin-Sonbong free economic zone and for the establishment of joint enterprises in the textile, timber-cutting, and fishing industries. They also reached agreement on the formation of a joint tourist company, as well as a joint transport company using helicopters and vehicles from Khabarovskiy Kray. Meanwhile, the provincial delegation agreed to study proposals involving the organization of oil refining at the Sungni Oil Refinery, joint manufacturing of trolley buses at a Pyongyang bus-manufacturing plant, and provision of telecommunication equipment for the modernization of lines of communication between Rajin and Khabarovsk.[24]

As Kuzenkov stated at a news conference, although most of the proposed projects required further study in order to evaluate their economic feasibility, it was possible to begin increasing trade between Khabarovskiy Kray and the DPRK and using more North Korean workers in construction and agriculture.[25]

The most impressive results were those stemming from the visit of the Primorskiy Kray delegation headed by Governor Nazdratenko, who had been invited by Chairman of the DPRK External Economic Committee, Lee Song Dae. Nazdratenko and Lee signed a memorandum concerning prospects for the development of trade and economic exchanges between Primorskiy Kray and the DPRK, highlighting cooperation in fisheries, construction, agriculture, and machine building as priorities. In the area of fisheries, they proposed to exchange catch quotas and to set up joint ventures, especially in the production of canned fish. The North Koreans also expressed interest in purchasing zinc concentrate, timber, and trotyl (an explosive for mining purposes). The parties agreed to consider the possibility of coordinated penetration into the markets of Southeast Asia, using heavy construction and mining machinery of the Primorskiy joint stock company Vostokremstroimash and the services of the (North) Korean Machine Import and Export General Society.[26]

During his talks in Pyongyang, Nazdratenko stressed that Primorskiy Kray commands a significant portion of Russian–North Korean trade and that provincial participation in Russian ties with the DPRK could be broadened by setting up joint ventures for providing energy to border regions. Some local Russian companies are interested in using the North Korean port of Rajin for the transit of Russian cargo, especially given the fact that Rajin is already connected to the Russian border by a wide-gauge railroad.[27]

But it is clear that the full range of mutually beneficial Primorskiy Kray–North Korean projects in trade and economic development cannot be realized without the radical improvement of political relations between Russia and the DPRK.

In August 1995, the Russian Ministry of Foreign Affairs (MFA) made yet another attempt to move Russian–North Korean relations out of their state of stagnation by proposing the negotiation of a new interstate treaty that would substitute for the 1961 Soviet-DPRK alliance treaty (then due to expire on September 10, 1996) after removal of the clause about unconditional Russian support of the DPRK in the event of the latter's military conflict with a third country. The Russian MFA told the North Korean side that "the 1961 Treaty . . . is obsolete and does not correspond to the new realities that have emerged in Russia and in Russian-Korean relations in Northeast Asia."[28] This proposal of the Russian MFA contradicted previous statements of Russian Deputy Foreign Minister Panov, who, in early 1994, had characterized the situation on the Korean Peninsula as "balancing on the brink of war." Hence, he found it expedient to state that "in accordance with the existing treaty between the USSR and the DPRK, Russia will render assistance to North Korea in the event of an unprovoked aggression against it, but will make a decision about [such assistance] only on the basis of its own analysis."[29]

The foreign news media viewed this statement as an "expression of support for Kim Il Sung's regime." But, later, the Russian MFA issued the following refinement:

In 1995, we shall decide whether we should prolong the existing treaty for a new term or conclude a new treaty with North Korea. Although only formally, the current treaty still remains in force at present. This means that Russia will render necessary assistance to the DPRK, if and when it decides that Pyongyang has become a victim of aggression. In addition, according to current legislation, the North Koreans will receive any military assistance only after a corresponding decision by the State Duma.[30]

A year later, the Russian MFA finally decided not to prolong the treaty of 1961 for another five-year term. Perhaps this shift was intended to fulfill the promise made by Russian President Yeltsin to South Korea during his official visit to Seoul in November 1992 and later reaffirmed by other Russian high-ranking officials.[31] One can only guess that the new DPRK leadership, headed by Kim Jong Il, did not insist on prolonging the treaty. Therefore, it becomes clear why the representatives of the DPRK MFA expressed their willingness to consider the new Russian draft of the treaty so quickly.

The new Russian draft of the treaty does not contain the article on security assistance that existed in the previous version. According to one official from the Russian MFA, the main provisions of the draft treaty do not go beyond the framework of analogous treaties that Russia has signed with Mongolia, Vietnam, and the Republic of Korea in recent years." The Russian MFA also supposes that "the obligation to provide military assistance is an attribute of an alliance treaty," whereas the current concept of the Russian foreign policy does not conceive of military alliances with other states, with exception of the CIS member states."[32]

In an opinion issued by Russian Deputy Foreign Minister Panov in mid-1995, he described the alliance treaty with the DPRK of 1961 as "a one-way treaty." He argued that "it was a burden for us, not a benefit. At present, we are pursuing a more realistic policy, without distinguishing 'potential enemies' or 'allies of our enemy' in the region. Without entering into any allied relations, we are nonetheless creating a security belt."[33] Panov noted that "the policy of the Russian MFA is aimed at promoting maximally friendly and comprehensive relations with the DPRK and the ROK. Today, our task is to intensify our economic ties with the DPRK and to launch a new mechanism of bilateral relations in this sphere."[34]

The approach adopted by the Russian MFA is extremely pragmatic: it is aimed at achieving economic gains while ignoring the political side of Russian–North Korean relations. According to Russia's Ambassador to North Korea Valery Denisov, "The [current] desire to broaden economic contacts with the DPRK testifies to the end of the period of 'democratic romanticism' [in the early 1990s], which was characterized by the disruption of the full range of Russian–North Korean ties. Today, Russia wants to restore these contacts."[35] But Pyongyang cannot be fully satisfied by Moscow's explicitly declared desire to confine this "restoration of ties" only to economic relations.

Moreover, the opinion of the Russian MFA is not shared by many factions within the Russian State Duma, especially the Communists and Liberal Democrats. In his talks with North Korean leaders, Duma Speaker Gennady Seleznyev (a Communist) who visited the DPRK as head of the State Duma delegation in August 1995, did not support the annulment of the 1961 treaty. On the contrary, during these talks, both sides advocated the idea of restoring cooperation between Russia and the DPRK up to the level "that could be found during the existence of the USSR, and then surpassing it."[36]

Vigorous discussions took place in the Russian news media and academic circles in connection with the MFA's proposal to substitute the 1961 treaty with a new one. For instance, in November 1995, the Institute of the Far East of the Russian Academy of Sciences sponsored an academic conference under the title of "Russian-Korean Relations: Past and Present," where opinions varied widely. A majority of scholars judged this proposal to be a step "corresponding to the spirit of our time" and as "a timely rejection of one of the relics of the Cold War." But others, like Yuri Vanin, expressed the apprehension that by rewriting the treaty Russia would destroy "one of the pillars upholding the military-political balance of power in Korea, which constituted a deterrent vis-à-vis the United States and the Republic of Korea, as well as the DPRK itself."[37]

Another analyst, B. Zanegin, maintained a similar view, arguing that "under current conditions, reconsideration of the legal foundations of Russian relations with the DPRK, as proposed by the Russian MFA, is likely to have a negative effect on the military-political situation in the region." Seoul and Washington, he argues, may again "revive their hopes to absorb the DPRK by force." Moreover, the Russian abandonment of the alliance with the DPRK "will finalize the complete squeezing out of Russian interests from the Korean Peninsula."[38]

During an official visit to Pyongyang from April 10 to 12, 1996, a Russian delegation headed by Deputy Prime Minister V. Ignatenko made another attempt to elicit the views of the North Korean leadership concerning the Russian proposal. Deputy Foreign Minister Panov and a representative of the Ministry of Foreign Economic Relations, G. Levchenko, were also members of the delegation.

Ignatenko transmitted to the North Korean leadership an oral message from Russian President Yeltsin to Kim Jong Il, who was away from Pyongyang during the visit of the Russian delegation. The message said that Russia was ready to activate the full complex of its relations with the DPRK on the basis of mutual respect, equality, and mutually beneficial cooperation, and was proposing to conclude a new treaty on the basis of friendly relations between Russia and the DPRK.[39]

Ignatenko held talks on trade and economic and scientific and technical cooperation with Vice-Premier of the DPRK Administrative Council Hong Song Nam. Both parties agreed to make some efforts aimed at elevating the economic and trade cooperation between the two countries back to the level of 1991.

During the visit, both sides agreed to set up a number of working groups in such important areas of cooperation as science and technology, timber cutting, light industry, and transportation and communication with the Russian Far East. One of the most promising areas was determined to be the intensification of regional ties, including the use of North Korean workers in agriculture, construction, and timber harvesting in Siberia and the Far East. Also, they discussed the possibilities for resuming cooperation in the industrial enterprises constructed with Soviet economic and technical assistance with the participation of third countries, such as the ROK.[40]

The most difficult part during the talks was the discussion of means for repaying North Korea's state debt to Russia, which in early 1996 exceeded 3.3 billion "foreign exchange rubles," and implementing mutual obligations by the enterprises of both countries.[41] For instance, North Korean owed $20 million to the Russian Far Eastern railroads. This sum had been accumulated in the past five years as a result of North Korea's seizure and use of Russian train cars on the territory of the DPRK. The situation had forced the Russian Far Eastern railroad administration to issue a directive in May 1996 forbidding the passage of trains with cargoes destined for the DPRK through the border station at Khasan.[42] According to the responsible secretary of the Russian delegation, V. Fitin, these negotiations did not achieve "any sufficient and satisfactory solution," but the issues "were discussed at the deputy prime minister level and an understanding was reached that both sides should continue to look for new approaches and solutions that will satisfy both parties."[43]

During the visit of the Russian delegation, Deputy Foreign Minister Panov held political consultations with the DPRK Foreign Minister, Kim Yong Nam, and his deputy, Lee In Gyu. The parties expressed their satisfaction at the pace of development of their bilateral relations, but they disagreed as to the ways of alleviating tensions on the Korean Peninsula. At the same time, the parties recognized that the mere fact that the first meeting of the intergovernmental commission, postponed since 1992, did convene and that political consultations did take place demonstrated that both countries were eager to develop mutually beneficial relations with each other.[44]

The visit ended with the signing of a protocol on the results of the work of the intergovernmental commission. But the protocol did not mention any time frame for the repayment of the North Korean debt. Nor did it mention a project involving the construction of a gas pipeline from Russia to the ROK via the territory of the DPRK. Nonetheless, the North Korean leadership promised to reconsider its position regarding repayment of the debt. In the opinion of both sides, good prospects exist for increased economic cooperation between the DPRK and adjacent Russian regions that have recently acquired considerable economic autonomy.

The results of political consultations were even less promising than the economic talks. The head of the Russian delegation failed to meet Kim Jong Il, so there was no official North Korean view provided regarding the Russian draft of the new interstate treaty. However, in unofficial conversations, high-ranking North Korean officials spoke about the "inevitability of war" on the Korean Peninsula,[45] hinting that the 1961 Soviet-DPRK treaty—if left in force—could play a positive role in preventing such a conflict in the future.

Judging by all signs, the North Korean leadership was not satisfied with the Russian draft treaty. It would be totally satisfied only if Russia made a decision similar to that of China,[46] by agreeing to prolong the 1961 treaty for another five-year term. It was obvious that the North Korean leadership had adopted a wait-and-see attitude in 1996, pending the results of the presidential elections in Russia. But their hopes did not materialize as Yeltsin was reelected and the Communists were not returned to power. Therefore, the North Koreans had to deal with the existing Russian government. Negotiations about a new framework agreement on cultural cooperation between the governments of Russia and the DPRK took place in Pyongyang in August 1996, representing a first step in that direction. According to an official from the Russian MFA who was involved in these talks, "We could sense some interest and initiative coming from the Korean side at these talks."[47]

Following the talks on cultural cooperation on September 3, 1996 (i.e., only one week before the official expiration date of the 1961 treaty), the North Korean side presented its draft of a new interstate treaty with Russia. According to Russian Deputy Foreign Minister Grigory Karasin, its text "in principle, was close to the Russian draft and was also aimed at improving relations between our countries and peoples."[48] One can suppose that, afterward, the parties would have discussed both drafts and begun to formulate the treaty text. However, the situation dramatically changed when, in September 1996, Russia began to ship armaments to the ROK, including the BMP-3 armored personnel carriers and T-80U tanks, which were being used as partial payment for Russian debts to South Korea.[49]

Under these circumstances, Pyongyang felt that its interests had been hurt and called Moscow's move "perfidious." The North Korean Central Telegraph Agency made the following comment about the Russian arms sales to the ROK: "Russia committed a crime instigated by the South Korean puppets and demonstrated that it was almost part of the camp of forces hostile to the DPRK. If Russia continues to move along the road of undermining peace and security, then we will have to settle our scores with it."[50] The Russian MFA made the argument that its actions were consistent with the principle that "both Korean states had the right to pursue a desire to enhance their own security" and that Russia "had supplied South Korea only with defensive weapons and in such quantities that would be insufficient to upset the military balance of power on the Korean

Peninsula...."[51] These statements, however, failed to calm the emotions of Pyongyang. As a result, North Korea decided to commit the desperate act of dispatching a submarine manned with infiltrators into South Korean territorial waters and begin other military preparations in the vicinity of the DMZ.[52]

It goes without saying that these actions were denounced by the United States and the ROK, and that both announced their intention to resume their joint Team Spirit war games. Afterward, the South Korean government announced its plan to increase its military budget by 12 percent in 1997 because of "the threat of armed invasion from the DPRK."[53] As one can see, the South Korean government—having secured the Russian refusal to prolong the 1961 Soviet-DPRK treaty—nonetheless failed to improve its position on the Korean Peninsula. On the contrary, Seoul realized that it had lost an important lever of putting pressure on the DPRK. Moreover, the South Korean government faced the necessity of increasing its defense spending.

The Murder of the South Korean Diplomat and Its Aftermath

The barbaric assassination of South Korean diplomat Choi Duk-kin on October 1, 1996, in Vladivostok,[54] which bore clear signs of participation by North Korean agents, compelled the population of Primorskiy Kray to reflect on the character of the activities being carried out by North and South Koreans on their territory. One of the theories advanced by the murder investigation involved possible participation by North Korean special forces pursuing retaliation against the South for the killing of North Korean crew members stranded by their submarine on South Korean soil. By October 22, 1996, the North Korean consulate in Nakhodka found it necessary "to deny groundless rumors spread by Primorskiy Kray newspapers" and demanded that "this artificial linking of the DPRK to the assassination be stopped immediately, because so far the Russian authorities investigating this murder case have not made any official statements."[55]

Correspondents from the Japanese and South Korean newspapers were the first to begin to discuss the theory about "the North Korean trail." They characterized Choi Duk-kin as a professional operative of the South Korean special services (i.e., the ROK Agency for National Security Planning), who legally collect intelligence about the DPRK. According to their information, at the murder victim's apartment, the police found some Russian internal police materials containing information about the circumstances surrounding the arrest of a son of a high-ranking North Korean politician, Pak Song Ch'ol, for illegal drug trafficking on the territory of Primorskiy Kray in 1994.[56]

The regional news media paid special attention to the method of Choi's assassination, which was very similar to that used previously in certain cases in Khabarovskiy Kray where, "in accordance with the Russian–North Korean intergovernmental agreement, a representative office of the DPRK's special services, headed by a major-general of the DPRK Ministry of State Security, still continues to operate."[57]

As of 1996, there were more than 600 North Koreans residing legally in Vladivostok. They were employed in construction and in seasonal jobs (such as agriculture). In the opinion of the Primorskiy Department of Internal Affairs and the Federal Counter-Intelligence Service, more than half of these workers were actually employees of the North Korean intelligence service and special forces. It is telling that a day after Choi's assassination, representatives of the Primorskiy Kray Immigration Service sent a document to the Russian MFA that stated that "a representative office of the North Korean Agricultural Committee in the Primorskiy Kray, located in the town of Artyem, has flagrantly violated Russian laws, and, therefore, the Provincial Immigration Service has to insist on its closure." This document also stated that although the purpose of that representative office was to develop cooperation and joint businesses in agriculture, in reality, it had grossly violated a number of rules regulating the stay of foreign citizens on the territory of Russia. For instance, in 1996 alone, this committee imported more than 250 citizens of the DPRK as "laborers" into Primorskiy Kray, abusing visa-free entry into the province and failing to process any of the necessary permits with either the Russian Federal Immigration Service or its provincial equivalent. According to the document, these newly arrived illegal immigrants "travel freely around the provincial territory without any control and are engaged in businesses incompatible with their status."[58] Clearly, the theory suggesting a political motivation behind the assassination of the South Korean diplomat is not without some basis in general North Korean behavior in the region.

The Russian and South Korean Ministries of Foreign Affairs exchanged official notes concerning the incident, assuring one another that "the assassination of the diplomat should not cause bilateral, interstate ties to deteriorate."[59] Nonetheless, the still-unsolved murder of the South Korean diplomat has undoubtedly had some negative effects on Russian–South Korean relations as a whole and on trade and economic relations between Primorskiy Kray and the ROK in particular, besides damaging Vladivostok's international reputation as a city. Some South Korean businessmen, fearful for their personal safety, have canceled trips to Primorskiy Kray. This, in turn, has undermined the investment activity of South Koreans in the Russian Far East. (Until recently, the ROK had been the leader among other nations in investing into the Russian Far Eastern economy and is only now resuming its active involvement in the region.)

Conclusion

One obvious conclusion from the analysis provided above is that developments on the Korean Peninsula and the status of inter-Korean relations have a direct effect on the security of Russia's Primorskiy Kray. As long as some deterrent measures were in place, primarily the 1961 Soviet-DPRK treaty, the balance of power on the Korean Peninsula was relatively stable. But the Russian self-withdrawal from the northern part of the Korean Peninsula and the dramatic reorientation of its foreign policy in the direction of South Korea—accompanied by Russian deliveries of the advanced military hardware to Seoul—has created a dangerous situation under which the DPRK leadership feels abandoned by everyone. The limited contacts between Pyongyang and Washington cannot compensate the DPRK for the loss of Russian support and serves instead only as a very weak consolation. Under these circumstances, the North Korean leadership has opted to pursue its own path toward self-preservation. Regretfully, as recent events have demonstrated, this path does not have a constructive character and is leading only to the aggravation of tensions.

The community of nations interested in peaceful resolution of the inter-Korean confrontation and in security for themselves should undertake a number of measures aimed at promoting the relaxation of tensions on the Korean Peninsula. First, these countries should recommend that Pyongyang and Seoul reject the use of force and other methods of confrontation inherited from the Cold War and instead concentrate their efforts on a search for mutually acceptable compromises. Second, the states interested in peaceful resolution of the inter-Korean disagreements should convene an international conference with the participation of the DPRK and the ROK, at which the latter two, first of all, should agree to refrain from staging military exercises and from importing all types of armaments into the Korean Peninsula. Third, and finally, Russia and China should pledge to refrain from allowing the transit of any armaments from third countries with a destination in either of the two Koreas. If such measures were adopted, some positive changes along the road toward the resolution of the Korean question might be achievable.

Part V

Unsettled Problems and Future Issues

17

The Renewal of Russian–North Korean Relations

James Clay Moltz

In the late 1980s, the Soviet government abruptly reversed the Korean Peninsula policy it had followed since 1948 of exclusive engagement with the Democratic People's Republic of North Korea (DPRK). By opening relations with South Korea in 1990, Moscow sent out a clear message that it would no longer support Stalinist regimes like Kim Il Sung's North Korea. This dramatic turnaround was a logical product of the reforms initiated by Soviet General Secretary Mikhail Gorbachev, whose primary aims in East Asia were to stabilize the Soviet Union's eastern borders and attract foreign investment for the development of the Russian Far East.

Consistent with his new goal of engaging Moscow's former adversaries in Northeast Asia (China, Japan, and South Korea), Gorbachev used this initiative to court investment, political relations, and even military ties with Seoul, while edging bankrupt North Korea to the sidelines. At the same time, the Soviet Union rapidly reduced and then eliminated its subsidies to Pyongyang for essential products like energy, foodstuffs, and military hardware, causing bilateral trade with the DPRK to plummet from $3.5 billion in 1988 to less than $100 million by 1995.[1] Meanwhile, trade with South Korea blossomed from a mere $290 million in 1988 to a robust $3.2 billion by 1995.[2] As these figures suggest, after December 1991, the newly noncommunist Russian Federation government under President Boris Yeltsin only accelerated this shift in foreign policy in favor of South Korea.

However, the Russian Federation's eagerness in its pursuit of a pro-Southern policy on the Korean Peninsula soon waned. Part of the problem related to

Moscow's inability to repay its $1.47 billion Soviet debt to South Korea, which resulted in a cutoff of government-to-government credits in the early 1990s. Meanwhile, despite promises made during the honeymoon period in the late 1980s, actual South Korean investments in the Russian economy turned out to be rather modest. South Korea's own financial crisis since late 1997 has further limited Seoul's willingness to channel new investments into the even more unstable Russian economy. These developments have caused grumbling in Moscow and have led to an active pro-Northern backlash in the Russian Far East, where economic interests have contributed to a renewal of pro-Pyongyang sentiments. Labor- and cash-poor enterprises in the region now hope to reinvigorate past policies of using cheap North Korean labor in local industries, while perhaps opening new markets for local manufactured goods that are unappealing to more sophisticated East Asian consumers.

The recent return to power in Moscow of old-style bureaucrats, communists, and enterprise directors has combined to form a political block of North Korean supporters eager to use Soviet-era contacts to make deals for Russian technology, as well as to show the world that Russia remains a crucial player in any future international peace settlement on the Korean Peninsula. The Kremlin has embraced the belief that the Russian Federation's sharp alienation of Pyongyang in the early 1990s was the reason that Moscow was left out of important international negotiations on the future of the peninsula, including the Agreed Framework.

As a result, there is a new emphasis in Russian policy toward the Korean Peninsula on pursuing a "balanced" policy between the two Koreas. In practical terms, this means reestablishing closer ties with Moscow's unrepentant former ally, North Korea. As the political tilt of the Yeltsin government continues to move in a conservative direction, this policy has gained momentum, both in Moscow and in the Russian Far East.

In light of these recent trends, this chapter examines the growing rapprochement in Russian–North Korean relations since 1994 from the perspective of political trends in Moscow, as well as the context of the lesser-known Russian Far Eastern factor. It argues that—in contrast to outside observers' tendency to view North Korea as a "dead end" state[3]—Moscow and the Russian Far East view the DPRK quite differently. In fact, many Russian analysts and officials see opportunities to renew Moscow's influence on the Korean Peninsula through the reestablishment of closer ties with a state that they view not as a renegade, as does much of the world, but as a very familiar old neighbor. Despite the dramatic differences that exist today between quasidemocratic Russia and Stalinist North Korea, this chapter shows how underlying historical, strategic, and regional economic forces are pushing the two sides back together again. The rise of representatives from Russia's security agencies to high positions in Moscow is likely to reinforce these trends. However, as discussed in the chapter's conclusion, there could be serious costs for Russia—and the international community—associated with Moscow's attempts to restore its close relationship with Pyongyang.

Changes in Russian Political Attitudes toward North Korea

Several recent factors have coincided to strengthen the North Korea "lobby" both in Moscow and in the Russian Far East. The first major influence on these developments has been the increasing weight of nationalist and communist factions in the Russian State Duma, which has heightened attention to North Korean concerns, putting pressure on liberal elements in the Foreign Ministry and the presidential staff that once favored the South. The strong showing of Vladimir Zhirinovsky and the Liberal Democratic Party of Russia (LDPR) in the December 1993 Russian Duma elections marked the first step in a growing anti-Western movement within the Russian government, whose influence was unaffected by the victory of Boris Yeltsin in the 1996 presidential elections. Several highly publicized trips by Duma members to Pyongyang, including one led by Communist Duma Speaker Gennadiy Seleznev in 1996,[4] have emphasized the point in Moscow that Russia needs to reevaluate its "failed" policies on the Korean Peninsula and seek to regain its influence, even at the expense of renewing aid to Pyongyang. At a press conference reviewing the results of his visit, Seleznev treated the pro-Northern shift in official policy as a fait accompli, going as far as to refer to his North Korean hosts as "our strategic partners."[5] A similar trip in February 1998 by Vice-Chairman of the Russian Communist Party Konstantin Nikolayev reiterated the desire of Russia's communists to maintain close ties with Pyongyang and the Kim Jong Il regime.[6]

Indeed, there have been strong calls in the Russian Duma, the Foreign Ministry, and even in academic circles arguing that the shift to the South in the late 1980s was a major policy blunder and that Russia should return to its "reliable" ally in the North, which is now the focus of considerable international attention. This interest, culminating in the signing of the U.S.-DPRK Agreed Framework in October 1994, effectively removed Moscow from its prior role as a power broker on the Korean Peninsula, causing waves of frustration and indignity in both the Russian Foreign Ministry and the Russian Ministry of Atomic Energy (Minatom).[7] Similar feelings were evoked by the 1996 U.S.–South Korean announcement of the "four-party" talks, which worked toward achieving a settlement on the Korean Peninsula (which included China, North Korea, South Korea, and the United States, but snubbed Russia). The move caused another backlash of irritation, as well as pressure to move to a more balanced policy on the peninsula. In reaction, Russia's Ambassador to Seoul Georgiy Kunadze branded the four-party proposal as a mere "propaganda exercise" by South Korea and the United States, and argued that any real proposal aimed at bringing Pyongyang to the bargaining table would have to include Moscow.[8] Russia's then–Deputy

Foreign Minister Alexander Panov again reiterated Moscow's proposal to convene a broad, multilateral conference to discuss means of reaching a lasting peace on the Korean Peninsula.[9] These sentiments were restated in a frosty meeting in Moscow between then–Foreign Minister Evgeniy Primakov and the South Korean foreign minister in May 1996. Afterward, a Russian journalist covering these events summarized the results of the meeting by commenting tartly: "Russian–South Korean relations are in a state of obvious stagnation."[10]

The Russian Foreign Ministry has largely supported the move toward closer ties with the North. In spring 1998, for example, the Russian embassy in Pyongyang held a reception to commemorate the forty-ninth anniversary of the agreement on economic and social ties between the two countries.[11] Russia's Ambassador Valery Denisov has also expressed his optimism about opportunities for broadening Russian ties and influence on the Korean Peninsula through more developed contacts with the North.[12] Russian officials have undoubtedly been cheered by the failure to date of the four-party talks to achieve any meaningful results toward a peace settlement on the peninsula, suggesting that a new venue for these negotiations might be more productive.

Russian–North Korean relations have also benefited recently from a significant setback in Russian–South Korean political relations—the fallout of the serious spy scandal that occurred in July 1998. After months of work by the Federal Security Service (FSB), Russian officials arrested South Korean embassy official Cho Son-u on charges of espionage.[13] In the course of his interrogation, Cho revealed a string of contacts in the Korean Section of the Russian Foreign Ministry, involving espionage activities dating back as far as five years. The spy charges against Cho ranged from working with Russian contacts to obtain scores of classified documents from the Russian Foreign Ministry on both North and South Korean issues, to working with South Korean organizations in the Russian Far East to try to establish a Korean autonomous region under Seoul's influence. As a result of the FSB investigation, Moscow removed several Russian Foreign Ministry officials from its Korea Section who were allegedly involved in the provision of secret documents to Cho.[14] The Russian government also deported Cho and another diplomat from the South Korean embassy in Moscow along with three diplomats from the South Korean consulate in Vladivostok on related charges. South Korea retaliated by deporting a Russian official from Seoul and threatening further action. Finally, the Russian and South Korean governments called a truce in late-summer 1998 to avoid further damage to bilateral ties, with the result that each side agreed to reduce its diplomatic presence in the other's capital. Although the crisis has now subsided, the long-term implications of the scandal are likely to set back relations for many years. Within the Russian Foreign Ministry, for example, it will now be dangerous to be seen as a "friend" of South Korea. Indeed, the ministry's China experts have been put in charge of the Korea

Section in order to ensure that there are no more breaches in security.[15] These points all work to the benefit of North Korean interests.

Pyongyang, meanwhile, has long been readying itself for just such an opportunity. It is clearly eager to benefit from the scandal and is hopeful of using it to catapult its own relations with Russia. North Korea continues to sponsor pro-Pyongyang conferences and publications in Moscow, in hopes of influencing public opinion and official views in its favor.[16] Despite its own food crisis, the North Korean government has even provided "goodwill" food packages to Russian war veterans, indicating the importance Pyongyang places on winning back Russian "hearts and minds."[17] In the Russian Far East, North Korea still maintains a bustling consulate in Nakhodka, whose compound was greatly enlarged in the early 1990s (although construction had begun in the early Gorbachev era).[18] The consulate continues to provide an active lobby for Pyongyang in the region and regularly purchases newspaper space for the publication of favorable propaganda regarding North Korean perspectives and developments.[19] The number of staff present at the consulate could be quickly increased if trends in Moscow and the region continue to develop in Pyongyang's favor.[20] Some Russian analysts espouse the view that Russia represents North Korea's best hope of assisting its gradual transition toward economic integration with the Pacific Rim, especially in the development of new market-oriented sectors of the North Korean economy, such as the Rajin-Sonbong free economic zone (FEZ).[21] It is worth noting that many of North Korea's economic managers received their training in the Soviet Union. As one Russian report observes:

> Russian visitors [to the DPRK] underline the fact that quite a few graduates of Soviet universities can be found now in responsible positions in the North: in the ministries, armed forces, at plants and factories, and in research institutes. They are especially friendly toward Russians.[22]

These residual connections provide North Koreans with a natural link to Russian enterprise managers, as well as with knowledge of how the system operates.

Within the Russian Far East especially, talk of reinvigorating such ties has met with increasing resonance among old-style agricultural and industrial enterprises, which look favorably on the government's growing reconsideration of its prior South Korea–centric policy. As Primorskiy Kray Governor Evgeniy Nazdratenko predicted after his 1995 visit to North Korea: "Resumption of ties with the DPRK may happen to be very profitable for the Maritime Territory."[23] Nazdratenko has frequently chastised the Yeltsin government for jeopardizing these relations in the past few years and has called on Moscow to reorient its policy regarding the two Koreas.

Similarly, after the replacement in the early 1990s of such reformist regional governors as Primorskiy Kray's Vladimir Kuznetsov and Sakhalin's Valentin Fyodorov with conservative, old-style bureaucrats (such as Nazdratenko), Far Eastern regional administrations have also tilted markedly toward Pyongyang. Whether this is simply a "market-based" solution aimed at acquiring new export customers and cheaper imports (including labor) for the region *or* instead indicative of a longer-term political shift and a new conservatism in dealing with East Asia is difficult to tell. In either case, however, a process of Russian "integration" into East Asia that takes place mostly with partners like North Korea will undoubtedly reverse some of the more optimistic trends in the direction of economic and political reforms in the Russian Far East.

Notably, these developments come in sharp contrast—and may partially respond—to the recent worsening of local Russian ties with China, until recently the most fruitful Russian trading partner in East Asia. While Moscow has courted Beijing assiduously due to its large-scale arms purchases and lucrative long-term deals in the nuclear power field, Russia's regional governors and their populations have bristled against the opening of the border to Chinese traders and illegal immigrants, whom they see as increasing crime and even threatening the sovereignty of the region. The overtures by Nazdratenko to North Korea, by contrast, have not raised similar public opposition. Despite Western perceptions, North Korea appears to local Russians (long used to close bilateral ties) as representing less of a threat than China does and, notably, a significant and neglected economic opportunity.

Economic Issues

When Russia cut off its subsidies to the North Korean economy in the late 1980s, the North Korean economy began a period of precipitous decline, from which it is still suffering today. Given its own economic crisis during the early 1990s, however, Russia was in no position to provide the assistance upon which Pyongyang had long relied. Moreover, in the early postcommunist political climate, it no longer had any geostrategic incentives in the region to do so.

Despite the warming of recent political relations, what remains of the Russian–North Korean trade relationship is still a "wilting vine" compared to old-style ties. Currently, Moscow's exports consist of 90 percent of total trade turnover, with imports from the DPRK making up only 10 percent.[24] Estimated figures for 1997 put bilateral trade turnover at a mere $60 million, and this figure dropped even further to an estimated $40 million in 1998.[25] A key problem in bilateral economic relations is the $4.32 billion debt owed by North Korea to Russia, largely for fuel, machinery, technology, and armaments provided in the years up to 1988.[26] This is a serious obstacle even for optimists who are pushing the two sides closer together.

At the regional level, however, the picture of economic relations is much more vibrant, especially in Primorskiy Kray. While Russian–North Korean trade is minor in national terms for Russia, 72 percent of this trade is conducted with Primorye.[27] Since 1994, several delegations of North Korean economic officials have toured the Primorskiy and Khabarovskiy regions, signing agreements in a variety of fields, including labor exchanges.[28] Many analysts argue that cheap labor is the main "product" that North Korean can offer Russia at this point. Governor Nazdratenko may especially value this commodity, since, according to Russian sources, this labor comes to the *kray* without charge, constituting partial repayment of North Korea's large debt.[29] This labor is a small gift from Moscow to the struggling Primorskiy region, which is itself reeling from cuts to local military industries and the Pacific Fleet.

Thus, at the regional level, the trade relationship looks considerably different. Indeed, some 15,000 North Koreans are still involved in logging operations in the Russian Far East's Khabarovskiy Kray, while an additional 4,500 North Koreans are working for agricultural, fishing, or construction enterprises in Primorskiy Kray and on Sakhalin Island.[30] After earlier complaints in the region over human rights violations in the North Korean camps, the Russian side successfully negotiated a revision of the bilateral timber agreement in 1995 that gives more favorable terms to local authorities. The new pact allows constant access by local officials to the camps (for the purpose of inspections) and, in economic terms, also increases the Russian portion of the harvest from 43 percent to 61.5 percent.[31] Thus, despite the gloomy data on current trade relations, many in the Russian Far East take a more optimistic view of current opportunities with North Korea.

Perhaps most outspoken on this score is Sergey Dudnik, speaker of the Primorskiy Kray Duma and former head of the Nakhodka Special Economic Zone (SEZ), close to the North Korean border. Dudnik sees potential for further cooperation in the lack of capacity in Russia to process existing timber harvests, which, he argues, could be overcome by utilizing North Korean mills. These enterprises were built by the Soviet Union, but many now are working only part-time shifts because of a lack of raw materials and power. Indeed, the Nakhodka SEZ and North Korea's Rajin-Sonbong FEZ signed a cooperative agreement in 1996 aiming at future joint activities of the two zones to explore these and other possibilities. To date, however, very little progress has been made, due in large part to a lack of financing on both sides of the border.[32]

At the government-to-government level, the 1996 negotiations on the renewal of the 1961 Soviet–North Korean Treaty of Friendship and Cooperation seemed to convince Pyongyang that it needed to begin to make some concessions on the economic front. To that end, North Korea compromised on a major transportation issue that had been blocking trade relations. This dispute centered around Pyongyang's seizure in 1991 of some 226 Soviet railcars and its failure to pay subsequently for a series of rail fees it owed in connection with Russian–North

Korean trade through the border station at Khasan. In spring 1996, the Russian side had actually closed the rail line, effectively isolating North Korea from the Russian market.[33] This threat and the fear of failure in the upcoming negotiations impelled Pyongyang to settle the dispute in September 1996 by promising to repay $26 million in back debts.[34]

Despite the still-tentative nature of the ongoing rapprochement, many Russian regional analysts are optimistic about working out a long-term cooperative relationship with Pyongyang. Moreover, many feel that this relationship is not only advantageous for Russia, but also crucial for the sake of broader Northeast Asian security. As one Vladivostok-based specialist argues, a future unified Korea will increase the "geostrategic position" of the southern Primorskiy Kray in stabilizing international contacts with the Korea Peninsula. In this context, he argues, "the need to reestablish and deepen multilevel connections with Pyongyang becomes clear."[35]

Cultural Issues

In the Russian Far East, there is a tense triangular dynamic among the three sides involved: Russia, North Korea, and South Korea. Relations before 1990 existed exclusively with North Korea, and a range of economic, Communist Party, and cultural relations flourished. But times have changed quickly, especially in the past few years, which have seen the formation of several new South Korean–funded cultural organizations, including the newly built $1.5 million Korean College, now affiliated with the Far Eastern State University, as well as the opening of a large South Korean business center, both in Vladivostok. Not surprisingly, within the academic and private business communities there has been a rise in pro–South Korean sentiments. Most South Korean organizations, however, are suspicious of any Russians working for them who maintain ties with North Koreans, thus adding an element of tension to these new relations.[36] On the other side, Pyongyang's growing fear of Russian political "contamination" is seen in its new visa regime, which now requires Russian citizens to obtain a visa before traveling to North Korea (not the case before 1991).[37]

Meanwhile, local government policy in Primorskiy Kray is moving back toward North Korea after a period of strongly pro–South Korean ties under the previous governor, Moscow-trained economist Kuznetsov. By contrast, Governor Nazdratenko is supported in his pro–North Korea policy by the strong anti-Chinese sentiments of the local population. He is also, as noted above, seeking a reliable (nonimmigrant) population of workers in labor-strapped Russian construction and agricultural enterprises in the region. Defections from North Korean camps are rare compared with the mass illegal immigration perpetrated in recent years by Chinese workers, as well as those on supposed "tourist" trips to Russia. Notably, many of these Chinese are actually ethnic Koreans whose families originally fled to China either after the Japanese occupation or, later, after the establishment of the DPRK. Many speak Mandarin badly and may begin to look to their ethnic Korean compatriots in the Russian Far East as role models for their own integration into Russian society.

Within this complex matrix of evolving relations, therefore, the place of Russian-Koreans in the Far East also plays an important role (and one often neglected by Western analysts). This long-standing Russian minority strengthens the already strong tendency to look to the two Koreas—as opposed to China or Japan—for closer trade and political relations. However, there are two distinct groups of minority Koreans in the Russian Far East, each with a very different trajectory.

The first group consists of the approximately 40,000 Koreans sent by the Japanese to Southern Sakhalin during World War II as forced laborers.[38] These people consider themselves Koreans and, even today, many do not possess Russian citizenship. During the Soviet period, even their offspring had to marry a Soviet citizen in order to be granted citizenship themselves.[39] Many are still seeking to return to South Korea (the area where most are from) with help from the Japanese government, which recently made some funds available for this purpose as compensation to these victims of the war. Unfortunately, the money is not enough and most of the original deportees will likely die on Sakhalin. Their offspring, however, are being folded into the Russian Federation and will continue to play a role in the regional politics of the Far East.

A second group consists of the surviving members and offspring of the 204,000 Russian-Koreans who had lived in the Russian Far East for generations until Stalin forcibly exiled them to Uzbekistan and Kazakhstan in 1937.[40] They speak an older Korean dialect and the Primorskiy Kray as their homeland, to which they are also seeking to return. Although rehabilitated in the early 1960s, by 1991 only about 8,000 had succeeded in moving back (out of a total of about 300,000 now, counting their children born in Central Asia). By 1996, however, some 16,000 had managed to return to the region.[41] The *kray* administration is supposed to assist them in relocating, but has not provided any help (according to local Russian-Koreans). However, some local farming areas (such as Ussuriysk) are attempting to assist returnees because of the excellent reputations Korean-run *kolkhozy* (collective farms) achieved in Central Asia during the Soviet period. Some returnees are also working for South Korean companies. There are even bigger plans to resettle Russian-Koreans in the region if the South Korean industrial park in Nakhodka is finally built. One problem in this plan is that many of the offspring of returning Koreans no longer speak Korean. A South Korean-funded program in Vladivostok, however, now provides free Korean language lessons to all comers.

It seems clear that the Russian-Korean population is bound to increase in the region. Ironically, it is also conceivable that North Koreans may eventually flee to Primorskiy Kray if the reunification of Korea occurs amid great social instability and distrust of Southern intentions. With this in mind, many local policy makers believe that ties with Pyongyang need to be renewed and put on a more stable footing to increase Russia's preparedness.

Notably, there remains an enduring feeling of closeness to North Koreans in the Russian Far East, which eight years of postcommunist reforms have not erased. In economic terms, Russian entrepreneurs and state enterprises find that they are more comfortable dealing with North Korean state enterprises than they are with private businesses from Japan, South Korea, or the United States. In addition, the plight of starving North Koreans, in the wake of several years of alternating floods and droughts, has resulted in the provision of food aid to Pyongyang, despite Russia's own economic difficulties. These points suggest that there may be broader support than assumed in the West for at least a more balanced policy between the two Koreas.

Risks of the New Rapprochement

Ironically, however, recent events have shown that the growing rapprochement with North Korean may also create new problems for a democratizing Russia, itself trying to shake off the legacy of communism. Indeed, if not managed carefully, closer ties may eventually backfire on supporters like Governor Nazdratenko and the pro–North Korean lobby in Moscow.

Reports of illegal activities conducted in the Russian Far East by North Koreans fill the regional press. These range from drug smuggling to money laundering to counterfeiting to poaching to small-arms trafficking to attempts to acquire nuclear materials. Many of the victims of these criminal activities are Russians. One recent border incident, for example, resulted in the confiscation of 10,000 counterfeit U.S. dollars from a Russian businessman returning from North Korea, where he apparently had been passed counterfeit dollars as part of his sale of Russian equipment to a North Korean enterprise.[42] In other cases, including several related to illegal activities in the North Korean trade office in Khabarovsk, North Korean diplomats used their immunity from prosecution to make deals with Russian firms and then skipped town still owing huge payments for undelivered Korean goods.

Of even greater concern are reports of North Korean attempts to acquire Russian help in developing nuclear weapons and delivery systems. Beyond the much-publicized interception of more than 60 Russian scientists and their families on a Pyongyang-bound plane in Moscow in 1992, several other cases have come to light. In January 1996, for example, seventeen North Korean agricultural workers were caught lurking around a nuclear submarine facility some 15 kilometers from the enterprise that they were supposed to be attached to.[43] Since these naval facilities house enriched uranium and other nuclear-related technologies, the potential threat from such activities (as well as the proximity of the North Korean border) cannot be treated lightly.

Although the Russian government remains concerned about these problems, it lacks sufficient wherewithal to fight many of them effectively. For this reason, it would seem that Russia should proceed carefully in its rapprochement with poverty-stricken and potentially unstable North Korea, so that its border officials, local police, and the local population will be able to deal better with expected problems in the context of their reduced, post-Soviet capabilities.

Another problem associated with Russia's growing rapprochement with North Korea is that it risks strengthening some of the very tendencies from Soviet-era political life that postcommunist Russia is seeking to overcome. One example is the disturbing continuity in old-style diplomatic cooperation between North Korean and the Russian Foreign Ministry when it comes to controlling the behavior of North Korean citizens. In spring 1996, a young North Korean shot three police pursuers while fleeing for political asylum inside the Russian embassy in Pyongyang. The Russian side (fearing a diplomatic incident) allowed North Korean security forces to enter the Russian compound and retrieve their compatriot, who was immediately shot.[44] A similar incident took place in May 1996, when three North Korean citizens were arrested in Vladivostok for attempting to board a flight to South Korea with forged documents. Russian border police brought them back to the North Korean border. Only after the first detainee was shot by North Korean officers as he reentered the DPRK did the Russian side take the other two into protective custody.[45] Russian authorities also have yet to solve the troubling case of the fall 1996 assassination of a South Korean embassy official, allegedly by North Korean agents. Together, these incidents suggest a willingness by at least some Russian officials to tolerate activities on the part of North Korea that most civilized countries would consider grounds for a diplomatic breach.

Other areas of ongoing Russian–North Korean cooperation have raised serious questions regarding Russia's long-term security interests. Some reports argue that Russia is continuing to provide Pyongyang with spare parts for its Soviet-era weapons,[46] while others claim that Moscow is still sharing satellite photographs of U.S. and South Korean force deployments with the North Korean regime (albeit now for a fee).[47] In October 1998, Russian customs officials in the Khasan region near the North Korean border apprehended five Mi-8T attack helicopters en route to the DPRK.[48] The culprits were allegedly part of a mafia group operating out of Khabarovskiy Kray, where they had purchased the helicopters from an impoverished Russian military unit at a drastically reduced rate. These trends bear watching as potential harbingers of a dangerous relationship between North Korea and the Russian military, especially if the Russian economy continues to decline.

Conclusion

Russia's move to a more balanced policy between the two Koreas, in the minds of many Russian government officials and even academic analysts, has been long overdue. After essentially removing itself from strategic developments on the Korean Peninsula during the mid-1990s, Moscow is seeking to reestablish its sway in the region and forge a new course toward a more sustainable long-term policy. Some of the efforts in this regard, especially considering the potentially highly unstable situation in North Korea do deserve credit. Moscow is becoming more concerned about the need to assist Pyongyang in maintaining its borders and in feeding its population. At the same time, there are also clearly self-interested Russian reasons for this policy, especially to avoid an uncontrolled breakup of North Korea, which could send millions of refugees into the neighboring Russian Far East. However, Russian attempts to play a role again in Korean Peninsula diplomacy will not come without a significant investment, both in material terms and in high-level political attention by the Russian government. Given the many problems in Moscow today, it is unclear what Moscow intends to draw upon for these resources—or whether it will be accepted at the table of regional talks if it does.

Whatever the outcome on the Korean Peninsula, current trends suggest that Moscow is beginning to "balance" between North and South, after several years of leaning heavily toward Seoul. Russia has been irritated by its exclusion from ongoing international discussions and sees strengthening its ties with the North as a means of restoring its role. At the regional level, there are even stronger forces pushing for economic and political rapprochement. Whether it wants to or not, therefore, it is unlikely that the international community can exclude Russia from reestablishing influence in the region. In the end, Russia's historic ties and shared border with the North, its own minority Korean population, and its willingness to recognize and pursue simultaneous ties with the South all constitute legitimate reasons for Moscow to be involved in any future regional settlement. If left out, moreover, Russia could also play a spoiler's role on the Korean Peninsula, especially if more conservative forces pushing for closer ties with Pyongyang regain power in Moscow.

Thus, it seems that the challenge facing the United States and South Korea is how to ensure that Russia's role is a positive one, rather than one that strengthens the very characteristics of the North Korean regime that led to the nuclear crisis in the first place. More active efforts at bringing Russia into the Agreed Framework may be one avenue, in return for greater Russian restraint on military cooperation with North Korea. A focused U.S. assistance program might also help beleaguered and underpaid Russian Far Eastern export control officials enforce regulations against illicit weapons transfers to Pyongyang. One positive note in this regard are the recent joint U.S.-Russian military exercises aimed at alleviating the impact of a possible political and economic collapse in North Korea. These and similar efforts would seem to represent the most prudent means of ensuring that Russia—a still-influential neighbor of North Korea—makes a useful contribution to the Korean Peninsula peace process, even if it cannot be brought immediately into the current four-party talks. The alternative, alienating Russia and creating conditions that may facilitate a new alignment between Moscow and an unreformed North Korea, is clearly not in the interests of the international community and could eventually set back efforts aimed at achieving a lasting peace in the region. For these reasons, the status and future character of Russian–North Korean relations remain subjects of considerable importance to the future of Korean Peninsula security.

18

The Korean Peninsula

From Inter-Korean Confrontation to a System of Cooperative Security

Alexander Zarubin

For decades, the Korean Peninsula has remained a subject of deep international concern. In fact, this tense region is probably the worst remaining legacy of the Cold War. Despite efforts to reduce arms elsewhere in the world, the North-South relationship frequently lingers on the brink of direct military conflict. The Korean situation, therefore, constitutes a hotbed of tension and instability in the Pacific region, with potentially global consequences should a conflict break out.

Moreover, the existing situation is an ongoing personal tragedy for millions of Korean people on both sides of the border. Both Pyongyang and Seoul declare reunification as their basic national aspiration and goal. Yet, since the Korean armistice in 1953, it seems that neither of the two Korean sides nor any external player on the scene has done anything to promote reunification. Though the Korean counterparts blame one another, the predominant attitude of the world community in the post–Cold War period is that Pyongyang is the more guilty of the two. There is no doubt that sometimes the behavior of North Korea in international affairs provides justifiable reasons for this kind of negative attitude. And, of course, there is widespread disgust with the North Korean regime's gross violations of universally accepted human rights.

Therefore, it is no wonder that many politicians and analysts argue that bringing down or "naturally dismantling" the North Korean regime is necessary in order to normalize the situation on the peninsula. However, further study of this issue suggests that the followers of the "dismantlement school" fail to take into account the range of highly undesirable developments that could be easily triggered if a political crisis or power vacuum developed in the North.

This chapter first discusses the perspectives of the various parties involved in the Korean Peninsula crisis, either as direct participants or as interested observers (due to their contacts with one side or the other). It critiques these views and suggests that each of them contains flaws that render their development of a reasonable solution to the crisis unlikely. With this in mind, the chapter then outlines another possible path to reunification, one focusing first on the all-important security dimensions. Using this crucial area as a point of departure, it shows how building trust through military exchanges may provide the best means of building substantive contacts between the two sides, thus creating a new dialogue on security concerns that is necessary for bringing a successful resolution to the crisis. Using the concept of "cooperative security," it shows that building trust between the two militaries may be the way out of the current impasse.

North-South Views on Reunification and Their Inherent Problems

A first question to ask is: Who is actually striving for reunification? While the two Korean governments are proclaiming that reunification is their principal goal, are they in fact articulating the true feelings of the population? Everyone agrees that the people in North Korea have no rights, procedure, or institution to express openly their feelings and opinions. Thus, it is very difficult to divine the actual attitudes of North Korean citizens regarding the idea of reunification. In South Korea, there are some associations and clubs openly supporting the idea, but it is difficult to find any influential public movement ardently in favor of reunification. To sum it up, the desire of reunification is quite alive among Koreans but mainly on the subconscious level and in terms of a national instinct. At present, therefore, reunification is not likely to result from the current policies of the two regimes as a result of "bottom-up" forces. Either there are no such possibilities (as in the North), or there is no practical emphasis on such movements active on the political scene (as in the South).

Due to human nature, the real problem may be the mutual fear of an unpredictable personal future, particularly among the top political leaderships. If reunification does take place sooner or later on the basis of the peaceful convergence of the two systems, the emergence of a new Korean society will make the dismantling of both existing establishments inevitable, which in turn will force the old bureaucracies on both sides out. This basic instinct of bureaucratic self-preservation outweighs any staunch devotion to Marxism in the North or any commitment to democracy and free-market values in the South. This self-preservation syndrome is not simply psychological and personal, but also a politically dominant factors in the behavior of the Korean political players. Therefore, they are led by a very pragmatic approach: meaningful reunification under the rule of only one of the existing regimes can be achieved solely through war.

No one (with the exception of North Korea) accuses South Korea of preparing for forcible reunification. On the other hand, those who blame North Korea for aggressive intentions have not put forward any more convincing arguments besides the "unpredictability of Pyongyang." Indeed, while Pyongyang is certainly the "enfant terrible" of the 1990s, its mischief is actually quite predictable. If the North wanted a military solution to the problem, war would have broken out a dozen times over before now. Nevertheless, a state of permanent "prewar" conditions has become the inter-Korean rule. This is where the real danger lies. There is always a risk of a nervous, uncontrolled reaction, including one from external powers.

How might the two states approach the prospects of a peaceful restoration of a united Korean nation from a more positive perspective? First of all, the term "reunification" should be replaced by the concept of "convergence." Yet, even in the case of Germany we see neither reunification, nor convergence, but rather a civilized assimilation. Convergence is only now beginning within a reunified Germany, and not without considerable troubles. There is no doubt that Pyongyang is afraid of such an uncontrolled "convergence" taking place on the Korean Peninsula. This is why the North is much more inclined to open itself to the United States and Japan rather than to its "fraternal" southern neighbor. The North believes that it can prevent American and Japanese ideological penetration, but it doubts its abilities to resist a Southern psychological and social expansion in the North, which will inevitably provoke an irreversible erosion of the communist regime.

But the South Korean government also has significant fears regarding convergence. Since the first days of its existence, the government in Seoul has been legitimately accused of corruption, totalitarianism, and antidemocratic actions. Many formerly high-placed officials remain under official investigation, and the government still resorts to authoritarian measures against strikers and other opponents. This is why the government fears that the result of convergence may be that a large part of the Southern population might choose a form of government closer to the North's.

Meanwhile, there are external players on the scene, including the United States, Japan, China, and Russia. It is important to understand how their policies influence the current "endgame" negotiations on the Korean Peninsula.

The Views of Outside Powers

Unfortunately, the policies of outside powers have not had a positive impact on the prospects for Korean reunification. This suggests that many of these states, deep down, may not actually favor reunification, due to its risks and similar fears of uncertainty regarding the outcome.

As for the United States, its media representatives frequently point out that neither North nor South Korea would be able to withstand the burden of reunification. Take, for example, the U.S. observers who stress that the German method of reunification is not suitable for Korea. What they especially underline is the fact that the gap in living standards and economic potential is much deeper and more difficult to overcome in the two Koreas than in Germany. Others warn that South Korea will lose its economic and political position as a rising industrial nation if reunification occurs, plunging it into decades of "recovery" with the North saddled to its back. Moreover, there are even some U.S. academics warning about alleged, secret Northern plans of underground communist resistance if a peaceful reunification does take place.

Meanwhile, the Japanese press is even more blunt and direct. Without tongue in cheek, its observers state that a strong united Korea will pose a direct threat to the security and national interests of Japan. This view is explained by their firm conviction that "traditional Korean nationalism has always had and will have a strong anti-Japanese orientation."

Current Chinese diplomacy points out that Beijing would also prefer the present status quo to an unpredictable reunification. Take, for example, one of the latest Chinese remarks on the subject: "The positive developments in China's relations with North and South Korea means that Beijing is consolidating its presence on the peninsula as a whole. This translates into new and strong guarantees for peace and stability in Korea." Between the lines, this quote makes it clear that China wishes to keep the current status quo on the peninsula. Beijing is openly worried about the possible geopolitical challenges to itself that might ensue as a result of actual reunification.

Russia, by contrast, is the only outside power that is not concerned by the prospects of Korean reunification. Russia does have some long-standing interests in the economic, political, and strategic spheres on the Korean Peninsula, but it does not foresee any grave damage to those interests as a result of reunification. Russia's position is that the terms, conditions, forms, and procedure of reunification are up to the Korean nation and the existing Korean governments to decide, as long as they are supported by their populations. Russia hopes that the reunification of the two countries will bring mutual benefit and prosperity to all Koreans and provide effective guarantees of human rights. Russia would like to ensure that a new united Korea brings a positive contribution to stability, security, and cooperation in the Asian-Pacific region. Of course, such a position assumes an ideal outcome, which may or may not occur as a result of actual events.

Given the policies of the various domestic and outside actors, it is not entirely surprising that the situation on the Korean Peninsula has become an entangled knot of contradictions. Quite a number of politicians, scholars, and observers believe that any attempt to untie this knot is hopeless. Therefore, some insist on the necessity of putting constant pressure on Pyongyang—the weaker and greater "evil" of the two sides—in order to accelerate the downfall of the Northern regime. In my view, such an approach will only lead to a dead-end. Thus, we must examine whether there is some potential in the existing situation for a mutually acceptable solution.

The Military Balance and Prospects for a Rapprochement in the Security Field

Glancing back at the history of inter-Korean confrontation, it can be observed that while the tension at times has been quite serious, it has also been well controlled. The presence of U.S. troops on the peninsula, it must be admitted, has generally had a positive effect in causing restraint by the two sides. These forces have held down Southern extremism and acted as a deterrent to the military adventurism of the North. At the same time, China, the former Soviet Union, and now Russia, Japan, and other nations must be recognized for their roles in trying to keep the situation under control. The North, with all of its risky actions and maneuvering, in the end always backed down from the brink of a real disaster. And this is not the only reason for cautious optimism in the Korean problem.

Optimists examining the potential for progress despite today's apparent stalemate can point to a number of developments:

- Despite the long period of tension and mutual confrontation, both Koreas developed relatively well economically. The South has practically leaped to the position of a new industrial state. While the achievements of the North are less impressive, it is worthwhile to note that the Democratic Peoples' Republic of Korea (DPRK) of 1998 is still far more developed than the new North Korea of 1945.
- Though the two Koreas have shown a tendency for confrontation and hostilities, neither of them has ever closed the door to détente for good. While they are far from a reconciliation, each understands the inevitability of establishing relations of mutual coexistence. Under Kim Dae-jung Seoul today is openly speaking about it.
- Pyongyang will always be Pyongyang in its public face, but its rhetoric is quite different from what it actually means and does. After all, there are already a number of formal and informal agreements between the North and the South that provide a rather sound base for an inter-Korean dialogue.

Given these points, the question arises: What can the world community do to help accelerate détente on the Peninsula?

What is clear for anyone who understands Asian politics is, first and foremost, a need for patience and tolerance, both toward the North and the South. In this regard, it should be viewed as a very dangerous situation if any state tries to interfere directly with the natural course of development of the North Korean regime. As Kipling wrote: "West is West, and East is East." Sometimes the West does not understand (or is not able to understand) that the East requires significantly more time for changes to occur that the West deems long overdue.

Not only international law, but common sense also demands that the future of the existing Northern system should be determined by the North Koreans themselves, no matter how indoctrinated they may seem. The short but rich postsocialist history in Eastern Europe shows all too clearly that indoctrination does not count for much when the situation becomes ripe. Interference, pressure, or attempts at isolation will bring only negative results. Similarly, a hurried and premature dismantling of the regime might bring unpredictable consequences, such as the possibility of the uncontrolled immigration of many tens of thousands of North Koreans to neighboring countries. There may be other, even worse implications.

An old Chinese proverb makes an important point relevant to North Korea today: "a frightened cat becomes a tiger." Too much pressure and interference could stimulate extremist forces within the Northern establishment or even help bring them to power. In regard to the nuclear program, we should not exclude the possibility that Pyongyang's unwillingness to open its nuclear facilities to international inspections could be explained by a quite different reason than the hostile intentions that the international community has convinced itself of. Kim Jong Il might not have any military nuclear program at all. In this case, the results of an international inspection would deprive his regime of the chance of holding the South hostage to Northern nuclear blackmail. Pyongyang may have doubts about the reliability of its former allies in a crisis, especially taking into consideration that it does not have close ideological ties with any state today.

While the South has become more and more involved in a wide scale of international relations, it must also be noted that the North is trying to find a "safe" way for such involvement. Yet, while Pyongyang may open up to the external world gradually, it will never accept the prospect of being opened up forcibly. Taking into consideration all the peculiarities of the regime, it might be observed that it is doing as much as it can in its own peculiar fashion to introduce an "open door" policy (albeit à la North Korea). That is, this means negotiations primarily with the United States and Japan, and an eventual (but not immediate) dialogue with the South.

Clearly, the above-mentioned processes are not developing smoothly. These policies are evolving behind a North-installed "iron curtain," which, unfortunately, keeps foreign observers from knowing what is really happening within the Northern ruling establishment, especially in its highest strata. But one can easily suppose that many of North Korea's actions may reflect a struggle of ideas and approaches, if not a direct power struggle among the Northern elite. It is probable that the fall 1996 "submarine incursion" is another example of such conflicts. These tendencies suggest that the most effective remedy in the North Korean case should be the careful, steady involvement of Pyongyang in international affairs, both global and regional.

Building a Positive Policy of Engagement

The first step in such a policy must be to help those in power in North Korea to elaborate an outlook that is broader than just its own country and broader even than the Korean Peninsula. This is why it could be very harmful if China and Russia were to drop the DPRK from their respective lists of allies (or, at least, friends). This does not imply, necessarily, a full military commitment, but more of a political and moral one. At the same time, in the author's view, there would be no damage caused to the Asian-Pacific region and global security if China and Russia supplied Pyongyang with some types of weaponry. Military ties, especially in this case, might be one of the ways of inducing North Korea to accept greater international obligations. In Russia's case, while the old treaty of friendship and cooperation signed between the DPRK and the Soviet Union in June 1961 must be modified and adapted to the new post–Cold War reality, a more limited Russian–North Korea treaty of friendship and cooperation might be very useful for drawing Pyongyang into more broad-based international cooperation. In this regard, the normalization of relations between North Korea and both the United States and Japan would provide another positive impetus. These ties, particularly in the economic realm, could be developed without damage to the two countries' respective ties with South Korea.

It can be observed from the history of tensions on the peninsula that the "stick" of deterrence has been used toward the North more than enough. It is time to try the "carrot" of mutually beneficial cooperation as a key to open the North to the external world. Stable security cannot be developed in Northeast Asia and in the Asian-Pacific region as a whole without a new level of intra-Korean ties. On the other hand, no real intra-Korean normalization can be achieved without a restructured system of Northeast Asian security. A significant contribution to such a system could be made by an intra-Korean security agreement with reliable international guarantees. In this way, one could conceive of a path in which the Northeast Asian security system and an intra-Korean security system could be constructed simultaneously. For such a development to occur, the first and most important step might be played by the initiation of direct contacts between the two Korean militaries.

Charting a Course for Direct North–South, Military-to-Military Contacts

To begin this process, a first step might be for Russia, the United States, China, and Japan to give cross-security guarantees to the North and the South. These should be provisional guarantees, meaning that they are valid only for a transitional, preunification period. Secondly, the United States, Russia, China, Japan, Canada, the DPRK, and the Republic of Korea (ROK) should start negotiations on establishing a collective security system in Northeast Asia with all the necessary institutions. At the same time, the North and the South could start negotiations on a bilateral system of cooperative defense on the peninsula to cover the pre-reunification period. During the negotiations, the Korean sides would have to come to an agreement on several important points:

- a common pan-Korean cooperative national security concept and common cooperative defense doctrine;
- a common understanding of requirements for troops, military equipment, material, and financial recourses needed to provide for the cooperative defense of the peninsula (which would provide a sound base for further agreements on bilateral reduction of military forces, confidence-building measures, etc.); and
- an inter-Korean institution for cooperative defense (possibly a joint military or defense committee with the functions of a joint command over troops detailed by the two sides for the task of cooperative defense).

A fundamental question, of course, arises over the issue of whether a political atmosphere exists between the two Korean sides to permit any such military rapprochement. This is a valid objection. But we might suppose that, as a first step, the U.N. Security Council could ask the North and South to create a joint military unit that could be used by the United Nations for peacekeeping or humanitarian operations. To make this option more palatable, it might be proposed that this unit be a three-sided one with the participation of one of the eventual Northeast Asian collective security members. The formation and training of such an intra-Korean peacekeeping contingent could provide the basis for the opening of a school for mutual confidence building.

It is no secret that the U.S. forces deployed in South Korea represent a highly sensitive subject in the North. But the proposed multilateral guarantees of security, the participation of the two Koreas in a collective security institution, and the gradual restructuring of the intra-Korean military relationship from confrontation to cooperative defense may help create new options to the U.S. military presence on the peninsula. Under the conditions suggested above, U.S. forces in the South could either eventually be incorporated into the military institutions of the Northeast Asian security system and the inter-Korean system of cooperative defense, or be withdrawn by the common agreement of the two Korean sides.

While, at present, the plan outlined above may seem only to be a poor fantasy, it is worth noting that many new organizations in international relations began with little more support than a simple plan by a few diplomats using their imaginations. The U.S.-DPRK Agreed Framework, certainly, would not have been believed even six months before its eventual formation. Similarly, if pursued with patience and prudence, the above cooperative security plan for the Korean Peninsula may provide the basic guidelines for moving beyond the hostile politics that dominate today's impasse in intra-Korean relations.

19

Russian Views of the Agreed Framework and the Four-Party Talks

Evgeniy P. Bazhanov

This chapter begins with an analysis of the main motives and goals of Russian policy toward the Korean Peninsula. It traces the evolution of the reaction of Russian diplomacy to North Korea's nuclear activities from 1991 through the signing of the Agreed Framework in 1994 and the establishment of the Korean Peninsula Energy Development Organization (KEDO) in 1995. The focus then turns to Russian commentaries on the four-party talks among the Democratic Peoples' Republic of Korea (DPRK), South Korea, the United States, and China toward a settlement in Korea. Presenting Russian arguments for including Moscow in this process. Finally, the chapter presents Russia's plan for the settlement of the Korean crisis. This research draws upon official Russian governmental documents, comments made by various Russian politicians, scholarly studies, and articles in the Russian news media.

Russia's Changing Policies toward North Korea

During General Secretary Mikhail Gorbachev's administration (1985–91), Moscow gradually distanced itself from the Soviet Union's lopsided and unconditional support for Pyongyang. The Soviet government began to reject the domestic and foreign policies of the North Korean leadership and, simultaneously, began a thaw in relations with Seoul.[1] When the anticommunist regime took over power in Russia at the end of 1991 under President Boris Yeltsin, these tendencies grew even more. The liberal democrats in the Kremlin had no emotional attachment but instead enmity toward the Stalinist regime in the DPRK. They expected its imminent collapse, and, therefore, reduced their contacts with the North to a minimum. They abruptly halted economic aid to the DPRK, causing Pyongyang to react to Moscow's actions in a similar manner. On the top of their indignation, the North Korean leadership was afraid that the Russian democrats might join the attempts by Seoul and Washington to undermine the DPRK regime from the inside.[2]

It was in this situation that Moscow's position in respect to Pyongyang's nuclear preparations was formulated. It was tough, almost uncompromising, and oriented toward Seoul's and Washington's approaches to this problem.

Russian policy on the nuclear question, which was the key problem in North Korean foreign relations from 1991 to 1994, in turn, contributed to the further deterioration of the entire spectrum of Russian–North Korean relations. By 1994–95, however, the Kremlin's foreign policy strategy had begun to change, due to the influence of domestic politics and geostrategic factors. Russian diplomacy became: less idealistic and more down-to-earth; less ideological (i.e., less anticommunist) and more pragmatic; less internationalist and more nationalist; less pro-Western and more pro-Eurasian; and oriented less toward the West than the East and the South. Greater priority was given to security concerns and Russia's remaining great power ambitions.[3]

Moscow has now begun to gain a better appreciation for the danger of destabilizing the Korean Peninsula. As a result, Russian diplomacy has resumed its efforts to play a role in the settlement of contradictions between Pyongyang and Seoul. The collapse of the Stalinist regime in the North does not now appear to be as inevitable in the near future as it used to be. Moreover, should it occur, it is believed that it might cause even greater threats to Russian security. New assessments of the Korean Peninsula situation are pushing Moscow to improve its relations with the DPRK and to pursue a more balanced policy on the peninsula.

Russia's great power ambitions, in turn, are also contributing to Moscow's new tilt toward North Korea. The Russian government is seeking to restore its influence and prestige in the Asian-Pacific region. Moscow is hoping to strengthen its ties with new partners and to revive its relations with former Soviet allies, which were so hurriedly and unreasonably abandoned in the past. The necessity of contacts with North Korea is supported by the argument that Moscow created Kim Il Sung's regime and spent a lot of time, effort, and money to develop it. In addition, analysts argue that while leaders come and go, popular memories and friendships endure and should be preserved.

Russia's renewed interest in the DPRK has been enhanced by the intensification of U.S. diplomacy toward North Korea. Russians view the situation as if the United States were trying to gradually convert Moscow's former ally to its side. Observers in Moscow see this tendency especially in the growing cooperation between the DPRK and the United States in the nuclear sphere and in the four-party talks for a settlement on the Korean Peninsula, which exclude Russian participation. Russian Ambassador to North Korea Valery Denisov has stated that the United States, without considering Russian interests, has launched a wide-ranging offensive aimed at increasing its influence in the northern part of the Korean Peninsula. In his opinion, the goal of this offensive is to make Washington the sole arbiter of the future of the peninsula.[4] He stresses that an active Russian role on the Korean Peninsula does not correspond to U.S. national interests.[5]

At the same time, as some Russian experts point out, the United States preserves its military-political alliance with South Korea in its original form and continues to exercise a dominant influence over its South Korean ally. As far as Moscow's prestige and influence in Seoul are concerned, they have been weakened recently precisely because the Kremlin has lost its influence in the North.

These Russian experts recall that in the 1980s, South Korea decided to accelerate the normalization of its relations with Moscow, supposing that the latter could exercise its influence on the North Korean leadership. But, as soon as the Kremlin offered Seoul diplomatic recognition, the South Koreans began to demand that Moscow end its military and other assistance to its ally in the northern half of the Korean Peninsula. When Russia essentially met this demand, the Southerners, instead of rejoicing, suddenly became disappointed with their new partner: If Moscow lost its levers of influence on Pyongyang, then of what use was it?

It is this kind of behavior that led Moscow to believe that its improvement of relations with the DPRK, along with other measures, could contribute to the restoration of confidence and respect toward Russia in South Korea. Renowned Orientalist and president of the Moscow State Institute of International Relations, (MGIMO) Alexander Torkunov stresses that only by exercising influence on both Korean states can Moscow "remain in the game" and retain a solid position with a future reunified Korea. The deterioration of its relations with the DPRK "limited Russian ability to exert influence on developments in the immediate vicinity of its borders."[6]

To some extent, economic considerations have added to the revival of Moscow's interest in a rapprochement with the North. Some experts argue that only through a renewal of ties can Moscow count on the repayment of North Korea's significant debt in the future. Moreover, without the DPRK's participation, it will be impossible to implement certain large-scale economic projects, such as the construction of a gas pipeline from Yakutiya to the southern part of the Korean Peninsula. Russia is also attracted on economic grounds to participation in the development of North Korean nuclear industry within the KEDO framework.[7]

One should also take into consideration that ideological barriers blocking ties with Pyongyang from the Russian side have also weakened recently. The Russian elite no longer feels disgusted at Kim Jong Il's regime, while the communists and some members of the nationalist opposition treat the North as their "favorite pet." Russian communists are attracted to Pyongyang on the basis of ideological solidarity, while the nationalists (and some communists) are attracted by the anti-American and anti-Japanese potential of the Stalinist regime. At the same time, the opposition speaks rather negatively about Seoul, blaming it for humiliating Russia, ignoring its interests, pursuing a pro-American policy line, and attempting to undermine stability in the North.[8]

In sum, one can observe that recent Russian approaches to the Agreed Framework between the DPRK and the United States and to the idea of the four-party talks have been formulated on the following new basis: Moscow has ceased to see its interests in Korea as congruent with U.S. interests; Moscow is watching Washington's activities on the peninsula with suspicion and jealousy; Russia's attitude toward its new partner Seoul has cooled; and Moscow is attempting to overcome the recent phase of "alienation" seen in relations with its old partner Pyongyang.

Russian Reaction to Pyongyang's Nuclear Activities (1991–94)

Pyongyang's nuclear activities were one of the factors that contributed to the deterioration of Russian–North Korean relations in the early 1990s. Russia welcomed the agreement on nuclear safeguards between the DPRK and the IAEA and the agreements reached between Pyongyang and Seoul regarding the nonnuclear status of the Korean Peninsula, the establishment of an intra-Korean Committee on Nuclear Control, and the initiation of international inspections of nuclear facilities in the North.[9] However, soon sharp disagreements concerning the Yongbyon facilities surfaced between Pyongyang and the Internatinal Atomic Energy Agency (IAEA). IAEA inspectors demanded access to these facilities, believing that separated plutonium was being stored there, whereas the North Koreans refused to provide them with access, arguing that these were military facilities outside the scope of IAEA jurisdiction.

Under active prodding from Washington and Seoul, Moscow attempted to persuade Pyongyang to agree to cooperate with the international nuclear agency in resolving outstanding issues. The North Korean side reacted to the Kremlin's efforts rather negatively. Pyongyang accused Moscow of "joint pressure with Washington, aimed at compelling the DPRK to make unilateral concessions."[10] Pyongyang stressed that it had "neither the capability nor the intention to develop its own nuclear weapons."[11] North Korean authorities expressed their "disappointment" at Boris Yeltsin's statement in Seoul that Moscow would "put political pressure on the DPRK if the latter failed to resolve the nuclear problem."[12]

Pyongyang was especially indignant at the coordination of actions between Moscow and Washington. Pyongyang protested the agreement of the Russian president to a joint statement in June 1992 with U.S. President George Bush regarding the nonproliferation of nuclear weapons on the Korean Peninsula.

Specifically, the U.S.-Russian statement said that "Only the DPRK's full implementation of all its obligations under the Non-Proliferation Treaty (NPT) and the joint intra-Korean declaration, including the IAEA controls, along with reliable and effective mutual nuclear inspections, will finally eliminate international concerns regarding the nuclear problem on the Korean Peninsula."[13] North Korean diplomats pointed out in conversations with Russian officials that "We do not like that you talk to us in the same manner as the United States does. We would like a Russia that is confronted with complicated problems itself to treat us more fairly."[14]

Also, Pyongyang expressed its resentment at the joint appeal by Boris Yeltsin and Bill Clinton calling upon the DPRK to abide by its IAEA safeguards agreement and to annul its decision suspending its membership in the NPT. But the Russian side again appealed to the North Korean leadership to change its mind, stressing that full adherence to the NPT corresponded to "the interests of the North Korean side itself, [as well as to] stability and security on the Korean Peninsula."[15] In response, Pyongyang accused the Kremlin of providing "lopsided support to the unjust demands of the United States" and suggested that instead Moscow should back the DPRK's decision to withdraw from the NPT.[16] North Korean newspapers published several commentaries denouncing Moscow for "joining the criminal pressure of the DPRK's foes regarding the nuclear question" and for "living by the brains of others and dancing to the tunes of American imperialism in advocating pressure and sanctions against the DPRK, which are aimed at finishing us off."[17] Russian government actions were explained by Moscow's desire to "receive huge sums of dollars in order to deflect the resentment of its population from the pernicious crisis and to prolong its rule."[18]

At the same time, the North Korean side moved to the counteroffensive against Moscow on the nuclear question. The DPRK Foreign Ministry began to make regular reprimands of Russia for its disposal of nuclear waste in the Sea of Japan. It stressed that Moscow had contaminated the environment and violated the international convention on the prevention of ocean dumping. Pyongyang demanded that Russia and the other nuclear powers be denounced and compelled to "make apologies before mankind" for "contaminating nature with nuclear waste, which endangered the lives of people."[19] The chairman of the DPRK's State Committee on Environmental Protection dispatched a letter to the U.N. Environmental Protection director in which he accused Russia of "enormously damaging the main fishing area" of North Korea as a result of the dumping of nuclear waste into the Sea of Japan. Pyongyang demanded that the United Nations intervene without delay and force the Kremlin to act in accordance with international norms.[20] The North showered Moscow with insults, saying that it was "incapable of resolving complex internal problems related to nuclear weapons, including the situation with the nuclear weapons stored in other CIS countries."[21] Pyongyang asserted that all these problems "posed real threats to peace in Asia and all over the world."[22] In its appeals to the United Nations, the DPRK's government reiterated that it was not the North Korean nuclear question that should have been placed on the U.N. agenda, but instead the behavior of the nuclear powers (including Russia), which implicitly threaten North Korea with nuclear weapons.[23]

Moscow rejected Pyongyang's claims as groundless and designed to deflect attention from the essence of the problem and raise issues totally unrelated to the question of North Korea's membership in the NPT. Russian Foreign Minister Andrei Kozyrev persistently stressed that Russia had used all possible avenues to prevent North Korea from withdrawing from the NPT. Kozyrev stated that as far as Russia was concerned, "It was a number one concern; maybe even a more important concern than for the United States, because the U.S. is located far away, whereas, for us, Korea is in the very immediate near-abroad."[24] In his opinion, "Even mildly speaking, the emergence of even a hint of a nuclear bomb in the immediate geographic proximity of the Russian Far East cannot generate anything but irritation."[25] Kozyrev reiterated that some other "nuclear candidates" might follow the example of North Korea, and, consequently, a large nuclear zone might emerge along the CIS borders to the East and to the South.[26] Moscow's concerns were reinforced by intelligence reports received through Russian channels. They demonstrated that Pyongyang had applied considerable efforts to develop its own nuclear, chemical, and biological weapons.[27]

Russia suspended cooperation with the DPRK in the sphere of nuclear energy: it refused to ship the equipment for the planned three-block reactor at Sinp'o, it ended the subsidized training of the North Korean nuclear scientists and engineers, and ended exchanges of delegations. At the same time, Russian diplomacy sought a compromise solution to the growing international crisis involving Pyongyang's alleged nuclear preparations. Moscow attempted to persuade Washington not to rush to impose sanctions against Pyongyang, to avoid the use of force, and to emphasize diplomacy. Russia welcomed the U.S.-DPRK negotiations in New York in spring 1993 that led to the North Korean decision to suspend its withdrawal from the NPT. Moscow declared its intention to join the nuclear powers' negative security guarantees toward North Korea.[28]

After the situation surrounding Pyongyang's nuclear policy flared up again, Russia proposed the convening of an international conference on security and the nonnuclear status of the Korean Peninsula. On March 24, 1994, the Russian Foreign Ministry issued a statement saying that it would make sense to try to settle the situation on the peninsula on a multilateral basis, since bilateral talks between Washington and Pyongyang seemed to be failing. Moscow proposed to include on the agenda of these multilateral talks questions regarding security guarantees for both Korean states and the denuclearization of the peninsula.[29] President Yeltsin insisted that the great powers should talk about sanctions against the DPRK only if the above-mentioned conference failed to fulfill its tasks. Moscow rejected the U.S. position on immediate sanctions. Russian experts believed unanimously that the position adopted by the Kremlin exerted some positive influence on subsequent developments of the situation around Pyongyang's nuclear preparations; in particular, it stimulated further U.S.-DPRK dialogue.[30]

Russia and the Agreed Framework

In October 1994, Washington and Pyongyang concluded the Agreed Framework. Moscow's reaction to this document was ambivalent. On the one hand, the Kremlin noticed the benefits of the Agreed Framework—that is, Pyongyang had agreed to remain within the NPT regime and, therefore, to abide by all its obligations under the treaty and under its IAEA safeguards agreement.[31]

On the other hand, Moscow stated that the Agreed Framework did not fully correspond to its interests. It stressed that, in compliance with the NPT, Moscow had frozen its cooperation with the DPRK in the peaceful use of nuclear energy (i.e., the 1985 agreement with Pyongyang on the Russian construction of light-water reactors [LWRs] in North Korea). But as a result of the Agreed Framework, the United States and South Korea were now becoming suppliers of the same technology to Pyongyang. Thus, Russia was in effect being "punished" for its consistent adherence to its obligations under the NPT. Consequently, Moscow concluded that the Agreed Framework could lead to a violation of one of the key principles embedded in all agreements on nuclear export controls, that is, that exporters should not use their "nonproliferation elements" to foster national commercial goals.[32]

Russian experts found other shortcomings in the U.S.-DPRK Agreed Framework as well. They pointed out that Pyongyang's implementation of its obligations under its agreements with the IAEA in fact was dependent upon the implementation of its agreements with Washington, and hence that IAEA inspections were to be postponed for at least five years. At the same time, the resolutions issued by the IAEA Board of Governors and the U.N. Security Council contained a demand that the DPRK first fulfill those obligations that ensued from its safeguards agreement with the IAEA. Moscow also expressed its apprehension that the Agreed Framework "could have a negative impact on the entire international system of nuclear nonproliferation controls, and, specifically, on the NPT extension conference in 1995, which would have to make a decision on permanent extension of the NPT."[33]

From a broader perspective, Russian officials grumbled that by signing the Agreed Framework with the DPRK, the United States "was seeking to squeeze Russia out of the DPRK and in essence has achieved definite results along this road."[34] One expert even stated that "although the DPRK's return to the NPT represented a positive achievement, it should not have been carried out at the expense of Russian national interests."[35]

In March 1995, an international consortium, KEDO, was set up in New York to supply and finance the transfer of LWRs to North Korea. Representatives of the United States, the Republic of Korea (ROK), and Japan formed KEDO's executive board of directors. They announced that South Korea was to supply LWRs made in the South to the DPRK and to contribute more than $2 billion for this purpose.

Russia argued that North Korea should use Russian reactors. Pyongyang welcomed this Russian proposal in principle, but the United States and South Korea categorically rejected it. Their argument was as follows: those who pay should supply the equipment. Because of this disagreement, Moscow decided against joining KEDO.[36]

Russian Attitudes toward the Four-Party Talks

Officials responsible for Russian policy in the Far East have frequently stated their views regarding the U.S.–ROK plan for four-party talks on a Korean Peninsula settlement with the DPRK and China. Russian Ambassador Denisov in Pyongyang has stressed that "As one can see, the Russian side has no place in this scheme. We cannot agree with this, because Moscow used to play and continues to play a positive role in the normalization of the situation in and around Korea. All attempts aimed at removing us from the settlement of a problem that directly concerns our interests, cannot be understood or accepted by us."[37]

Russian official circles do not express any doubt that the problem of Korean unification must be and will be resolved by the Korean people itself. At the same time, Moscow stresses that "the problem of unification of Korea has an international aspect, too; history has determined that Russia, China, the United States, Japan, and the United Nations, indirectly or directly, are involved."[38] Sometimes, they state this thesis more directly, without diplomatic niceties. Specifically, they argue that "no matter how negatively the Koreans view this circumstance, in any case, China, the United States, Japan, and Russia are fully determined to participate and will take part in the process of a settlement on the Korean Peninsula."[39]

Russian government experts and scholars share the opinion that since 1945 the fate of Korea has been in the hands of the Soviet Union and the United States, and that it was the Americans who, using U.N. mechanisms, became the initiators of the de jure formalization of the division of Korea in 1948. At the same time, they admit that the Soviet Union, by providing substantial assistance to the North, bears some share of responsibility, because the decisions adopted at the December 1945 Moscow meeting of the foreign ministers of the Soviet Union, the United States, and Great Britain, about the formation of a provisional all-Korean government were never implemented. However, Moscow mostly reacted to prior "initiatives" by Washington in this regard.[40]

Russian experts argue that, at the initial stage, the Korean conflict was a civil war, which eventually dragged in the Soviet Union, the United States, China, and the United Nations. On May 27, 1953, an armistice agreement was signed. Moscow did not put its signature on this document, but neither did Seoul. One cannot infer from this circumstance that neither Russia nor the ROK have anything to do with the Korean settlement (although this is Pyongyang's position).[41] Russia believes that although the armistice is obsolete, it is not as outdated as to warrant considering the use of force.[42] Russian official circles believe that "abandonment of the armistice agreement is in fact a way to undermine the armistice and its supporting mechanisms, including the recognition of the demarcation line and the demilitarized zone."[43]

Arguing in favor of Russian participation in the Korean settlement, far from presenting only historical arguments, experts are seeking to convince Seoul of the utility and importance of having Moscow as a partner in this respect. In particular, they point to the congruence of the national interests of Russia and the ROK. They say that Seoul should take into consideration the fact that no other country has national interests so congruent with those of the ROK as does Russia. In the area of security, the ROK and Russia do not pose and will not pose any threat to one another in the foreseeable future. Both states are interested in promoting peace and stability in Northeast Asia. In the political sphere, Russia is perhaps the only power sincerely hoping for Korean unification. This is so not because Russians are the nicest people, but because of the fact that a strong Korea will balance out Japan and China. As far as Tokyo and Beijing are concerned, for a number of well-known reasons, they are unlikely to want to see the emergence of a powerful unified Korea on the map of the world. In the event of the elimination of the division of the Korean nation, even the Americans may lose—because their troops are likely to be asked to leave the peninsula.

Russian analysts stress that in an economic sense, Russian and South Korean interests are also mutually complementary. The ROK is in need of Russian raw materials, whereas Russia is interested in ROK manufactured goods of various kinds. So far, both sides have not been satisfied with the state of their bilateral economic ties. But stabilization of the domestic situation in Russia (which is already forthcoming) is likely to contribute to a real boom in Russian–South Korean cooperation. There are all the objective preconditions for Russia to become the main destination for South Korean investment.[44]

Moscow supposes that Seoul's interest in Russian participation in the Korean settlement has declined because of its disappointment at the diminished power and influence of the Kremlin. They note that opinion makers in the South Korean capital advance the thesis that Russia is no longer the same "superpower" it was five to ten years ago. Moscow's position has gotten considerably weaker, especially in Asia. If this is the case, then one can ignore Russia or, at least, not count on its effective contribution in the Korean settlement.

Russian experts contend that they cannot agree with such arguments. They underscore that even in the past, the Soviet Union did not have a solid position in the Asian-Pacific region. Moscow was in conflict or confrontation with a majority of countries in this region: from huge China to miniature Singapore. In fact, the Soviet Union was not present in the Pacific area, except for its navy. Soviet military might was great and frightening. This military machine has not disappeared yet, and Russia still possesses thousands of deadly nuclear missiles deployed in silos, on aircraft, and at sea, as well as powerful conventional forces. But the new democratic Russia does not want to be feared. Nor does it want to threaten anybody. Moscow's efforts are aimed at something completely different: that is, developing friendly and mutually beneficial contacts with all states of the Asian-Pacific region. In this regard, democratic Russia has achieved far more than the Soviet Union already.

Russian relations with China have changed from hostility to being so intimate that other Asian countries are already beginning to talk about another "alliance" of the two giants. At the same time, Moscow has succeeded in overcoming the suspicions of Taiwan and in smoothing its relations with the ASEAN countries, Australia, and New Zealand. Past U.S.-Russian confrontation in the Pacific is gone. Moscow has also made considerable progress in overcoming its alienation from Seoul. It looks as if the only serious mistake of the nascent Russian diplomacy has been to have reduced its ties with Pyongyang. But the Kremlin is intent on gradually correcting its relations with even the DPRK, so that Russia should not be discounted by any standpoint.[45]

The analysis presented here of the statements made by Russian officials and by analysts in the Russian news media suggests that, in essence, Moscow proposes to use three principles as guidelines for the resolution of the Korean problem.

The first principle is to recognize the fact that six states should participate in the settlement on an equal basis: the ROK, the DPRK, China, the United States, Russia, and Japan. Whether Koreans like it or not, the Korean problem includes inseparable internal and external aspects. No matter how justified the demands that other countries not interfere in the affairs of the Korean Peninsula may be, China, the United States, Russia, and Japan must take part in the settlement. Any attempt to sidestep or block altogether the participation of any one of these six states will cause the delay and interruption of the peace process itself.

The second principle is for the six powers to approve the normalization of relations among all of them. Seoul does not like it that Washington and Pyongyang have increased their mutual dialogue, while North Korea is seeking to ignore the South. One can fully understand the indignation of the Southerners. Nonetheless, the ROK should show some reserve because the DPRK is the weaker party, increasingly lagging behind South Korea in economic, social, and military respects. Otherwise, Pyongyang would never have agreed to a meaningful dialogue and reconciliation with Seoul. First of all, Pyongyang wants to solicit international support and find its own balancing points on the world stage. It does not really matter that the North Korean leadership is obsessed with the idea of isolating the South. It will certainly not succeed in this goal anyway. It is ridiculous even to think that the United States might "fall in love" with the communist regime in the DPRK and, for this reason, might sacrifice its ally on the southern half of the peninsula. Of course, it will not. But, as far as Pyongyang is concerned, it may start to open up and transform itself in the process of rapprochement with the United States and Japan. As a result, it may become ready for a full normalization with the ROK.

In the same vein, Russia should not be concerned that the United States is "winning the game" versus Russia in the DPRK. In turn, China should not impede the normalization of North Korean–Japanese relations or, perhaps, envy the progress in Russian-ROK relations. It is also time for South Korea to learn how to react soberly to the changing situation in relations between Moscow and Pyongyang. It is hoped that the Kim Dae-jung government possesses such understanding.

The third principle is for all the members of the six to reject (if they are currently pursuing them) policies of undermining the domestic regimes of the other five states of this group. Recently, both Seoul and Washington have persistently reiterated they will refrain from interfering in the domestic affairs of the DPRK and not seek to foster the downfall of its communist regime. Nonetheless, Pyongyang has some grounds to doubt these statements. Even a few years ago, when the communist regimes of Eastern Europe were crumbling, the South Korean elite openly sought to crush its opponents in the northern part of the peninsula. After the disintegration of the communist order in the Soviet Union, democratic Russia also looked forward to seeing the DPRK crumble. In 1993, the United States seriously considered a plan for delivering military strikes against North Korea's nuclear facilities. It goes without saying that the fears of North Korean leaders will not disappear quickly, especially given the rapidly deteriorating conditions in the DPRK.

Russia's Plan for a Settlement on the Korean Peninsula

On March 24, 1994, Moscow introduced its own plan for a Korean settlement.[46] Its core was to convene an international conference that would allow the creation of a new mechanism for the comprehensive settlement of problems on the Korean Peninsula.

The DPRK, ROK, permanent members of the U.N. Security Council (Russia, the United States, the Peoples' Republic of China [PRC], France, the United Kingdom—since the latter two are also nuclear powers, their participation is important from the nuclear nonproliferation standpoint), Japan, as well as the U.N. secretary general and the IAEA director general, would be invited to take part in the conference. Representatives of certain other states and international organizations concerned, for instance, ASEAN, might be invited to participate as observers, as well.

Tentative Agenda for the Conference

1. Improving relations between the North and the South.
2. Substituting the armistice with a peace structure.
3. Adopting confidence-building measures on the Korean Peninsula.
4. Creating a nonnuclear Korean Peninsula and a zone free from all types of weapons of mass destruction.
5. Normalizing relations between the states participating in the conference (i.e., between the DPRK and the United States and between the DPRK and Japan).

It is assumed that on the eve of the conference, the United States and Japan would announce their diplomatic recognition of the DPRK. In the course of the conference, they could discuss means of resolving concrete bilateral problems with the aim of normalizing relations between the DPRK and the United States, and the DPRK and Japan.

Tentative Conference Proceedings

The conference would take place at the level of foreign ministers, who would agree on the agenda and conference proceedings, as well as create working groups on the particular sections of the agenda.

One working group would discuss ways to improve relations between the DPRK and the ROK. The subject of its discussions would be the full range of proposals already introduced by both sides. Its main goal would be to elaborate a range of measures aimed at developing a lasting, multifaceted dialogue between the North and South. As a first step, the parties could reach an agreement specifying how they are going to implement the provisions of existing agreements on reconciliation, nonaggression, cooperation, and exchanges between the North and the South. In particular, they would need to deal with such issues as the establishment of direct telephone links between the military commands of both countries, the advancement of economic cooperation, the restoration of interrupted transportation lines, the establishment of postal links, and the unification of separated families. Also, they should consider how to implement their agreement on the formation of a standing commission on military matters and a commission on cooperation and exchanges.

It would also be expedient if they agreed to have regular summit meetings and talks between the heads of governments and at the ministerial level. Agreement on propaganda policy would also be helpful.

In addition, this group could start a discussion of principles that should underpin the unification of Korea. Such talks could rely on the principles already outlined in the joint statement of the North and the South, dated July 4, 1972. In accordance with this statement, the unification of Korean should be achieved independently and without foreign interference, by peaceful means, and on the basis of "national consolidation." Along the lines of the agreement reached by Pyongyang and Seoul in December 1972, they could resume the operations of the Coordination Committee, whose functions would include the creation of conditions for the peaceful unification of Korea, broad exchanges between political parties, public organizations, and private individuals of the North and the South, as well as cooperation in the economic and cultural areas. The committee would set up five subcommittees covering political, military, diplomatic, economic, and cultural issues. The committee could be assigned to study proposals of the North and the South and to outline an action program aimed at the unification of Korea.

The DPRK is known to have introduced the idea of creating a confederation of the North and the South in the form of the Democratic Confederate Republic of Koryo. According to this idea, the North and the South would preserve their existing social-political systems but form a unified national government that should include, on equal terms, representatives of the DPRK and the ROK. At the same time, the North and the South would pursue regional self-government under the general guidance of the national government. A Standing Confederate Committee, formed within the Supreme National Confederate Assembly and consisting of an equal number of representatives of the North and the South, would serve as the unified government. This unified government would deal with the problems of external relations and defense, coordinate the activities of the two regional governments (of the North and South), and provide for the overall economic, social, and cultural development of the country, as well as for cooperation between the North and South. The regional governments, within the limits of preserving their national interests, could pursue policies independent from one another but, at the same time, could make efforts aimed at eradicating the differences existing between two halves of Korea.

The ROK's unification proposals are also well known. In 1989, ROK President Roh Tae-woo introduced the concept of a "Korean community." That concept envisaged the formation of a Council of Presidents of the two Korean states and the formation of a Council of Ministers, composed of the prime ministers of the North and the South and ten ministers from each side. Their priority task would be finding a solution for the problem of unification of the 10 million members of separated families and to relax the military-political confrontation between the two Koreas. His concept also envisaged the formation of a Representative Council composed of 100 parliamentarians drawn equally from the South and the North. Its mission would be to draft a constitution for a unified Korea.

Proposals advanced by the North and the South differ substantively from one another according to the forms and stages of unification. However, this fact does not mean that one could not use them to elaborate a compromise concept acceptable to both parties.

Agreements reached by the North and the South would be "approved" by other participants of the conference, who would then undertake an obligation to guarantee their implementation.

A second working group would be formed to deal with the substitution of the armistice with a new peace structure. Not only representatives of the countries that signed the armistice agreement but also representatives of the member states from the Neutral Nations Supervisory Commission (NNSC) that supervises the implementation of the armistice agreement could take part in the work of this working group. It is natural that representatives of the United Nations and of other countries participating in the conference would take part in it as well.

It is clear that the annulment of the armistice agreement will require the revocation of the relevant U.N. resolutions. In particular, the question of the use of the U.N. flag and attributes by U.S. troops stationed in the ROK would also have to be resolved.

As an intermediary step toward creating a new peace structure on the Korean Peninsula, one could consider the question of substituting U.S. troops stationed near the demilitarized zone (DMZ) with troops from "neutral nations." It is useful to remember that in 1987 Pyongyang proposed the formation of special military units drawn from the military personnel of NNSC states (Poland, Czechoslovakia, Sweden, and Switzerland) to supervise the situation within the DMZ. On this basis, one could make decisions about the gradual dismantlement of military facilities near the DMZ and the withdrawal of troops from the vicinity of the DMZ further into the rear areas.

A third working group could analyze possible confidence-building measures for the Korean Peninsula. This working group could concentrate its efforts on elaborating ways for gradual implementation of comprehensive measures aimed at confidence building in the military sphere, as well as on the radical reduction of troops and armaments. Such measures might include: (1) the invitation of observers to military exercises; (2) the banning of military exercises involving troops exceeding a certain numerical limit; (3) the exchange of data regarding the armed forces of both sides; and (4) the formation of joint groups for the exchange of views on the military situation on the peninsula.

At the same time, it would be useful to discuss the question of reducing the U.S. military presence on the Korean Peninsula. The United States is known to have repeatedly planned reductions in its military presence in the ROK. For

Russian Views of the Agreed Framework and the Four-Party Talks 233

instance, the Nixon Doctrine called for the gradual withdrawal of U.S. troops stationed on the southern part of the peninsula. One U.S. division was indeed withdrawn, but the withdrawal of all army troops planned for 1976 did not occur. After its defeat in the Vietnam War, the United States returned to its policy of maintaining a military presence in the ROK.

During the Carter administration, the U.S. government developed a five-year plan for the withdrawal of the U.S. troops from the ROK. The U.S. planned to reduce their number by 6,000 men by the end of 1978. After the completion of the five-year program, only 13,000 U.S. troops, mainly Air Force units, would have been left in the ROK. However, these plans were not implemented because of opposition from the ROK, the U.S. military, and the right wing of the U.S. political elite.

The Bush administration also planned a three-stage reduction of U.S. troops in the ROK. During the first stage (which was to last three years), U.S. forces were to be reduced by 7,000; during the second stage (from three to five years), depending on the progress achieved in the dialogue between the DPRK and the ROK, a considerable further reduction was to be made; and during the third stage (from five to ten years), troops would be cut to the level of a minimum containment force.

Thus, the question of reducing the U.S. military presence in the South is not new, and the U.S. side has already studied its parameters and possible stages in considerable detail.

A fourth working group at the proposed conference would seek to develop means to secure the nonnuclear status of the Korean Peninsula and establish a zone free of weapons of mass destruction. While the North Korean "nuclear problem" has been frozen, it has not been resolved. The IAEA continues to monitor North Korea's declared facilities, but is waiting for further implementation of the Agreed Framework in order to receive access to two undeclared facilities. For these reasons, the goal of securing a nonnuclear status for the Korean Peninsula is still a very pressing one. Of course, the idea is not new. The Soviet Union was in fact the first state to advance the idea of establishing a nuclear-weapon-free zone on the Korean Peninsula. The Soviet Union proposed that nuclear powers undertake the following obligations:

1. not to undertake any steps (or induce any other states to undertake any steps) that violate (or might lead to the violation) of the nonnuclear status of the zone;
2. not to use (or threaten to use) nuclear weapons against any participant in the treaty on the nonnuclear zone;
3. not to provide any assistance in the development, manufacture, or acquisition of nuclear weapons, or in training of troops of the participant states in the treaty in how to handle nuclear weapons;

4. to refrain from the direct or indirect transfer of nuclear weapons or any other nuclear explosive devices to states party to the treaty;
5. not to deploy or store means of delivery of nuclear weapons on the territory of the nonnuclear zone; and
6. to abide by a ban on the transit of nuclear weapons through the territory of the nonnuclear zone, provided that the ban does not infringe on the freedom of navigation of the open seas, the right to peaceful passage via territorial waters, and the regime of the straits used for international navigation.

An international conference on a Korean settlement could consider these obligations and conceivably could finally resolve the question of nuclear weapons on the Korean Peninsula.

The main goal of this working group would be to identify means of implementing the U.S.-DPRK Agreed Framework and the joint North-South declaration on denuclearization of the Korean Peninsula, including their pledge not to test, manufacture, possess, import, store, or deploy nuclear weapons. For these purposes, the working group might elaborate a series of measures aimed at:

1. assisting in the implementation of the Joint North-South Declaration on Denuclearization of the Korean Peninsula;
2. assisting in the implementation of the Agreed Framework between the DPRK and the United States;
3. securing the nuclear powers' guarantees for the nonnuclear status of the Korean Peninsula;
4. promoting KEDO; and
5. assisting in the resolution of the spent fuel problem and the supply of heavy fuel oil to the DPRK.

Within the framework of this working group, the United States and the DPRK could reaffirm their adherence to their commitments made in Geneva in October 1994. The United States could confirm the absence of nuclear weapons in South Korea and would commit itself not to reintroduce them into the ROK in the future. The United States, the ROK, and the DPRK could express their readiness to open the military facilities in both Koreas to international inspection with the aim of certifying the absence of nuclear weapons there.

This working group might also consider the question of a possible ban on the manufacture, deployment, and acquisition by the North and South of chemical and biological weapons, as well as long-range missiles. The DPRK and the ROK could jointly ratify the Chemical Weapons Convention and join the Missile Technology Control Regime.

Finally, a fifth working group could focus on the normalization of relations between the states participating in the conference (i.e., between the DPRK and the United States and the DPRK and Japan). This working group could consider questions related to the resolution of the problems impeding the full normalization of bilateral relations between the DPRK and the United States and between the DPRK and Japan.

Since all of these groups would be working under the "roof" of the same international conference, they could convene joint meetings of several working groups for the consideration of overlapping problems and to search for comprehensive solutions. Recommendations drafted by the working groups would be submitted to the collective body of the participating foreign ministers for their final approval on the behalf of their governments.

In conclusion, given the lack of significant progress in the ongoing four-party talks (which currently exclude Russia), the analysis presented above suggests that Russia's plan for a multilateral conference on Korean Peninsula issues could provide a valuable new forum for stimulating movement toward an eventual peaceful settlement of the North-South conflict.

20

Pyongyang's Stake in the Agreed Framework

Alexandre Y. Mansourov

Juch'e ideology has always played an important role in the policy-making process in North Korea's political system. It represents a unified worldview for policy makers and outlines the realm of desirable goals and legitimate means to achieve them. It elaborates fundamental principles and criteria guiding policy makers in their choices. The *juch'e* framework places concrete policy questions and dilemmas of the day into the broader socioeconomic and political context of the country. Also, *juch'e* plays a considerable role in setting policy-making priorities and offering indisputable justifications for policy choices, including in the foreign policy arena.

At the same time, *juch'e* ideology has to coexist in the minds of policy makers with the pressing demands of *realpolitik* and the constraining legacy of historical and institutional traditions. *Realpolitik* thinking compels North Korean leaders to constantly reinterpret and broaden the *juch'e* framework in order to custom tailor it to new realities, thereby allowing themselves greater ideological flexibility in dealing with pressing new problems. The historical legacy, however, serves as an anchor of tradition and inertia rooted in historical and institutional memories that constrain any fundamental ideological change and rapid policy innovation.

In this chapter, I will discuss how *juch'e* ideology, political realism, and historical traditions combine to shape the strategic calculations of and define the stakes for the North Korean leaders who have emerged at the helm of the Democratic Peoples' Republic of Korea (DPRK) since the death of its founder, Kim Il Sung. In particular, the chapter will analyze how these strategic policy guidelines are likely to affect future North Korean negotiating behavior in regards to the implementation of the 1994 Agreed Framework.

Kim Jong Il's Strategic Calculations

Since Kim Il Sung's death in July 1994, the ideologues of the Workers' Party of Korea (WPK) have been busy striving to redefine the great leader's place in Korean history, account for his historical legacy, and identify the challenges and new missions facing his heir, Kim Jong Il, on the eve of the twenty-first century within the context of the unified *juch'e* ideology.

From a historical perspective, Pyongyang's mythologists credit Kim Il Sung with the following accomplishments: the liberation of Korea from the Japanese colonial yoke in 1945; the "defeat" of both the U.S. imperialists armed with nuclear weapons *and* their South Korean puppets during the Korean War; and the postwar economic rehabilitation and rapid construction of socialism in the North from the late 1950s to the early 1990s. They place Kim Il Sung, who founded the socialist state in the North and ruled it with an iron fist for almost half a century (from 1945 to 1994), on par with the widely recognized mythical father of the Korean nation, the founder of the Old Choson, King Tangun, who is said to have lived more than 5,000 years ago.[1]

But despite this impressive record of achievement, they assert that Kim Il Sung left three historical tasks unaccomplished: (1) raising the living standards of his people through increased investment in agriculture, light industry, and foreign trade by reversing years of stress on the military and heavy industries; (2) officially settling scores with the United States and Japan by concluding peace treaties and normalization accords with both; and (3) reunifying the entire Korean Peninsula on the terms of *juch'e* ideas. It is the mission of his son and successor, Marshal Kim Jong Il, to complete these Herculean tasks.[2] In order to meet these challenges, Kim Jong Il must display not only the virtues and leadership skills of a second Kim Il Sung, he must prove himself to be better than Kim Il Sung.

According to the WPK's propaganda, the Agreed Framework—billed as the first "landmark feat of Kim Jong Il's brilliant leadership" in post–Kim Il Sung foreign policy—goes a long way toward advancing the above-mentioned strategic policy aims. Therefore, it must be faithfully upheld and meticulously and expeditiously implemented.[3] In general terms, a number of Kim Jong Il's associates appear to harbor some hope that the implementation of the Agreed Framework and related agreements is likely to create favorable international conditions for domestic economic recovery and industrial modernization by literally *refueling* the North Korean economy, unlocking international financial resources, and attracting technological expertise, all of which were previously unavailable (due to Pyongyang's bad credit reputation in international markets).[4]

More specifically, Kim's coterie entertains some hope that the nearly $5 billion—or a capital investment equal to roughly 20 percent of the DPRK's gross national product—that is to be provided by the Republic of Korea (ROK), Japan, and others to finance the construction of the two light-water reactors (LWRs), should contribute significantly to its economic modernization and structural adjustment plans. They believe that the annual supply of 500,000 tons of heavy fuel oil for the period of ten years, agreed to in the Agreed Framework, should guarantee the satisfaction of the DPRK's minimum energy needs until the time when, purportedly, it will be able to get back on its feet.[5] The two 1,000-megawatt (MW) LWRs scheduled to be put in operation in the year 2003 are seen as a decisive factor in overcoming the country's chronic energy bottlenecks and in reequipping the DPRK's nuclear energy industry at low cost (and with a safer and more sophisticated technology). The modernization and dramatic expansion of North Korea's communications network (including its telephone, cable, computer, and satellite contacts), transportation sector (its port facilities, airport facilities, highways, and railroads), and service sector (its banking, insurance, health care, and entertainment fields) are believed to be among many other potential beneficiaries of the successful implementation of the Agreed Framework.[6] Lastly, Pyongyang hopes that the easing of the U.S. economic embargo against the DPRK is likely to increase its external trade and induce more direct foreign investment to flow into the Rajin-Sonbong Free Trade Economic Zone, which is billed as a future "Singapore of Northeast Asia."[7] It goes without saying that North Korean economic policy planners believe that, in the long run, all these measures are likely to lead to the betterment of the living standards of the North Korean people, the foremost declared domestic goal of Kim Jong Il's administration. Consequently, they hope that these measures will boost the legitimacy of Kim junior's rule, while strengthening their own positions in power. This is why one can expect very strong support for continuous and good faith implementation of the nuclear accords from the economic bureaucracy of the North.[8]

Secondly, Kim Jong Il and his associates appear to believe that implementation of the LWR project should be instrumental to furthering the full-scale normalization of U.S.-DPRK relations, including the conclusion of a peace treaty.[9] The Agreed Framework laid out a comprehensive agenda for improving bilateral relations: from mutual diplomatic recognition to security guarantees to political consultations on other issues of mutual interest to broader economic cooperation.

Pyongyang was quick to notice the beneficial consequences of the conclusion of the 1994 Geneva and 1995 Kuala Lumpur nuclear accords. Although the United States is still officially in a state of war with the DPRK, notwithstanding the July 1953 armistice agreement, it has agreed to recognize the DPRK as a sovereign and independent state and to deal with it as a legitimate negotiating partner. Washington also issued negative security assurances that it would no longer threaten Pyongyang with nuclear weapons, as long as the latter remained within the NPT regime. The Clinton administration suspended its highly provocative "Team Spirit" military exercises (thereby reducing the perceived security threats facing the North) and implicitly acquiesced to North Korean insistence on separating the U.S.-DPRK dialogue from the North-South dialogue. Although heavily armed, trigger-happy combat troops still stare off at each other along both sides of the 38th Parallel—ever ready to unleash an overwhelming premeditated attack—the United States, ROK, and Japan have decided to transfer highly sensitive nuclear technology to their foresworn enemy in the North. While the overburdened North Korean regime seems to be slowly approaching its economic self-destruction, the United States—evidently abandoning its long-held strategic goal of eliminating the North Korean state—appears to be intent on orchestrating a "soft landing" for its North Korean foes by pushing its allies to bail out Pyongyang, while providing humanitarian aid to the flood-stricken population in the North.[10] Sensing a new opportunity in the ambiguous U.S. stance toward the North, the latter seems to be striving to engage its former enemy in multichannel diplomacy aimed at full normalization of interstate relations.

Ultimately, one can envision three conceivable scenarios for future U.S.-DPRK relations in light of the implementation of the Agreed Framework and its enabling agreements, assuming that the DPRK retains its full sovereignty and independence for the foreseeable future. The first scenario contends that if the implementation process proceeds as scheduled in the early twenty-first century, a reborn and possibly (but not necessarily) reformed DPRK, newly equipped with sophisticated Western nuclear reactors, will be fully recognized by the international community of nations, including the United States and ROK (with which it will by then have signed peace treaties). However, although there appears to be enough political will on all sides to see to it that the Geneva agreement and all subsequent protocols are fully and duly implemented, this first scenario still has plenty of opponents in Pyongyang, Seoul, and Washington, and contains more wishful thinking than political realism at present.

The second scenario supposes that various incidents[11] and miscalculations (for instance, unwarranted mutual threats and provocations, and Western failures to deliver funds or equipment for the LWR project or heavy fuel oil) might spiral out of control and derail the implementation of the Agreed Framework. Then, the LWR transfer project might be aborted altogether. Consequently, Pyongyang is likely to resume its indigenous unsafeguarded nuclear activities, proving that those experts who assert that the North Korean nuclear freeze has already reached the point of no return are wrong. In turn, this is sure to lead to the escalation of tensions on the Korean Peninsula and, possibly, to the brink of a full-scale shooting war between the Korean People's Army and U.S. and ROK forces. In order to boost its weakening credibility and counter the arguments of those in the West who believe that, as the economic situation deteriorates or as Pyongyang opens itself up in the course of the implementation of the nuclear accords, North Korea will become more vulnerable, the DPRK leadership leaked to the press its own apocalyptic version of this worst-case scenario:

With operating nuclear power plants on the Korean peninsula singled out as prime targets of a first strike and retaliatory action in kind, a second Korean War would produce tens of Chernobyl-like meltdowns and leave millions of lives lost and tens of thousands wounded with deadly radioactive fallout raining on all parts of the Asia-Pacific region. The world's booming economies of Northeast Asia would end up in ruins.[12]

The warning from Pyongyang to the West is clear: do not try to take advantage of our difficulties, or we will do more damage to you than you will do to us. Hopefully, the international community will never have to call this bluff.

The third scenario is neither rosy nor pessimistic; it is also less dramatic. One might call it a "muddling through" alternative. There will be twists and turns, and ups and downs in an uphill battle for the unprecedented transfer of the LWRs to the North, sponsored by the Korean Peninsula Energy Development Organization (KEDO). But the North Koreans are willing to see to it that, at the end of the day, they will get what they bargained for.[13] They seem to believe that time is on their side because their leaders can discount the need for immediate political payoffs, as well as any political liabilities arising from the package deal and focus on its long-term value. For instance, they do not have to worry about electoral politics or public opinion, as their Western counterparts do. Besides, relatively speaking, their part of the bargain is not particularly exacting. After all, what they seem to believe they committed themselves to do is just to maintain a nuclear freeze and demonstrate good faith in implementing negotiations, and almost nothing else. In contrast, the Western side committed itself to exert its sweat to design, manufacture, install, and pay for the LWRs (in the form of a long-term, interest-free loan), while delivering heavy fuel oil in the meantime.[14]

In other words, in this low-cost, high-payoff game, as long as the overall integrity of the Agreed Framework remains intact from Pyongyang's point of view, its leaders seem to be prepared to be patient and flexible enough to accommodate unavoidable changes in the schedule and modalities of the implementing agreements, in order to achieve their ultimate goals of political survival, economic modernization, and diplomatic normalization.[15] Of course, the big question highlighted by the economic difficulties now being experienced in Seoul and Tokyo since late 1997 is how much the West can push, and how far Pyongyang is prepared to bend over backward, in order to accommodate the changes in the implementation schedule and terms of cooperation proposed by the West. Barring an outright Western provocation of state-threatening proportions or a unilateral suspension of the LWR project by the United States or any of its allies for political or security reasons (both of which are rather unlikely), for the time being, it would seem that a dramatically weakened Pyongyang is willing to compromise with KEDO considerably, albeit not without a good fight.

Kim Jong Il and his associates seem to believe that the implementation of the

Agreed Framework could be conducive to the *juch'e*-led improvement of intra-Korean relations.[16] This may appear to contradict Pyongyang's persistent opposition to Washington's demand that it open a direct North-South dialogue as a precondition for further improvement of the U.S.-DPRK bilateral ties. But it is not. Pyongyang is not opposed to the expansion of inter-Korean ties per se, but it is opposed to its hurried timetable and certain specific routes for such a rapprochement, as well as a Washington-advocated linkage between the U.S.-DPRK and ROK-DPRK direct talks. In other words, it wants to promote inter-Korean exchanges, but later rather than sooner, and on its own terms.[17] This is quite understandable, given its current extremely weak bargaining position.

In addition, one can attribute this reluctance of the North to open a genuine direct dialogue with the South to a number of other factors, including: (1) the painful historical legacy of intra-Korean relations in general; (2) imbalanced geopolitical realities between two Koreas; (3) the fact that the DPRK's ROK and U.S. policies are formulated by different, unrelated, and highly compartmentalized bureaucracies and controlled by very different political interests that do not necessarily overlap[18]; (4) the fact that agenda items on the U.S.-DPRK negotiating table are very different from those on the ROK-DPRK negotiating table; and (5) the fact that the stakes involved in normalizing relations with Seoul and Washington are very different from the standpoint of national survival.[19] This last point stems from the fact that even after fully normalizing its relations with Washington, Pyongyang will still be faced with a mortal threat to its national security emanating from Seoul. Only after normalizing its relations with the South and reaching a new modus vivendi in the context of peaceful coexistence between the two Korean states, will North Korea's national security be strengthened significantly and its state sovereignty and political independence guaranteed. This is why in terms of strategy, intra-Korean relations will always dominate the foreign policy-making agenda in Pyongyang, despite the fact that tactically North Korean leaders may reveal at times more interest and determination in improving their relations with the United States or Japan.

Furthermore, from an ideological standpoint, the North Korean–style communist worldview dictates that the United States, the "superpower bastion of global imperialism," now more than ever, will exercise absolute control over its "puppet regimes," including South Korea. Therefore, if the DPRK wants to resolve its disputes with the South, it has to go to Washington first and discuss and settle them with the United States. Only then, after securing the colonial master's support can it seek to engage the ROK in direct talks. Pyongyang's hard-headed logic is simple: there is very little utility in any talks with the puppet until his master's consent is secured. Consequently, North-South relations cannot improve until there is a bilateral understanding in relations between the DPRK and the United States.[20]

From a realist point of view, however, Pyongyang cannot help noticing that

Washington is trying to listen attentively to Seoul's wishes and incorporate them as a loyal ally into its policy toward the North. Therefore, North Korean policy makers do not pass up any opportunity to drive a wedge between Seoul and Washington, in order to capitalize on their growing differences and to play one off against the other.[21]

Nonetheless, some senior North Korean officials appear to believe that, in the medium term, expanding North-South cooperation and forthcoming exchanges of nuclear experts, technicians, engineers, and other personnel related to the LWR transfer project, as well as intensive intra-Korean diplomatic bargaining within the KEDO-DPRK negotiating setting, is likely to generate plenty of opportunities for intra-Korean teamwork and the accumulation of good will at the grassroots level, as well as creating a favorable atmosphere for building mutual trust, understanding, and better direct working relations between representatives of both Koreas at a high political level.[22]

Although initially Pyongyang was suspicious of KEDO's mission and purposes and was reluctant to give up its newly established channel of direct communication on all nuclear-related matters with the United States, it later came around to recognize KEDO as a legitimate negotiating partner and accepted its principal role in all negotiations on the implementation of the Agreed Framework.[23] Moreover, as KEDO's membership has grown,[24] Pyongyang has appreciated its role more and more, apparently, because it has seen the latter as an indication of increasing international support for its nuclear policy and the Geneva accords.[25] In addition, the DPRK has come to view KEDO as an institutional setting offering an opportunity for direct contacts on a daily basis with the South Korean government on major issues of North-South relations at very low political cost in terms of publicly compromising its anti-South propaganda stance.[26]

In addition, Pyongyang, probably, feels that Sinp'o (the site of the future construction of the LWR plant, also referred to as Kumho by South Koreans) is going to become a major battleground and a great symbol in a decades-long competitive struggle for paramount legitimacy on the peninsula between two opposite Korean lifestyles and ways of governance. Despite evident apprehensions that a massive influx of foreign personnel and ideas might undermine domestic stability and expose the falsity of some of the myths legitimizing the North Korean regime, its leaders appear to be confident that they will be able to set up an adequate net "to keep the mosquitoes out while the window is open." And the underlying calculus is, of course, that they have more to gain than to lose on the developmental, technological, and ideological fronts vis-à-vis the South on their own home turf than abroad.[27]

Moreover, one of Pyongyang's political calculations is that sooner or later the ROK government might be compelled to abolish or at least substantially amend the National Security Law (NSL), which bans all nongovernmental contacts with the communist North and has long been one of Pyongyang's policy goals. At the very least, one can suppose that the Kim Dae-jung government will have to adjust its evolving NSL enforcement policy somehow to the reality (and domestic political implications) of several thousand KEDO-hired ROK contractors. This will include construction, maintenance, service, and nuclear personnel employed by the Korean Electric Power Company (KEPCO).[28] They will be shuttling back and forth regularly between the LWR construction site at Sinp'o and their homes in the South and maintaining contacts with their North Korean counterparts at the worksite outside the ROK government's surveillance and control. The Kim Dae-jung government has already indicated its willingness to expand all sorts of exchanges with the North as part of its new "sunshine policy," which suggests the possibility of such progress. In Pyongyang's traditional view, this kind of policy change in Seoul would remove one of the major obstacles to greater inter-Korean integration.[29]

In the long run, greater inter-Korean cooperation achieved through the implementation of the Agreed Framework along with a new peace mechanism on the Korean Peninsula are likely to contribute to fostering favorable political, economic, and international conditions for the establishment of an independent, peaceful, democratic, and reunited Korea. In a major break with the official party line, some unorthodox minds in Pyongyang, apparently enjoying Kim Jong Il's backing, now go as far as to admit that "whether a post-reunification Korea will go socialist or capitalist is a matter to be determined by future generations."[30] They argue that what matters most for the new Great Leader today is "the reassertion of traditional Korean values throughout the Korean Peninsula, which is the core of the *juch'e* philosophy of Marshal Kim Jong Il and the will of the late President Kim Il Sung."[31] In this light, mutual exchange of Korean technological expertise, predominantly Korean funding and management of a nuclear energy project involving construction and operation of two Korean-version LWRs on Korean soil by Korean specialists for use by the Korean people is likely to boost a sense of "Koreanness" among all participants. This process could, in turn, reflect the triumph of the idea of self-reliance and lead to greater awareness of national unity and the power of *juch'e* by all Koreans living on both sides of the 38th Parallel. In turn, this rapprochement ultimately will serve Kim Jong Il's goals of legitimizing his rule, strengthening public sup-port for his policies, and projecting himself as an all-Korean leader capable of peacefully reunifying the entire peninsula under the banner of *juch'e* ideas, the idea of self-reliance, and making North Korea "an affluent Switzerland or Sweden of Asia."[32]

Conclusion

To sum up, existing evidence shows that despite some short-term concerns and difficulties, in the long term, Pyongyang is keenly interested in the successful realization of the provisions of the Geneva Agreed Framework. The agreement serves its goals of: (1) economic modernization and regime survival; (2) broader international recognition and normalization of diplomatic and trade relations with advanced Western countries, including the United States and Japan; and (3) improvement of North-South relations along the road of the greater spread of *juch'e* throughout the Korean Peninsula.

Notes

Chapter 1
Russia, North Korea, and U.S. Policy toward the Nuclear Crisis

1. Leon V. Sigal, *Disarming Strangers: Nuclear Diplomacy with North Korea* (Princeton, N.J.: Princeton University Press, 1998), 254.
2. Michael J. Mazarr, *North Korea and the Bomb: A Case Study in Nonproliferation* (New York: St. Martin's Press, 1995), 239.
3. Don Oberdorfer, *The Two Koreas: A Contemporary History* (Reading, Mass.: Addison-Wesley, 1997), 411.
4. Among these studies, the edited volume by Young Whan Kihl and Peter Hayes (*Peace and Security in Northeast Asia: The Nuclear Issue and the Korean Peninsula*, M.E. Sharpe, 1997) provides some of the most substantive discussions of issues of relevance to the solution of the ongoing North Korean nuclear crisis, particularly in the field of alternative energy provision. Another is the new book by Chuck Downs (*Over the Line: North Korea's Negotiating Strategy* [Need City: American Enterprise Institute, 1999]), a rich history of international negoatiations involving the North Korean government.
5. In the nonproliferation field, North Korea's behavior remains highly problematic. For example, at the April–May 1997 NPT Preparatory Committee meeting in New York, the North Korean delegation asked the Secretariat to be seated among the observer states, not the member states. Only when the Secretariat refused to grant this request did the DPRK delegation relent and accept seating among the member states in the treaty. Shortly thereafter, however, the members of the delegation left the meeting and did not return for the duration of the two-week conference. (Author's interview with 1997 NPT Preparatory Committee President Pasi Patokallio [Finland], April 1997, Monterey, California.)
6. Associated Press (Seoul), April 26, 1999.

Chapter 2
A Technical History of Soviet–North Korean Nuclear Relations

1. "Report on the work of the Soviet specialists' team in the DPRK on contract no. 9559/5 for the period 1963–65" (in Russian), State Committee on Atomic Energy (GKAE), USSR Council of Ministers, Moscow, 1965.
2. Ibid.
3. Ibid.
4. Ibid.

5. "Contract for the construction of the Object 9559" (in Russian), GKAE, USSR Council of Ministers, Moscow, 1962.
6. A. A. Schischkin, "Certificate of Supply of Nuclear Fuel for the Nuclear Research Reactor in the DPRK" (in Russian), Tekhsnabexport, 1993.
7. Ibid.
8. Ibid.
9. Ibid.
10. Exchange of letters between the governments of the Soviet Union and the DPRK.
11. Itar-Tass (Beijing), June 18, 1998.

Chapter 3
Nuclear Institutions and Organizations in North Korea

1. See "Report on the work of the Soviet specialists' team in the DPRK on contract no. 9559/5 for the period 1963–65" (in Russian), State Committee on Atomic Energy, USSR Council of Ministers, Moscow, 1965.
2. See *Documents and Materials of the 5th Congress of the WPK* (in English) (Pyongyang: Foreign Languages Publishing House, 1970); also *Documents and Materials of the 6th Congress of the WPK* (in English) (Pyongyang: Foreign Languages Publishing House, 1980).
3. See *Geology of Korea* (in English) (Pyongyang: DPRK Academy of Sciences, 1995).
4. *Rodong Shinmun* (Pyongyang), March 14, 1988, 1.
5. N. E. Bazhanova, *Vneshneekonomicheskie svyazi KNDR* (Foreign economic relations of the DPRK) (Moscow: Nauka Press, 1993), 84–100.
6. Author's personal notes.
7. Ibid.
8. A torr is a unit of measure equivalent to 1/167 of an atmosphere.
9. *Trade & Economic Ties between the DPRK and Industrially Developed Capitalist Countries in 1970–1977: A Collection of Statistical Data* (in Russian) (Moscow: Commerce Research Institute, Ministry of Foreign Trade of the USSR, 1978), 17, 27.
10. "Confidential Memorandum from V. Kryuchkov, Director of the KGB, to the CPSU Central Committee" (in Russian), February 22, 1990, as published in *Izvestiya* (Moscow), June 24, 1992.
11. Ibid.
12. See interview with V. Belous, retired major-general of the KGB, *Segodnya* (Moscow), August 26, 1994.
13. David E. Sanger, "North Korea Site an A-Bomb Plant, U.S. Agencies Say," *The New York Times*, August 17, 1998, A1.

Chapter 4
A Political History of Soviet–North Korean Nuclear Cooperation

1. *Otnosheniya Sovetskogo Soyuza s Narodnoy Koreyey 1945–1980. Dokumenty i materialy* (Relations of the Soviet Union with the People's Korea 1945–1980. Documents and materials), Moscow, 1981, 87–90.
2. Ibid., 109–10, and 139–43.
3. Ibid.
4. *Obyedinenniy institut yadernykh issledovaniy* (United Institute for Nuclear Research) (Dubna, 1994), 4–5.
5. *Sovetskiy entsyklopedicheskiy slovar* (Soviet encyclopedic dictionary) (Moscow, 1983), 911.
6. *Finansoviye Izvestiya* (Moscow), November 10, 1995.
7. Ibid.
8. S. G. Nam, *Obrazovaniye i nauka KNDR v usloviyakh nauchno-tekhnicheskoy revolutsii* (Education and Science of the DPRK under the Conditions of Science and Technology Revolution) (Moscow: Nauka Press, 1975), 48.

Notes

9. *Nerushymaya druzhba* (Unbreakable Friendship) (Moscow: Nauka Press, 1971), 106.
10. Nam, *Obrazovaniye i nauka KNDR*, 54.
11. *Control Figures of the Seven-Year Plan of Development of the People's Economy of the DPRK for 1961–1967* (in Korean) (Pyongyang, 1961), 32.
12. Nam, *Obrazovaniye i nauka KNDR*, 54.
13. *Segodnya* (Moscow), August 21, 1994.
14. Author's interview with the DPRK Deputy Minister of Atomic Energy Park Hyon-gyu, March 27, 1987, Pyongyang.
15. *Komsomolskaya Pravda* (Moscow), July 13, 1994.
16. *Rodong Shinmun* (Pyongyang), November 10, 1970.
17. *Komsomolskaya Pravda* (Moscow), July 13, 1994.
18. Kim Jong Il, *On the Further Development of Science and Technology* (Pyongyang: Foreign Languages Literature Publishing House, 1989), 3.
19. Ibid., 7–8.
20. See "Agreement between the Government of the USSR and the Government of the DPRK in the Construction of a Nuclear Power Plant in the DPRK" (in Russian), December 26, 1985, Moscow.
21. Author's interviews with Russian nuclear engineers involved in the project.
22. *Rodong Shinmun* (Pyongyang), April 22 and 24, 1987.
23. Author's interview with Park Hyon-gyu, March 27, 1987, in Pyongyang.
24. Ibid.
25. *Nezavisimaya Gazeta* (Moscow), September 20, 1994.
26. *Izvestiya* (Moscow), June 24, 1994.
27. "Noviy vyzov posle 'kholodnoy voyny': rasprostraneniye oruzhiya massovogo unichtozheniya" (The new post–Cold War challenge: proliferation of weapons of mass destruction), in *Doklad SVR* (The FIS Report), Moscow, 1993, 92–93.
28. *Komsomolskaya Pravda* (Moscow), July 13, 1994.
29. *Korea Herald* (Seoul), January 16, 1994.
30. *Segodnya* (Moscow), August 26, 1994.
31. "Dogovor o nerasprostranenii yadernogo oruzhiya. Problemy prodleniya," (Treaty on Non-Proliferation of Nuclear Weapons. Problems of its extension), in *Doklad SVR* (The FIS Report), Moscow, 1995, 26.
32. *Komsomolskaya Pravda* (Moscow), July 13, 1994.
33. Ibid.
34. *Rossiyskaya Gazeta* (Moscow), December 11, 1992.
35. *Moskovskiye Novosti* (Moscow), April 14, 1993; *Novaya yezhednevnaya gazeta*, April 1, 1993.
36. *Segodnya* (Moscow), October 21, 1994.
37. *Kommersant* (Moscow), no. 14, 1992.
38. *Komsomolskaya Pravda* (Moscow), July 13, 1994.

Chapter 5
Economic Aspects of the North Korean Nuclear Program

1. *Korea and the World: Key Indicators 1995* (Seoul: Korea Foreign Trade Association, 1996), 96; also, *South Korea, North Korea Country Profile 1996–1997* (London: The Economic Intelligence Unit, 1997), 3, 6.
2. "A Perspective of the Democratic People's Republic of Korea," in *Trade and Investment Complementarities in North East Asia* (New York: United Nations, 1996) 111.
3. *South Korea, North Korea Country Profile 1996–1997*, 3, 6.
4. *Korea and the World: Key Indicators 1995*, 17.
5. "A Perspective of the Democratic People's Republic of Korea," 106–7.
6. Ibid., 108.

7. Ibid., 109.
8. See Bank of Korea website (*http://www.bok.or.kr*).
9. *Democratic People's Republic of Korea, 1992 Report* (New York: United Nations Development Program, April 1994), 29.
10. Ibid., 32.
11. See Bank of Korea website (*http://www.bok.or.kr*).
12. "The North Korean Economy: Challenges and Prospects," in *The Korean Economy at a Crossroads* (Westport, Conn.: Praeger Press, 1994), 169–87.
13. Ibid.
14. See Bank of Korea website (*http://www.bok.or.kr*).
15. *Korea and the World: Key Indicators 1995*, 96.
16. Ibid., 38.
17. Ibid., 96.
18. Ibid., 43.
19. See Bank of Korea website (*http://www.bok.or.kr*).
20. Ibid.
21. Ibid.
22. *BIKI*, No. 128, October 31, 1996, 6.
23. See Kim Jong U, "DPRK's Policy on Development of Rajin-Sonbong Area," website of *The People's Korea* (*www.korea-np.co.jp/pk*).
24. Ibid.
25. See "Enforcement Regulations for Foreign Investment Business and Foreign Individual Tax Law" (approved by the Administrative Council of the DPRK on February 21, 1994), website of *The People's Korea* (*www.korea-np.co.jp/pk*).
26. *BIKI*, No. 128, October 31, 1996, 6.
27. See "Update of Foreign Investment Situation in Rajin-Sonbong," website of *The People's Korea* (*www.korea-np.co.jp/pk*).
28. Nautilus Institute, "Estimated Energy Supply/Demand Balances, Democratic People's Republic of Korea (DPRK)," Estimated Summary Energy Balance for 1996 (Appendix), Berkeley, California, July 1997, A1–18.
29. "A Perspective of the Democratic People's Republic of Korea," 114.
30. For specific details of these shipments see *KEDO Annual Report 1995* (New York: KEDO, 1995), and *KEDO Annual Report 1996/1997* (New York: KEDO, 1995).
31. See David Von Hippel, "Global Dimensions of Energy Growth Projections in Northeast Asia," Nautilus Institute website (*www.nautilus.org*).
32. *North Korean Nuclear Problem* (Seoul: Hanul Academy Publishing Co., 1995), 83.
33. See *Japan Atomic Energy Industrial Newsletter*, July 16, 1998, 3; also *IAEA Daily Press Review*, September 9, 1998 (citing *Enerpresse* article of September 9, 1998).
34. Ibid., 36.
35. "Pyongyang and Hyundai Agree [sic] Nine Economic Projects," website of *The People's Korea* (*www.korea.-np.co.jp/pk*).
36. John Burton, Michiyo Nakamoto, and Mark Suzman, "N Korea test-fires missile aimed towards Japan: Launch before talks on sanctions seen as bargaining tactic," *Financial Times* (London), September 1, 1998, 1.

Chapter 6
The North Korean Energy Sector

1. K. V. Martynov, *Koreya* (Moscow, 1970), 71.
2. *Koreiskaya Narodno-Demokraticheskaya Respublika* (The Democratic People's Republic of Korea) (Moscow, 1985), 74.
3. *Central Yearbook of Korea* (in Korean) (Pyongyang, 1971), 138.

Notes 249

4. *Promyshlennost KNDR: Ekonomicheskiye osnovy, sovremenniy uroven i struktura* (Industry in the DPRK: Its economic bases, current level, and structure) (Moscow, 1977), 71.
5. See Kim Il Sung, *Collected Works* (in Korean), vol. 12 (Pyongyang, 1992).
6. *Economic Dictionary* (in Korean), vol. 2 (Pyongyang, 1995), 307.
7. Ibid., 308.
8. Kim Il Sung, *Summary Report of the Central Committee of the Korean Workers' Party to the 6th Party Congress* (Pyongyang, 1980), 63–64.
9. *Istoriya Koreyi* (A History of Korea), vol. 2 (Moscow, 1974), 132, 187.
10. G.V. Gryaznov, *Stroitelstvo materialno-tekhnicheskoy bazy sotsializma v KNDR* (The construction of the material-technical base for socialism in the DPRK) (Moscow, 1979), 99.
11. *Elektricheskiye stantsii-Energeticheskoye khozyaistvo za rubezhom* (Electric power stations—the energy economy abroad), Moscow, no. 4, 1960, 24.
12. *Collection of Statistical Data on the Development of the People's Economy in the DPRK (1946–1960)* (in Korean) (Pyongyang, 1961), 117.
13. *Elektricheskiye stantsii*, 3.
14. *Central Yearbook of Korea* (in Korean) (Pyongyang, 1961), 161.
15. *Koreya* (Korea, North Korean journal published for the Soviet Union), no. 48, 1960, 10.
16. Gryaznov, *Stroitelstvo materialno-tekhnicheskoy bazy*, 101.
17. *Koreiskaya Narodno-Demokraticheskaya Respublika* (The Democratic People's Republic of Korea) (Moscow, 1975), 63.
18. Gryaznov, *Stroitelstvo materialno-tekhnicheskoy bazy*, 103.
19. See *Aktualniye problemy Koreiskogo poluostrova* (Current problems on the Korean Peninsula) (Moscow, 1996), 29–45.

Chapter 7
Economic Factors and the Stability of the North Korean Regime

1. E. Bazhanov (ed.), *Materiali seminara po problemam Severnoi Korei* (*Proceedings of Seminar on the Problems of North Korea*), held at the Rand Corporation in Santa Monica, Calif., on 5–7 February 1996, and at the Institute of Contemporary International Problems (hereinafter referred to as ICIP), Moscow, on May 10–11, 1996 (Moscow: Nauchnaya Kniga, 1996), 30–31, 70.
2. E.Bazhanov (ed.), *Koreiskaya problema na sovremennom etape: Materiali rossisko-yuzhnokoreiskogo simpoziuma* (The Korean problem in the current era: Proceedings of a Russian-South Korean symposium held in Moscow) (Moscow: ICIP, 1996), 18–19.
3. For example, see "Unusual phenomena on the Paektu Mountain" (editorial), *Rodong Shinmun* (Pyongyang), July 10, 1996, 1.
4. *Rodong Shinmun* (Pyongyang), October 7, 1998, 1.
5. *Rodong Shinmun* (Pyongyang), September 10, 1998, p. 1.
6. E.Bazhanov (ed.), *Materiali seminara*, 82–85.
7. Ibid., 88, 93–94.
8. See A. Ivanov, "Kim-mladshy zaigrivaet s podchinennymi" (Kim junior seeks favors from his subordinates), *Obshaya Gazeta* (Moscow), no. 22, 1995, 5.
9. E. Bazhanov (ed.), *Materiali seminara*, 102–120.
10. Ibid., 121–28.
11. For 1990–95 figures, see data in *The DPRK Report*, no. 2 (July–August 1996), written by the ICIP (posted on the Center for Nonproliferation Studies website at *http://cns.miis.edu*). For 1996 and 1997, see Bank of Korea statistics (*http://www.bok.or.kr*).
12. A. Torkunov, "Problemi bezopasnosti na Koreiskom poluostrove" (Problems of security on the Korean Peninsula), in *Diplomatichesky Ezhegodnik, 1996* (Diplomatic Yearbook, 1996) (Moscow: Nauchnaya Kniga, 1996), 373.
13. Ibid., 374.

14. L. Mlechin, "Golod v Severnoi Koree?" (Famine in North Korea?), *Izvestiya* (Moscow), November 20, 1995, 3.
15. *The DPRK Report*, no. 2 (July–August 1996) (*http://cns.miis,edu*), also Bank of Korea statistics, Seoul, August 1998.
16. Ibid.
17. L. Mlechin, "Golod."
18. See statistics of the Japan External Trade Organization (JETRO), Tokyo, 1997 and 1998 (partial); also, *The People's Korea* (Tokyo), August 1998, 8.
19. Torkunov, "Problemi," 374.
20. For details of the North Korean economic model see Natalya E. Bazhanova, *Vneshneekonomicheskie svyazi KNDR* (Foreign economic relations of the DPRK) (Moscow: Nauka Press, 1993), 3–7.
21. Ibid., 32–34, 39–40, 44.
22. See "Up the Banner of the Motherland's Defense!" (editorial), *Rodong Shinmun* (Pyongyang), September 15, 1996, 1.
23. N. Bazhanova, *Between Dead Dogmas and Practical Requirements* (in Korean) (Seoul: Korean Economic Daily Publishers, 1992), 19–91.
24. See JETRO, "Report on the DPRK's External Trade," Tokyo, August 1998, 2.
25. Ibid.
26. N. Bazhanova, *Ekonomika Severnoi Korei: Makroaspekti* (The North Korean economy: macroeconomic aspects) (Moscow: ICIP, 1995), 5.
27. Ibid., 6.
28. Ibid.
29. V. Shetinin, "Sovremennoe mirovoe khozyaistvo" (The contemporary world economy), in *Diplomatichesky ezhegodnik, 1996* (Diplomatic Yearbook, 1996), (Moscow: Nauchnaya Kniga, 1996), 65.
30. Ibid.
31. Bazhanova, *Ekonomika Severnoi Korei*, 8.
32. Bazhanov (ed.), *Koreiskaya problema*, 26–28.
33. Bazhanova, *Ekonomika Severnoi Korei*, 10.
34. Bank of Korea, "Results of an Estimation of North Korea's GDP in 1997," June 15, 1998; cited in FBIS-EAS-98-217 (online version).
35. *Materiali sitanalysa "Severnaya Koreya segodnya"* (Proceedings of a situational analysis on the topic "North Korea Today"), ICIP, Moscow, January 10, 1996, (unpublished), 24.
36. Ibid., 26.
37. Bazhanova, *Vneshneekonomicheskie svyazi KNDR*, 159–60.
38. Ibid., 160.
39. Ibid., 40–42.
40. Bazhanova, *Ekonomika Severnoi Korei*, 10.
41. On these ideas, *The People's Korea* (Tokyo), November 1997, 8.
42. See *The People's Korea* (Tokyo), November 1997, 8.
43. See *The People's Korea* (Tokyo), October 1998, 1, 8.
44. See Han O-ryol, The Main Economic Strategy, *Rodong Shinmun* (Pyongyang), May 20, 1996, 3.
45. Reuters (Seoul), "Korean Tycoon Returns from Mercy Trip to North," June 23, 1998.
46. Shetinin, "Sovremennoe mirovoe khozyaistvo," 6.
47. Bazhanova, *Between Dead Dogmas*, 88–89.
48. Bazhanov (ed.), *Materiali seminara*, 123.
49. See Eugene Bazhanov and Natasha Bazhanov, "The Evolution of Russian-Korean Relations," *Asian Survey* 34 (September 1994), 796–798.
50. This table was composed by Dr. Alexandre Y. Mansourov on the basis of materials obtained from the Committee for the Promotion of External Economic Cooperation of the DPRK.

Notes

Chapter 8
The Natural Disasters of the Mid-1990s and Their Impact on the Implementation of the Agreed Framework

1. The health care and school systems partially broke down in the flood-stricken areas. The damage was so profound that the government drew up plans and began to promote a shift to home-based health care and traditional medicine. It also moved the beginning of the school year from September 1 to April 1 since most of the children in the flood-stricken areas, homeless or not, could not attend destroyed or unheated schools from September until early spring, in any case. See *Rodong Shinmun* (Pyongyang), February 5, 1996, 1.
2. See "Preliminary Findings of the U.N. Disaster Assessment Mission to the DPRK, the U.N. Department of Humanitarian Affairs," *The People's Korea* (Tokyo), September 30, 1995, p. 8.
3. Ibid.
4. KCNA (Pyongyang), July 29, 1996.
5. DPRK Ministry of Foreign Affairs statement, Pyongyang, August 7, 1996.
6. *ROKNRC Report* (Seoul), September 13, 1996.
7. DPRK Ministry of Foreign Affairs statement, Pyongyang, August 7, 1996.
8. KCNA (Pyongyang), August 25, 1998.
9. See *IAEA Newsbriefs*, vol. 11, no. 1 (70) (January–February 1996) and vol. 12, no. 1 (74) (January–February 1997).
10. See *IAEA Newsbriefs*, vol. 10, no. 3 (69) (November–December 1995).
11. Ibid.
12. As everywhere in flood-stricken areas, in the Kuryong River valley local authorities faced insurmountable difficulties in clearing out the debris and repairing damaged infrastructure because they did not have heavy construction machinery needed to lift and remove heavy objects, nor did they have enough construction materials required for rebuilding. For reference, see *The People's Korea* (Tokyo), March 2, 1996, 6.
13. See *IAEA 1995 and 1996 Annual Reports* (Vienna: IAEA, correspondingly January 1996 and January 1997).
14. See *IAEA Newsbriefs*, vol. 11, no. 1 (70) (January–February 1996); vol. 11, no. 4 (73) (November–December 1996); vol. 12, no. 3 (76) (July–August 1997).
15. See *IAEA 1995 and 1996 Annual Reports*.
16. See *IAEA Newsbriefs*, vol. 12, no. 3 (76) (July–August 1997).
17. Eric Schmitt, "Washington Spat Threatens to Halt Disposal of North Korean Nuclear Fuel," *The New York Times*, April 20, 1996.
18. In 1995, the U.S. Congress appropriated $13.6 million for the treatment of spent fuel rods in North Korea. In 1996, the DOE asked for $25 million; but in its foreign aid bill the U.S. Congress decided to allocate again only $13 million, which President Clinton threatened to veto on these grounds. For details on the U.S. Congressional politics on this issue, see Schmitt, "Washington Spat Threatens to Halt Disposal of North Korean Nuclear Fuel."
19. *IAEA 1996 Annual Report* (Vienna: IAEA, 1997).
20. See "Statement of the IAEA Director-General to the IAEA Board of Directors," Vienna, December 1997.
21. Phone interview with senior KEDO official, March 10, 1999.
22. See *The People's Korea* (Tokyo), September 9, 1995, 1.
23. See *The People's Korea* (Tokyo), September 16, 1995, 8.
24. See *The People's Korea* (Tokyo), February 17, 1996, 6.
25. For example, in South Hwanghae Province 180 bridges, including ten railroad bridges, were washed away. In North Pyongan Province, where the Yongbyon Nuclear Research Complex is located, more than 150 bridges, including seven railroad bridges, were washed away by floods. See *The People's Korea* (Tokyo), February 17, 1996, 6.
26. See *The People's Korea* (Tokyo), February 27, 1996, 6.
27. Ibid.

28. According to Bernie Krishner, despite the fact that he was accompanied by FDRC officials, rented all ten trucks from the government, and paid $40 for gasoline per truck, he still had to give a bribe as a goodwill gesture to almost every top local official whose town he passed by on his way to the destination where he planned to distribute his donations. Mr. Krishner's presentation to the Korea Society in Washington, D.C., June 1996.
29. Ibid.
30. Estimates of the daily rice allotment vary from 600 grams a day (an official DPRK government figure), to 450 grams a day (an ROK government figure), to 260 grams a day (a U.N. figure). For comparison, see *The People's Korea* (Tokyo), March 30, 1996, 6, and *The Vantage Point* 19 (Seoul) (March 1996), 44–45.
31. Trevor Page, presentation to the Asia Society, fall 1996, New York.
32. Author's interview with a Russian diplomat who had just returned from a trip to the DPRK in September 1996, Moscow.
33. *The People's Korea* (Tokyo), March 30, 1996, 6.
34. *The People's Korea* (Tokyo), February 3, 1996, 6.
35. I am grateful to Dr. Stephen Linton for this insight.
36. See KCNA (Pyongyang), August 25, 1998.
37. Ibid.
38. Out of $15 million requested, North Korea received less than $5 million, mostly in-kind donations. See World Food Programme, *North Korea Update* (United Nations: New York, February 1996), 1.
39. Author's interview with an official at the North Korean Embassy in Moscow, February 1996.
40. On June 6, 1996, after clearing it with the DPRK government, the United Nations issued a second appeal for international humanitarian assistance to Pyongyang in the amount of $43.7 million. See World Food Programme, *North Korea Update* (United Nations: New York, June 1996), 1.
41. Author's interview with a Russian diplomat who had just returned from a trip to the DPRK in September 1996, Moscow.
42. On September 6, 1995, the Korean Central News Agency reported that "upon orders issued by Marshal Kim Jong Il, the DPRK government rapidly mobilized army, navy, and air force units to evacuate people in imminent danger and to rebuild the flood-damaged infrastructure." See the KCNA's statement in *The People's Korea* (Tokyo), September 16, 1995, 8.
43. Gen. Gary Luck, commander of the U.N. Forces in Korea, was quick to interpret this move as an indication of increasingly belligerent ambitions in Pyongyang. See General Luck's testimony before the National Security Appropriations Committee of the U.S. House of Representatives on March 16, 1996.
44. Author's interview with a Russian diplomat who had just returned from a trip to the DPRK in September 1996, Moscow.
45. Of course, the West is fearful of the exactly opposite scenario. That is, policy makers in Washington and Seoul suffer from recurrent nightmares featuring the KPA lashing out against the South in total desperation. See Kim Tae-seo, "Pyongyang Is Dreaming of a Unified Korea under Its Flag," *The Vantage Point* 19 (Seoul) (March 1996), 9–11.
46. It is no wonder that since late 1995, some foreign observers have contended that the military was gaining an upper hand in the decision-making process in Pyongyang. See "Does Kim Jong Il Really Have Control Over the Military?" *The Vantage Point* 19 (Seoul) (April 1996), 1–8.
47. For example, see "Dominant Military Influence," *The Vantage Point* 19 (April 1996), 3–4.
48. See the Constitution of the DPRK, Pyongyang, September 1998.
49. KCNA (Pyongyang), September 12, 1998.
50. See *The People's Korea* (Tokyo), September 16, 1995, 5.
51. I am grateful to Dr. Stephen Linton for sharing these insights with me.
52. Author's interview with a senior Russian diplomat who was in charge of Korean affairs at the Ministry of Foreign Affairs of the Russian Federation, May 1996, Moscow.

Notes

53. See "Dominant Military Influence," *The Vantage Point* 19 (Seoul) (April 1996), 4.
54. See *The People's Korea* (Tokyo), March 30, 1996, 6.
55. See KCNA (Pyongyang), October 5, 1998.
56. See Carol Ciacomo, Reuters, May 31, 1996.
57. Attentive observers could not miss a startling statement made in January 1996 by one of the DPRK's Foreign Ministry officials, who openly suggested that the North's powerful military was getting irritated at the course of the ongoing attempts by the Foreign Ministry officials to raise humanitarian aid for the flood-stricken victims through U.N. channels and threatened to pull the plug on this humanitarian dialogue. See *The New York Times*, January 17, 1996. Surely, in the North Korean political world, the Foreign Ministry would not have dared to issue this kind of statement without obtaining prior clearance from Kim Jong Il.
58. *The People's Korea* (Tokyo), August 5, 1995, 2.
59. However, these rumors may have been spread by South Korean intelligence officials, as U.S. officials suggest that his ex-wife remains in residence, as previously, in Moscow.
60. See KCNA (Pyongyang), August 25, 1998.
61. U.S. House Representative Bill Richardson's news conference in Seoul, Reuters, May 29, 1996.

Chapter 9
Nuclear Blackmail and North Korea's Search for a Place in the Sun

1. Author's conversations with senior Soviet diplomats in Pyongyang in 1989.
2. Author's interview with a senior diplomat from the Soviet embassy in Pyongyang who participated in these talks (name withheld by request).
3. See *Rodong Shinmun* (Pyongyang), January 2, 1995, 1.

Chapter 10
Military-Strategic Aspects of the North Korean Nuclear Program

1. Author's interviews with Ministry of Atomic Energy (Minatom) officials who participated in cooperative nuclear projects with the DPRK (names withheld by request), September–November 1995, Moscow.
2. Ibid.
3. Ibid.
4. Russian Ministry of Defense experts on international security matters (names withheld), interview with author, November 1995.
5. Ibid.
6. USSR Ministry of Defense Archives for 1975–78 (specific file location and information withheld).
7. Ibid.
8. "The DPRK's Missile Program," background material prepared by the Department on Relations with the Communist and Workers' Parties of the Socialist Countries, Central Committee of the Communist Party of the Soviet Union, August 1994, 10 (unpublished).
9. Ibid., 11.
10. Ibid., 12.
11. Ibid., 13.
12. Author's interviews with experts at the Russian Ministry of Defense (names withheld by request), August 21, 1996, Moscow.
13. Ibid.
14. Ibid.
15. Ibid.
16. Ibid.
17. V. Zvetkov, "Novye rakety v KNDR?" *Krasnaya Zvezda* (Moscow), March 5, 1995, 5.
18. Ibid.

19. *Jane's Intelligence Review Yearbook*, 1995, 134.
20. D. Ivanov, "Severnaya Koreya okhotitsya za voennymi sekretami," *Nezavisimaya gazeta*, January 12, 1994, 4.
21. *Asahi Shimbun* (Tokyo), December 14, 1993, 8.
22. *The Independent* (London), February 10, 1994, 12.
23. Ibid.
24. Russian officials (names withheld), interview with author, during 1995 and 1996.
25. See Nicholas D. Kristof, "North Koreans Declare They Launched a Satellite, Not a Missile," *The New York Times*, September 5, 1998, A5; and Korean Central News Agency (Pyongyang), "Successful Launch of First Satellite in DPRK," September 4, 1998.
26. ITAR-Tass (Tokyo), August 21, 1998, citing Western predictions on the eve of the test.
27. ITAR-Tass (Moscow), September 15, 1998.
28. *Jane's Intelligence Review*, vol. 7, no. 4, 189–90.
29. "De Facto," broadcast on Russian television, Channel 11, August 17, 1996.
30. Ibid.
31. Experts interviewed in "De Facto" broadcast, Ibid.
32. Greg J. Gerardi and James A. Plotts, "An Annotated Chronology of DPRK Missile Trade and Developments," *The Nonproliferation Review* 2 (Fall 1994), 67.
33. *Ot reformy k stabilizatsii: Vneshnaya, voennaya, ekonomicheskaya politika Rossii* (Moscow: Promyshlenniy Vestnik Rossii, 1995), 137–53.
34. See, for example, A. Panov, "Mezhdunarodnye aspekty problemy obedineniya Korei" (unpublished paper), Institute for Contemporary International Problems, Moscow, July 1996, 13.
35. *Rodong Shinmun* (Pyongyang), March 12, 1993, 1.
36. Xinhua (Beijing), October 23, 1998.
37. Dong Sung Kim, "China's Policy Towards North Korea and Cooperation Between South Korea and China," *The Korean Journal of International Studies* 25 (Spring 1994), 44.

Chapter 11
Leadership Politics in North Korea and the Nuclear Program

1. *Rodong Shinmun* (Pyongyang), October 6, 1997, 1. It is worth noting that the WPK Central Committee elected Kim Jong Il to the position of general secretary of the whole party, not only of the WPK Central Committee as had been the case with his father.
2. *Rodong Shinmun* (Pyongyang), September 6, 1998, 1.
3. See Kim Jong Il, "To Respect the Senior Generation of Revolution Is the Highest Moral Duty of Revolutionaries" (in English), Foreign Languages Literature Publishing Company, Pyongyang (December 1995).
4. The Tenth SPA changed the official title of these men from vice presidents of the DPRK to honorary vice presidents of the SPA Presidium. This change may reflect, on the one hand, the fact that the institution of the presidency of the DPRK was abolished for all practical purposes, and, on the other hand, the fact that although their practical influence is waning, they were still awarded with sinecures at the SPA for symbolic, sentimental, and familial reasons, as well as for the sake of continuity between the two regimes. It is also worth noting that the Tenth SPA dropped Kim Byong Sik and added Jou Muu Sop to the list of honorary vice presidents. See *Rodong Shinmun* (Pyongyang), September 7, 1998, 1.
5. Lee Ul-sol is a member of the National Defense Commission.
6. The Tenth SPA elected Yang Hyong Sob vice president of the SPA Presidium. One should not interpret this change in his status as a demotion. Rather, the status of the SPA Presidium and its president was elevated by the new constitution, lending stature to Yang as well.
7. The Tenth SPA elected Kim Yang Nam as president of the SPA Presidium, now arguably the second-most powerful civilian position in the country. In this sense, Kim Yong-sam's career and institutional role resemble those of former Soviet Foreign Minister Andrei Gromyko, who was the last head of the Soviet state apparatus.

Notes

8. Kim Yong Sun is in charge of informational affairs with the WPK Central Committee.
9. The Tenth SPA appointed Han Song Ryong as chairman of its Budget Committee.
10. Pak Nam-gi was elevated by the Tenth SPA to chairman of the State Planning Commission.
11. Yon Hyong Muk served as the premier of the Administrative Council in the early 1990s. Then, he was demoted and dispatched to head the local party and government organs of one of North Korea's most underdeveloped provinces (Ryonggang). However, in 1997, he was brought back from political oblivion (despite his old age) and propelled to membership in the highest organ of state power, the National Defense Commission.
12. Kim Dar Hyon, a former deputy premier of the Administrative Council, is currently the head of the party and government organs in North Homgyong province.
13. In this context, it is worth noting that two defense ministers, O Jin U and Choe Gwang, have already passed away during Kim Jong Il's relatively short rule.
14. The 10th SPA appointed Cho Myong Nok the first vice-chairman of the National Defense Commission, thereby making him the second man in the line of command after Kim Jong Il, and, arguably, higher than Defense Minister Kim Il Chol.
15. Kim Yong Ch'un is also a member of the National Defense Commission.
16. The new DPRK Constitution changed the DPRK's Defense Committee into the National Defense Commission. It broadened its powers and reduced its membership to ten people, including its Chairman Kim Jong Il.
17. See *Minju Choson* (Pyongyang), June 18, 1996, 1.

Chapter 12
North Korea's Decision to Develop an Independent Nuclear Program

1. E. Bazhanov, *SSSR i Aziatsko-Tikhookeansiy Region* (USSR and the Asian-Pacific Region) (Moscow: Znaniye Press, 1991), 31.
2. See speeches by Soviet General Secretary Mikhail Gorbachev in Vladivostok in 1987 and in Krasnoyarsk in 1988 regarding new directions in Soviet policy toward Asia.
3. Bazhanov, *SSSR i Aziatsko-Tikhookeansiy Region*, 31–32. See also E. Bazhanov, "Druzhba i sotrudnichestvo mezhdu narodami SSSR i KNDR" (Friendship and cooperation between the peoples of USSR and the DPRK), *Problemy Dal'nego Vostoka*, no. 3, 1986.
4. E. Bazhanov, "Soviet Policy towards South Korea under Gorbachev," in I.G. Chun, ed., *Korea and Russia: Towards the 21st Century* (Seoul: Sejong Institute, 1992), 81.
5. Ibid., 82–83.
6. N. Bazhanova, *Vneshne-ekonomicheskie svyazi KNDR: V poiskakh vykhoda iz tupika* (Foreign economic ties of the DPRK: In search of an exit from the dead-end) (Moscow: Vostochnaya Literatura, 1993), 134.
7. See Bazhanov, "Soviet Policy towards South Korea," 84–85.
8. The All-Russian Center for Preservation of Contemporary Documents (VTsKhSD) (Moscow), File 8, List 6, Item 205, 130–31.
9. N. Bazhanova, *Between Dead Dogmas and Practical Requirements* (in Korean) (Seoul: Korean Economic Daily Publishers, 1992), 218–20.
10. Such predictions could be found in at least fifty to sixty articles published in major Soviet newspapers throughout 1990. See, for instance, *Komsomolskaya Pravda*, April 10, 1990. Other materials are cited in E. Bazhanov and N. Bazhanova, "Soviet Views on North Korea," *Asian Survey* 31 (December 1991), 1124, 1135–38.
11. *Rodong Shinmun* (Pyongyang), April 10, 1990, 2.
12. *Rodong Shinmun* (Pyongyang), June 12, 1990, 2.
13. Correspondence between the communist parties of the Soviet Union and the DPRK, VTsKhSD, File 8, List 6, Item 153, 60–62.
14. N. Bazhanova, "Nikakikh starshikh i mladshikh" (Neither older nor younger brothers), *Pravda* (Moscow), August 6, 1990, 4.

15. For a list of articles, see E. Bazhanov, *Kitay i Vneshniy Mir* (China and the outside world) (Moscow: Mezhduranodnie Otnosheniya, 1990), 301.
16. Ibid., 303.
17. See, for example, *Kulloja* (Pyongyang) 1986, no. 9; cited in Ibid., 304.
18. See Bazhanov, *Kitay i Vneshniy Mir*, 305.
19. Interview by Kim Il Sung with the editor in chief of the Japanese political journal *Sekai* (in Korean), published by the Korean Central News Agency, Pyongyang, 1985, 18.
20. *Rodong Shinmun* (Pyongyang), June 23, 1986, 1.
21. See *Rodong Shinmun* (Pyongyang), July 14, 1987, 1.
22. Ibid.
23. *Rodong Shinmun* (Pyongyang), November 10, 1989, 1.
24. B. Barakhta, "Novie voennie plany" (New military plans), *Pravda* (Moscow), September 24, 1989, 4 (citing *Ton'a Ilbo* (Pyongyang), September 23, 1989.
25. VtsKhSd, File 8, List 6, Item 39, 6.
26. Bazhanov, *Kitay i Vneshniy Mir*, 305.
27. N. Bazhanova, "Vneshnaya politika Severnoi Korei" (The foreign policy of North Korea) (Institute for Contemporary International Problems, Moscow, [unpublished paper]), 20–21.
28. Ibid., 23.
29. *Rodong Shinmun* (Pyongyang), April 16, 1988, 4.
30. *Rodong Shinmun* (Pyongyang), July 12, 1989, 3.
31. See A. Torkunov and E. Ufimtzev, *Koreiskaya problema: noviy vzglyad* (The Korean problem: A new perspective) (Moscow: Ankil, 1995), 163.
32. Ibid., 164.
33. Ibid., 190.
34. Ibid., 191.
35. *Rodong Shinmun* (Pyongyang), October 18, 1984, 4.
36. Kim En Gu, "The Japanese Imperialists Will Be Crushed" (in Korean), *Kulloja* (Pyongyang), no. 6, 1986, 20.
37. See Bazhanova, *Vneshne-ekonomicheskie svyazi KNDR*, 26.
38. See Ibid., 30.
39. Torkunov and Ufimtzev, *Koreiskaya problema: noviy vzglyad*, 109.

Chapter 13
North Korea and the Nuclear Nonproliferation Regime

1. Kim Hak-jun, *Nuclear Age and the Future of Korea* (in Korean) (Seoul 1994); Lee Sam-song, *The Nuclear Issue on the Korean Peninsula and U.S. Foreign Policy* (in Korean) (Seoul, 1984); Kim Yong Jeh, "North Korea's Nuclear Program: Problems and Prospects," *Korea Observer* (Seoul) (Autumn 1994), 317–40; Kim Dae-jung, "A Discussion of the North Korean Nuclear Question with Prof. Scalapino," *The Korean Problem: Nuclear Crisis, Democracy, and Reunification* (Seoul, 1994); and *Yadernye vzryvy v SSSR* (*Nuclear Explosions in the USSR*) (First ed.) (Moscow, 1993).
2. See Gary T. Gardner, *Nuclear Nonproliferation: A Primer* (Boulder, Colo.: Lynne Rienner Publishers, 1994).
3. The NPT allows signatories to deposit instruments of ratification in Moscow, Washington, or London.
4. *Rodong Shinmun* (Pyongyang), June 22, 1986.
5. See Young Whan Kihl, Chung-in Moon, and David I. Steinberg, eds., *Rethinking the Korean Peninsula* (Washington, DC: Georgetown University, Asian Studies Center, 1993), 121.
6. *Sino-Soviet Affairs* (Seoul) 18 (Winter 1994/1995), 144.
7. Quoted in *Problemy Dal'nego Vostoka* (Moscow), no. 4, 1995, 26.
8. *Yaderniy faktor kak istochnik novykh konfliktnykh situatsiy i ugroz* ... (The nuclear factor as a source of new conflict scenarios and threats) (Moscow, 1994), 71.

9. On the evolution of the Chinese nuclear program, see John Wilson Lewis and Xue Litai, *China Builds the Bomb* (Stanford, Calif.: Stanford University Press, 1988).
10. *Yaderniy faktor*, 74.
11. Ibid., 70.
12. *Problemy Dal'nego Vostoka* (Moscow), no. 4, 1995, 25; "On co-operation between the PRC and South Korea in Nuclear Energy," *Nuclear News*, 1994, no. 15.
13. Itar-Tass Bulletin, September 26, 1994.
14. *Sbornik deystvuyushchikh dogovorov i soglashenii* (Handbook of current treaties and agreements), Moscow, issue 26, 45–49.
15. Estimate based on known civilian trade figures for 1980–93 and estimated value of known missile sales. For more on these transactions, see Greg Gerardi and James A. Plotts, "Annotated Chronology of DPRK Missile Trade," Monterey Institute of International Studies, November 1, 1994.
16. See the current archive of the Asian-Pacific Research Center, Diplomatic Academy of the Russian Foreign Ministry, August 15, 1994.
17. Archive of the Secretariat of the KGB of the USSR, File no. 132, vol. 4, 143, "OS," *Izvestiya* (Moscow), June 24, 1994.
18. S. Levchenko, "Ya ushel ot nenavistskoi sistemy" (I fled the hated system), *Vechernyaya Moskva* (Moscow), November 24, 1995.
19. See Bill Gertz, "U.S. Intelligence: NK Could Have Nukes," *The Washington Times*, December 2, 1993, 3; Stephen Engelberg and Michael Gordon, "Intelligence Study Says NK Has Nuclear Bomb," *The New York Times*, December 26, 1993, 1; and Kim Young Jeh, "North Korean Nuclear Program: Problems and Prospect," *Korea Observer* (Seoul) (Autumn 1994), 320–21.
20. *Kompas* (Moscow), no. 92, June 21, 1994, 14.
21. Ibid., 11.
22. For more on the U.S. preparations, see Leon V. Sigal, *Disarming Strangers: Nuclear Diplomacy with North Korea* (Princeton University Press, 1997), 155.
23. *The Washington Times*, June 16, 1994.
24. *Mirnoe sotrudnichestvo v Severo-Vostochnoi Azii* (Peaceful cooperation in Northeast Asia) (Moscow, 1995), 59.
25. U.N. Security Council, official documents, June 22, 1994.
26. *Sbornik deystvuyushchikh dogovorov i soglashenii*, 45-49.
27. ITAR-TASS Bulletin, September 22, 1994.
28. David E. Sanger, "North Koreans Agree to Survey of Atomic Sites," *The New York Times*, February 16, 1994, A1.
29. Allison Smale, "IAEA Asks U.N. to Intervene on North Korea Nuclear Dispute," Associated Press, March 21, 1994.
30. ITAR-TASS Bulletin, September 22, 1994.
31. *Rodong Shinmun* (Pyongyang), April 16, 1994.
32. ITAR-TASS Bulletin, April 17, 1994.
33. *Rodong Shinmun* (Pyongyang), June 22, 1994.
34. *Mirnoe sotrudnichestvo*, 36.
35. Ibid.
36. "Clinton: U.S. May Press for N. Korea Sanctions," Reuters, June 2, 1994.
37. *Izvestiya* (Moscow), July 18, 1995.
38. *Rodong Shinmun* (Pyongyang), June, 27, 1994.
39. ITAR-TASS Bulletin, June 14, 1994.
40. *Rodong Shinmun* (Pyongyang), June, 27, 1994.
41. *Izvestiya* (Moscow), June 24, 1994.
42. Quote in *Mirnoe sotrudnichestvo*, 36.
43. See the current archive of the Asian-Pacific Research Center, Diplomatic Academy of the Russian Foreign Ministry, March 16, 1992.

44. Byun Dae-Ho, *North Korean Foreign Policy* (Seoul 1991), 249–58.
45. Nuclear Energy International, 1995, no. 490.
46. *Korean Central News Agency Bulletin* (Pyongyang), September 22, 1994, and December 26, 1994.
47. *Pyongyang Times*, November 4, 1995.
48. "Japan Plans to Launch Spy Satellites by 2002," Associated Press, November 6, 1998.

Chapter 14
North Korea's Negotiations with the Korean Peninsula Energy Development Organization (KEDO)

1. Author's conversation with KEDO Deputy Executive Director Itaru Umezu, June 12, 1996, New York.
2. See Office of the South-North Dialogue, *South-North Dialogue in Korea*, Ministry of National Unification, Seoul, ROK, January 1996, no. 63, 91.
3. Gary Samore from KEDO and Kim Yong-ho from the ROK delegation played an especially constructive role in these discussions, according to KEDO sources.
4. See an interview of the DPRK MFA spokesman with the KCNA reporter on September 28, 1995, cited in *The People's Korea* (Tokyo), October 7, 1995, 1, 8.
5. See Office of the South-North Dialogue, *South-North Dialogue in Korea*, 91.
6. See *The People's Korea* (Tokyo), September 30, 1995, 8.
7. Ibid.
8. See *The New York Times*, September 28, 1995, 3.
9. *The People's Korea* (Tokyo), October 28, 1995, 1.
10. See *The Joongang Daily News* (Seoul), November 6, 1995, 1.
11. See *The People's Korea* (Tokyo), December 2, 1995, 1.
12. Author's interview with KEDO Deputy Executive Director Ambassador Choi Young-jin, April 1996, New York.
13. See *KEDO Annual Report, 1996/1997*, July 31, 1997, 23–29.
14. Author's conversation with KEDO Deputy Executive Director Itaru Umezu, June 12, 1996, New York.
15. The Preliminary Works Contract, or the PWC, is a fixed-price, lump-sum turnkey contract, based on the so-called Orange Book model and designed to cover preliminary site preparation work, as well as the installation of temporary construction offices and warehouses, housing, welfare, and medical facilities for the personnel performing the work on the site. Its scope of work is estimated to be approximately twelve months, $45 million, and about 1,200,000 cubic meters of grading to complete. When the Prime Contract is signed, it is supposed to supersede the PWC, and the PWC scope of work will be included within the scope of the Prime Contract. See *KEDO Annual Report, 1996/1997*, July 31, 1997, 8.
16. In July 1998, the Executive Board agreed to extend the terms of the PWC for two months, until October 15, 1998. The PWC extension was intended to allow work on the LWR project to continue smoothly without a break. It also enabled 130 construction personnel working at the Sinp'o site to complete the KEDO office space and the facilities for heavy-equipment repair, electric generation, water treatment, and the warehousing of materials and explosives.
17. See *KEDO Annual Report, 1996/1997*, 6.
18. Ibid.
19. See *KEDO Annual Report, 1996/1997*, Articles I and II, 23.
20. See Ibid., Articles IV and VI, 24–25.
21. See Ibid., Article II, 23.
22. See Ibid., Article IV, 24; and *KEDO Press Release*, December 16, 1995, New York.
23. Author's conversation with Dr. Mitchell B. Reiss, June 12, 1996, New York.
24. See *KEDO Press Release*, May 23, 1996, New York, 1.
25. See *KEDO Press Release*, June 15, 1996, New York, 1.

Notes

26. See *South-North Dialogue in Korea*, Office of the South-North Dialogue, Ministry of National Unification, Seoul, ROK, December 1996, no. 64, 29–30.
27. See *The People's Korea* (Tokyo), June 1, 1996, 1.
28. See *Protocol between the Korean Peninsula Energy Development Organization and the Government of the DPRK on the Juridical Status, Privileges and Immunities, and Consular Protection of the Korean Peninsula Energy Development Organization in the DPRK*, KEDO Annual Report 1996/1997, 30–34.
29. See *Protocol between the Korean Peninsula Energy Development Organization and the Government of the DPRK on Transportation for the Implementation of a Light-Water Reactor Project*, Article 3, *KEDO Annual Report 1996/1997*, 38–41.
30. Ibid.
31. See *Protocol between the Korean Peninsula Energy Development Organization and the Government of the DPRK on Communications for the Implementation of a Light-Water Reactor Project*, Articles 5, 6, 7, *KEDO Annual Report 1996/1997*, 35–37.
32. Author's interview with Dr. Mitchell B. Reiss, February 19, 1999, New York.
33. See *Protocol between the Korean Peninsula Energy Development Organization and the Government of the DPRK on Site Take-over, Site Access, and Use of the Site for the Implementation of a Light-Water Reactor Project*, Articles 3, 4, 5, in *KEDO Annual Report 1996/1997*, 45–50.
34. See *Protocol between the Korean Peninsula Energy Development Organization and the Government of the DPRK on Labor, Goods, Facilities, and Other Services for the Implementation of a Light-Water Reactor Project*, Articles 3, 6, 7, in *KEDO Annual Report 1996/1997*, 42–44.
35. Ibid.
36. See *The People's Korea* (Tokyo), June 22, 1996, 1.
37. See, for instance, *The New York Times*, September 19, 21, 22, 1996, 1, 6.
38. This view was corroborated by Choe In Hwa, the DPRK chief delegate at the fourth and fifth protocols' negotiations, in his initial remarks to a KEDO official, on September 18, 1996, New York, which he later retracted when he fell back onto the official line stating that the whole incident was just a fabrication and provocation by the South.
39. Author's interview with an official from the U.S. Department of State, April 1998, New York. Among others, Mark Minton, former director of the Office of Korean Affairs at the U.S. Department of State, deserves considerable credit for his skillful shepherding of both Korean sides toward a mutually acceptable compromise on this issue.
40. See *The New York Times*, December 23, 1996, 1. Inside KEDO, analysts viewed North Korea's shift in attitude as a major concession.
41. See *KEDO Annual Report 1996/1997*, 45–50 and 42–44.
42. See *KEDO Annual Report 1996/1997*, 5, 7, 51–52.
43. One of the unyielding demands placed on KEDO by Pyongyang is that the site survey teams be headed by a U.S. citizen all the time. Obviously, this demand was politically motivated, given the North Korean sensitivities about "direct and central" South Korean participation in the LWR transfer project. So far, KEDO has been able to overcome Seoul's predictable opposition to this arbitrary restriction erected by its "little brother." Author's conversation with a senior KEDO official, June 12, 1996, New York.
44. See *KEDO Annual Report 1996/1997*, 7–8.
45. See *The People's Korea* (Tokyo), August 28, 1995, 1.
46. Author's interview with one of the American participants, September 10, 1995, New York.
47. See *The People's Korea* (Tokyo), November 18, 1995, 1.
48. Author's interview with one of its American participants, November 23, 1995, New York.
49. See *The People's Korea* (Tokyo), January 20, 1996, 6.
50. See KEDO Executive Board's decision no. 96-3, March 19, 1996.
51. KEDO purchased two container loads of equipment to be used in these surveys, including three sets of boring machines and one seismic analyzer, which were delivered from Pusan to Rajin by a

vessel of Tongyong Shipping Company and then transported to the Sinp'o area by land. See *South-North Dialogue in Korea*, Office of the South-North Dialogue, Ministry of National Unification, Seoul, ROK, December 1996, no. 64, 87.
52. On October 28, 1997, former U.S. State Department official Desaix Anderson replaced Ambassador Bosworth to become the second executive director of KEDO. Ambassador Bosworth subsequently was named U.S. ambassador to South Korea.
53. On April 28, 1998, Tae Shik-lee was appointed as the new ROK deputy director of KEDO.
54. On December 21, 1997, Masaaki Ono was appointed as the new Japanese deputy director of KEDO.
55. See *The People's Korea* (Tokyo), March 30, 1996, 1.
56. See *The People's Korea* (Tokyo), May 18, 1996, 8.
57. On March 16, 1996, the commander of the U.S. forces in Korea Gary Luck told the National Security Appropriations Committee of the U.S. House of Representatives that North Korea's disintegration was not a question of if but when, and that it may take the form of a military attack against the South (see Reuters from Washington, March 16, 1996, Clarinet). This comment raised eyebrows and provoked a defensive response from Pyongyang. On March 29, 1996, First Vice-Minister of the People's Armed Forces Kim Gwang-jin threatened to take action against what he called South Korean provocations and stated that "at present, the question is not whether or not a war will break out on the Korean peninsula, but when it will be unleashed" (see *Vantage Point* 19 (Seoul) (April 1996), 50.
58. Author's conversation with KEDO Deputy Executive Director Itaru Umezu, June 12, 1996, New York.
59. See *The People's Korea* (Tokyo), August 10, 1996, 8.
60. On a number of occasions, Russian government officials made formal representations to the DPRK government in regard to its "improper use" of the results of the Sinp'o site survey conducted by Russian specialists in the late 1980s–early 1990s. Moscow's position is that unless and until Pyongyang has paid up in full for those site surveys (which cost between $4–5 million) to the Russian organizations concerned, all materials related thereto remain the property of the Russian side. Therefore, the DPRK's sharing of these materials with third parties, including the KEDO, violates relevant Russian–North Korean contracts, infringes upon the Russian intellectual property rights, and violates international law. Some Russian experts believe that this could become an issue of contention between the Russian Federation and KEDO, not to mention the DPRK, sometime down the road. Author's interview with Oleg Agaphontsev, deputy chairman of the State Duma Working Group on North and South Korea, January 1997, Moscow.
61. See *South-North Dialogue in Korea*, Office of the South-North Dialogue, Ministry of National Unification, Seoul, ROK, December 1996, no. 64, 88.
62.. *KEDO Annual Report 1996/1997*, 5, 6, and 8.
63. See *KEDO Annual Report 1997/1998*, 7.
64. Ibid.
65. Ibid.

Chapter 15
China and the Korean Peninsula: Managing an Unstable Triangle

1. On this point, see Larry A. Niksch, "Comprehensive Negotiations with North Korea: A Viable Alternative for a Failed U.S. Strategy," *Korea and World Affairs* (Seoul), vol. 18 (Summer 1994), 259.
2. Quoted in Ibid.
3. On this point, see Leonid P. Moiseyev, "Kitay i Koreyskiy Poluostrov: Liniya na Status-Kvo" (China and the Korean Peninsula: Pursuit of the *Status Quo*), paper presented at the conference on Security on the Korean Peninsula, Russian Diplomatic Academy, Moscow, November 21, 1996.
4. "Foreign Ministry News Briefings," *Beijing Review*, December 28–January 3, 1999, 8.

Notes

5. Thomas J. Christensen, "Chinese Realpolitik," *Foreign Affairs* 75 (September/October 1996), 37.
6. As Moiseyev notes, "China's position was clearly demonstrated in connection with the South Korean proposal to condemn the DPRK at the U.N. Security Council in light of the incident with the North Korean submarine on September 18, 1996." (See Moiseyev, "Kitay i Koreyskiy Poluostrov.")
7. "Foreign Ministry News Briefings," *Beijing Review*, February 15–21, 1999, 8.
8. Moiseyev, "Kitay i Koreyskiy Poluostrov."
9. Cited in Ibid., drawing from *China's Diplomacy Yearbook* (in Chinese), Beijing, 1996, 25.
10. Korean Central News Agency, "Reception given to mark 37th anniversary of signing of DPRK-China treaty," July 10, 1998, reprinted on the Korean Central News Agency website (http://www.kcna.co.jp).
11. For example, according to Moiseyev, the Chinese Navy sent two missile destroyers to the DPRK immediately following the release of the revised U.S.-Japanese security guidelines. (See Moiseyev, "Kitay i Koreyskiy Poluostrov.")
12. AP (Beijing), "China Repatriates over 100 North Korea Refugees," as reported in *Vantage Point* (Seoul) 22 (January 1999), 46.
13. Korean Central News Agency (Pyongyang), July 10, 1997.
14. Ibid.
15. Moiseyev, "Kitay i Koreyskiy Poluostrov."
16. The authors gratefully thank Marcus Noland for providing these figures, which are corrected IMF "mirror statistics" after discounting arbitrary estimates added by the IMF for estimated cargo, insurance, and freight charges.
17. *Korea Herald*, March 3, 1999.
18. *The Korean Economic Weekly*, October 28, 1996, in FBIS-EAS-96-207 (October 28, 1996).
19. Xu Baokang and Wang Linchang, *Beijing Review*, November 30–December 6, 1998, 6.
20. On this point, see Moiseyev, "Kitay i Koreyskiy Poluostrov."
21. Cited in Ibid.
22. AP (Taipei), "Taiwan to Open an Office in North Korea," January 2, 1999, as reported in *Vantage Point* (Seoul) 22 (January 1999), 48.
23. Agence France Presse (Taipei), "North Korea May Open Direct Air Links with Taiwan," January 4, 1999, as reported in *Vantage Point* (Seoul) 22 (January 1999), 49.

Chapter 16
The Korean Peninsula and the Security of Russia's Primorskiy Kray

1. *Istoriya Kitaya s drevneyshikh vremen do nashikh dney* (A history of china from ancient times up to the present days) (Moscow, 1974), 205.
2. *Kompas* (Compass, Itar/Tass report), November 28, 1994, no. 182, 58.
3. Larisa V. Zabrovskaya, *Politika Tsinskoy imperii v Koree* (The policy of the Qing Empire in Korea) (Moscow, 1987), 109.
4. *Istoriya Koreya*, 413.
5. Author's interview with P. D. Pak, who served at the border check point at the Khasan Junction of Primorskiy Kray in the 1950s (September 1996, Vladivostok).
6. *Ocherki istorii Primorskoy organizatsii KPSS* (Notes on the History of the Primorskiy Kray Party Organization of the CPSU) (Vladivostok, 1971), 372, 419.
7. A photograph, which depicts L. I. Brezhnev and Kim Il Sung shaking hands at this summit, was published in *AiF* (*Argumenty i Fakty* weekly), no. 25, 1994.
8. After the communist-led unification of Vietnam, Kim Il Sung reportedly arrived in China in May 1975 and requested that the Chinese government provide military assistance to the DPRK for a drive to unify Korea. But his request is said to have been rejected. The Chinese reportedly argued that they did not want to see a repetition of the Korean War of 1950–53. Unlike the Soviet leadership, apparently, Chinese leaders did not hide from their population Kim Il Sung's

visit to Beijing, or the motives for their rejection of Kim's request. See Chung Jin-wui, "Future Developments in North Korea's Relations with China and the Soviet Union," *The Korean Journal of International Affairs* (Seoul) 11, No. 2, 1980, 107.
9. *Izvestiya* (Moscow), March 26, 1994.
10. *Vladivostok* (Vladivostok), November 5, 1994.
11. *Asiya-Kur'er* (Asian courier, Itar-Tass report), April 7, 1994.
12. *Kommersant'-Daily* (Moscow), July 17, 1996, no. 120.
13. Ibid.
14. Ibid.
15. *Globus* (Itar-Tass report), no. 43, October 27, 1995, 51–52.
16. *Vladivostok* (Vladivostok), August 13, 1996.
17. Some officers of the Russian Pacific Fleet came to the conclusion that the United States used such naval exercises to elaborate its operational plans for invasion of the DPRK, both from the east and the west, in the event of escalation of the situation on the Korean Peninsula. They also feared that Russia might be compelled to take part in such a hopeless action, which might inevitably lead to the initiation of a guerrilla struggle by the Northerners.
18. *Pul's planety* (Pulse of the planet, Itar/Tass report), September 6, 1996.
19. Ibid.
20. A. Valiyev, "K poezdke V. Zhirinovskogo v KNDR" (On the visit of V. Zhirinovsky to the DPRK), *Kompas*, no. 159, 1994, 13.
21. A. Valiyev, "Vizit rossiyskogo diplomata otkryvaet novie perspectivy v otnosheniyakh Moskvy i Pkhen'yana" (Visit of Russian diplomat opens new prospects for the relations between Moscow and Pyongyang), *Kompas*, no. 151, 1994, 5.
22. Ibid., 7.
23. Later on, the DPRK agreed to accept the ROK-made LWRs and the U.S. humanitarian aid, leaving the Russian proposal without further consideration.
24. *Ekonomicheskay zhizn' Dal'nego Vostoka* (Economic life of the far East, weekly bulletin), Vladivostok, no. 12, 1995.
25. Ibid.
26. A. Valiyev, "Primortsy Rossii vozrozhdayut svyazi c KNDR" (Russia's Maritime population revives ties with the DPRK), *Kompas*, no. 21, 1995, 27–28.
27. Ibid., 26.
28. *Izvestiya* (Moscow), September 8, 1995.
29. *Izvestiya* (Moscow), March 31, 1994.
30. *Izvestiya* (Moscow), September 8, 1995.
31. L. V. Zabrovskaya, *Rossiya i Respublika Koreya: ot konfrontatsii k sotrudnichestvu (1970–1990-e gg.)* (Russia and South Korea: from confrontation to cooperation [1970–90s]) (Vladivostok: Institute of History, 1996), 36–37.
32. *Izvestiya* (Moscow), September 8, 1995.
33. A. Varlamov and A. Kirillov, "Kak zakrepit'sya na aziatsko-tikhookeanskoy arene" (How can we strengthen our position in the Asia-Pacific Region?), *Dipkur'er* (Diplomatic courier), no. 1, 1996, 42.
34. Ibid.
35. *Sovetskiy Sakhalin* (Yuzhno-Sakhalinsk), November 30, 1995.
36. *Pul's planety* (Itar-Tass report), April 12, 1995.
37. *Sovetskiy Sakhalin* (Yuzhno-Sakhalinsk), November 30, 1995.
38. *Pravda* (Moscow), October 10, 1995.
39. *Pul's planety* (Itar-Tass report), April 12, 1995.
40. *Pul's planety* (Itar-Tass report), April 17, 1995.
41. In 1991, the official exchange rate of the USSR Bank of Foreign Trade equaled one U.S. dollar to sixty-four "foreign exchange kopecks."
42. *Vladivostok* (Vladivostok), April 3, 1996.

Notes 263

43. *Pul's planety* (Itar-Tass report), April 15, 1996.
44. Ibid.
45. Ibid.
46. L. V. Zabrovskaya, *Russia and the Republic of Korea*, 40–41.
47. *Pul's planety* (Itar-Tass report), September 25, 1996.
48. *Vladivostok* (Vladivostok), September 11, 1996.
49. *Kommersant'-Daily* (Moscow), no. 164, October 1, 1996.
50. Ibid.
51. *Kommersant'-Daily* (Moscow), no. 165, October 2, 1996.
52. *Pul's planety* (Itar-Tass report), September 25, 1996.
53. *The Korea Herald* (Seoul), September 30, 1996.
54. *Vladivostok* (Vladivostok), October 3 and 5, 1996.
55. *Vladivostok* (Vladivostok), October 25, 1996.
56. *Vladivostok* (Vladivostok), October 10, 1996.
57. *Vladivostok* (Vladivostok), October 5, 1996.
58. Ibid.
59. Ibid.

Chapter 17
The Renewal of Russian–North Korean Relations

1. L. V. Zabrovskaya, *Rossiya i Respublika Koreya: ot konfrontatsii k cotrudnichestvu (1970–1990-e gg.)* (Russia and the Republic of Korea: from confrontation to cooperation [1970–90s] (Vladivostok: Institute of History, Archaeology, and Ethnography of the Peoples of the Far East, 1996), 12.
2. Ibid., Table 1, 45.
3. Ironically, these recent developments have largely overlooked or discounted by the Western news media. See, for example, James Brooke, "In Big Shift, Russia Courts South Korea," *The New York Times*, July 3, 1996, A4. The article makes no mention of the significant rapprochement developing with the North, replaying old arguments from the late 1980s focusing purely on Moscow's economic incentives in the region, without considering the political and strategic dimensions of the new Russian policy.
4. Konstantin Eggert, "Gennadiy Seleznev udostoitsya vstrechi c Kim Chen Irom kak chastnoe litso" (Gennadiy Seleznev is honored with a private meeting with Kim Jong Il), *Izvestiya* (Moscow), May 25, 1996, 3.
5. Aleksandr Platkovskiy, "Severokoreytsev prizyvayut sobirat podnozhniy korm, chtoby spastis' ot goloda" (North Koreans are told to gather grass, in order to save themselves from starvation), *Izvestiya* (Moscow), June 5, 1996, 1.
6. "Russian guest here," Korean Central News Agency website (*http://www.kcna.co.jp*), February 16, 1998.
7. As Samuel S. Kim observed, Moscow's frustration stemmed from "... the realization that the Russians had made a cardinal mistake when they put all their eggs into the South Korean basket at the expense of credibility and leverage in Pyongyang." (Samuel S. Kim, "North Korea in 1994: Brinkmanship, Breakdown, and Breakthrough," *Asian Survey* 35 [January 1995], 23).
8. Kunadze's remarks quoted in David Cho, "Russian Envoy Says He's 'Not Impressed' by 4-Way Talks Proposal," *Korean Herald*, April 21, 1996 (Northeast Asian Peace and Security Network [NAPSNet] Daily Report, Berkeley, Calif., Nautilus Institute), April 22, 1996.
9. Yonhap, "Russia Insists Only on 8-Party Conference," Korean Times (NAPSNet Daily Report, April 17, 1996), 1.
10. Vladimir Abarinov, "V Moskve chitayut, chto uchastie Rossii v koreyskom uregulirovanii neobkhodimo" (Moscow considers Russian participation in Korean settlement essential) *Segodnya* (Moscow), May 6, 1996, 2.

11. "Russian ambassador gives party," Korean Central News Agency website (*http://www.kcna.co.jp*), March 19, 1998.
12. Author's interview with Valery Denisov, then–ambassador-designate to North Korea, Moscow, December 6, 1995. (Ambassador Denisov took up his post as ambassador in Pyongyang in August 1996.)
13. The rest of this paragraph relies on several accounts of the incident that appeared in the Russian press. See Igor Korotchenko, "Arestovan agent yuzhnokoreyskoy razvedki" (South Korean intelligence agent arrested), *Nezavisimaya gazeta* (Moscow), July 7, 1998, 1; Igor Ilyushin, "Menya zovut Bond. Cho Son U 'Bond'"(My name is Bond: Cho Son-u Bond), *Vladivostok*, July 10, 1998, 5; Igor Korotchenko, "Iz Rossii vysylayut pyaterykh yuzhnokoreyskikh pazvedchikov" (Russia deports five South Korean intelligence agents), *Nezavisimaya gazeta* (Moscow), July 21, 1998, 2.
14. At least one of these officials is still being held and may be tried on charges of treason.
15. Author's discussion with Alexandre Mansourov, based on the latter's discussion with Russian Foreign Ministry officials, September 1998.
16. See conferences described by Russian analysts in *The DPRK Report*, no. 3 (September–October 1996), website of the Center for Nonproliferation Studies, Monterey Institute of International Studies (*http://cns.miis.edu*).
17. Remarks by Dr. Vladimir Li, conference on "Conflict Prevention in Korea," Russian Diplomatic Academy, Moscow, December 4, 1997.
18. Author's interview with Vyacheslav Lee, president, Vladivostok Social-Cultural Center of Koreans, Vladivostok, January 25, 1996.
19. For example, see "Velikiy rukovoditel' tovarishch Kim Chen Ir vedet k pobede koreyskuyu revolyutsiyu" (The Great Leader Comrade Kim Jong Il is leading the Korean revolution to victory), *Vladivostok*, September 9, 1998, 5.
20. By contrast, the South Korean consulate in Vladivostok consists of just one small wing on one floor of a multiple-use office building, even though the South has a far greater trade volume with Russia. Whether this is any indication of pessimism on the part of Seoul is difficult to tell, and may simply be part of "slow but steady" approach to Russia, which guards it in case of future reversals.
21. See Russian views expressed in *The DPRK Report*, no. 3 (September–October 1996), website of the Center for Nonproliferation Studies, Monterey Institute of International Studies (*http://cns.miis.edu*).
22. See comments by recent Russian visitors to the DPRK in *The DPRK Report*, no. 9 (September–October 1997), website of the Center for Nonproliferation Studies, Monterey Institute of International Studies (*http://cns.miis.edu*).
23. *Kommersant* (Moscow), March 30, 1995, A2.
24. Itar-Tass, "Snova Stena?" (A new barrier?), *Zavtra Rossii* (Vladivostok weekly), April 19–26, 1996, 11.
25. See website of the South Korean Ministry of Unification (*http://www.unikorea.go.kr*); also figures provided by Alexandre Mansourov for 1998, based on his interviews with Russian Foreign Ministry officials, October 1998.
26. "NK's Foreign Debts Reach $11.5 Bil. In '95," *Korean Times*, May 18, 1996, 2, in NAPSNet Daily Report, May 20, 1996.
27. Author's interview with Alexander Kislitsyn, Vladivostok academic specialist on North Korea, Institute of History, Vladivostok, January 27, 1996.
28. Ibid.
29. Author's interview with Russian Consul General (and former Primorskiy Kray governor) Vladimir Kuznetsov, San Francisco, Calif., March 14, 1996.
30. Larisa Vyacheslavovna Zabrovskaya, "KNDR i Politika Rossii (1994–1998 gg.)" (The DPRK and Russian Policy [1994–1998]), *Rossiya i ATR* (Vladivostok), no. 3, 1998, 87.
31. Oleg Kryuchek, *Segodnya* (Moscow), March 11, 1995, 3

Notes

32. Other factors include the almost complete lack of construction and infrastructure at Rajin-Sonbong. Russian visitors from Nakhodka to the North Korean SEZ in January 1996 were shocked at the almost complete lack of facilities of any type. (Author's interview with Svetlana Vikhoreva, deputy chairman of development and marketing, Nakhodka SEZ, in Nakhodka, January 24, 1996.)
33. "V KNDR—Vagony ne vypuskat'?" (In the DPRK: Will they release the trains?) *Vladivostok*, April 11, 1996, 3.
34. AP-Dow Jones News Service (Seoul), "Rail Traffic Resumes Between N. Korea, Russia after Debt Paid," September 12, 1996, in *NAPSNet Daily Report*, September 12, 1996.
35. Gennadiy Syromyatnov, "Rossiya i KNDR: segodnya i zavtra" (Russia and the DPRK: Today and tomorrow), *Krasnoe Znamya* (Vladivostok), July 12, 1995, 1.
36. Several Russian-Koreans confirmed to the author that North Koreans are in fact actively engaged in spying on South Korean activities in Primorskiy Kray. (Author's interviews, Vladivostok, January 1996.)
37. Author's interview with Vyacheslav Lee, president, Vladivostok Social-Cultural Center of Koreans, Vladivostok, January 25, 1996.
38. For more on this group, see Anatoliy Kuzin, *Dal'nevostochnie Koreytsy: Zhizn' i Tragediya Sud'by (The Russian Far Eastern Koreans: Life and the Tragedy of Fate)* (Yuzhnosakhalinsk: "LIK" Publishers, 1993), 198–206.
39. Information provided by Nina Kim, a Russian-Korean staff member at the Republic of Korea State Center for Education in Vladivostok and a second-generation descendant of the Sakhalin-Korean diaspora. (Author's interview in Vladivostok, January 26, 1996).
40. For more on this group, see B. D. Pak, *Koreytsy v Sovetskoy Rossii (1917–konets 30-kh godov)* (Koreans in Soviet Russia [1917–the end of the 1930s]) (Moscow and Irkutsk: Diplomatic Academy and Irkutsk Pedagogical Institute, 1995).
41. Ibid.
42. Interfax, "V Primorskom krae tamozhenniki iz'yali u predprinimatelya 10 tysyach fal'shivykh dollarov" (Primorskiy Kray customs agents grab 10,000 counterfeit dollars) *Vladivostok*, June 27, 1996, 1.
43. NTV (Moscow), January 30, 1996; in FBIS-SOV-96-022 (January 30, 1996).
44. Information provided to author by a Russian academic who spoke with Russian embassy officials after the incident.
45. AFP (Tokyo), "Two North Koreans Held by Russia," *Moscow Tribune*, May 28, 1996, 5.
46. See Vladimir Abarinov, "Moskva blokiruet peregovory po eksportomu kontrolyu" (Moscow blocks discussions on export controls), *Segodnya* (Moscow), April 4, 1996, 1.
47. "NK Receives Spy Satellite Photos of South from Russia: Defector," *Korean Times*, June 25, 1996, 1; also "Defector Says Russia Supplies Data on South Military Facilities to North," *Korean Herald*, June 25, 1996, 3 (both from *NAPSNet Daily Report*, June 25, 1996).
48. "Posadili boevuyu eskardil'yu: rossiyskie vertolety prednaznachilis' dlya prodazhi v Severnuyu Koreyu" (A military squadron is grounded: Russian helicoptors destined for sale to North Korea), *Vladivostok*, October 9, 1998, 1. See also follow-up report on the subsequent investigation published on the *Vladivostok News* website *(http://vn.vladnews.ru)*, "Agents stop chopper sale to N. Korea," October 21, 1998.

Chapter 19
Russian Views of the Agreed Framework and the Four-Party Talks

1. For more details, see Natalya Bazhanov, "North Korea and Seoul-Moscow Relations," 328–46, and Evgeniy Bazhanov, "Soviet Policy towards South Korea under Gorbachev," 61–109, in Il Yung Chung (ed.), *Korea and Russia* (Seoul: Sejong Institute, 1992).
2. For details, see Evgeniy Bazhanov and Natalya Bazhanov, "Soviet Views on North Korea," *Asian Survey* 31 (December 1991), 1123–38.

3. For more details, see Evgeniy Bazhanov, *Russia's Changing Foreign Policy* (Cologne, Germany: BIOST, 1996), no. 30, 15–36.
4. V. Denisov, "Russia on the Korean Peninsula" (in Russian), *Segodnya* (Moscow), November 4, 1994, 4.
5. V. Denisov, "Excommunication from Integration" (in Russian), *Nezavisimaya Gazeta* (Moscow), August 26, 1994, 4.
6. A. Torkunov, *Problems of Security on the Korean Peninsula* (Moscow: MGIMO, 1995), 14.
7. For more details on Russian economic interests in the DPRK, see N. Bazhanov, *The Macroeconomy of the DPRK* (in Russian) (Moscow: Institute for Contemporary International Problems [ICIP], Diplomatic Academy, 1995), 35–40.
8. See Evgeniy Bazhanov, "Russia and North Korea," *Report of the Seminar on "Relations Between the DPRK and the Great Powers"* (Seoul: Kim Dae-jung Peace Foundation, November 26, 1996), 12–14.
9. A. Torkunov and E. Ufimtsev, *Koreiskaya problema: noviy vzglyad* (The Korean problem: A new perspective) (Moscow: Ankil, 1995), 182.
10. E. Bazhanov (ed.), *Problems of Maintaining Security in Northeast Asia*, Seminar materials (Moscow: ICIP, June 10, 1996), 8.
11. Ibid., 9.
12. Ibid.
13. *Rossiyskiye Vesti* (Moscow), June 19, 1992, 1.
14. Torkunov and Ufimtsev, *The Korean Problem*, 183.
15. "Statement of the Russian MFA" (in Russian), *Rossiyskiye Vesti*, March 13, 1993, 1.
16. "Unjust Move," *Rodong Shinmun* (Pyongyang), March 20, 1993, 1.
17. Kim Yok, "Base Crime Committed by a Number of Neighbor Countries," *Informational Bulletin*, (Pyongyang), Central Telegraph Agency of Korea, April 9, 1993, no. 83, 6.
18. Ibid.
19. "Statement of a Spokesman of the DPRK's Ministry of Foreign Affairs," *Rodon Shinmun* (Pyongyang), April 12, 1993, 1.
20. "Information," *Rodong Shinmun* (Pyongyang), April 30, 1993, 1.
21. Lee Hwan-il, "Who Poses a Threat to Peace?" *Rodong Shinmun*, May 4, 1993, 3.
22. Ibid.
23. "Statement of a Spokesman of the DPRK Ministry of Foreign Affairs," *Rodon Shinmun*, April 16, 1993, 1.
24. Torkunov and Ufimtsev, *The Korean Problem*, 185–86.
25. Ibid., 186.
26. "Interview with A. Kozyrev" (in Russian), *Izvestiya* (Moscow), June 18, 1994, 1.
27. See *The New Challenge After the Cold War: Proliferation of Weapons of Mass Destruction* (in Russian) (Moscow: Foreign Intelligence Service, 1992), 22–24.
28. "Statement of a Spokesman of the Russian Ministry of Foreign Affairs" (in Russian), *Rossiyskiye Vesti*, June 14, 1993, 2.
29. "Statement of a Spokesman of the Russian Ministry of Foreign Affairs" (in Russian), *Rossiyskiye Vesti*, March 24, 1994, 1. For a more detailed description of these proposals, see Denisov, "Russia on the Korean Peninsula."
30. Torkunov and Ufimtsev, *The Korean Problem*, 183.
31. Bazhanov (ed.), *Problems of Maintaining Security*, 24–25.
32. Ibid., 26.
33. Ibid., 27.
34. E. Bazhanov (ed.), *Nuclear Nonproliferation in the Asian-Pacific Region* (in Russian) (Moscow: Nauchnaya Kniga, 1996), 42.
35. *Materials of the Internal Conference on the Problems of Asia-Pacific Region* (Moscow: Russian Ministry of Foreign Affairs, November 1994), 74.

36. Bazhanov, *Nuclear Nonproliferation*, 45.
37. E. Bazhanov (ed.), *Problems of Maintaining Security in Northeast Asia*, 23.
38. Ibid., 14.
39. *Materials of the Internal Conference on the Problems of Asia-Pacific Region*, 75.
40. Ibid., 76.
41. Bazhanov (ed.), *Problems of Maintaining Security*, 15.
42. Ibid., 16.
43. Ibid.
44. E. Bazhanov, "Moscow and a Settlement on the Korean Peninsula" (in Korean), *Seoul Shinmun*, September 12, 1996, 6.
45. Bazhanov (ed.), *Problems of Maintaining Security*, 42–44.
46. "Statement of a Spokesman of the Russian Ministry of Foreign Affairs" (in Russian), *Rossiyskiye Vesti*, March 24, 1994, 1.

Chapter 20
Pyongyang's Stake in the Agreed Framework

1. *Rodong Shinmun* (Pyongyang), January 2, 1995, 2.
2. Ibid.
3. *Rodong Shinmun* (Pyongyang), February 15, 1995, 2, and February 19, 1996, 2.
4. Author's interview with a high-ranking DPRK diplomat in Moscow (name withheld by request), August 1996.
5. 150,000 metric tons of heavy fuel oil were delivered to the DPRK between October 1994 and October 1995. In 1996, KEDO delivered to the DPRK 42,000 tons in January; 85,000 tons in March; 44,000 tons in May; 38,000 tons in June; 59,000 tons in July; 66,000 tons in August; 62,500 tons in September; and 103,500 tons in October. The total cost (including commodity, freight, interest, and demurrage charges) reached about $67.4 million. In 1997, KEDO delivered to the DPRK 43,500 metric tons in February; 42,500 tons in April; 42,000 ton in May; 82,000 tons in June; 61,000 tons in July; 70,000 tons in August; 70,000 tons in September; 45,000 tons in October; and 44,000 tons in November. The total cost approached $67.7 million. In 1998, KEDO delivered to the DPRK 86,000 metric tons in February–March; 22,000 tons each in April, May, and June; 64,000 tons in July; 94,000 tons in August; 95,000 tons in September, and 95,000 in October (or 500,000 tons at a cost of $49.8 million). See *KEDO Annual Report, 1996/1997* (New York: KEDO, 1997), July 31, 1997, 9–11, as well as author's personal notes from discussions with KEDO officials in 1998, New York and *KEDO Annual Report 1997/98* (New York: KEDO, 1998), July 31, 1998, 9–11.
6. Author's interviews with a high-ranking DPRK diplomat (name withheld by request), November 1996, New York.
7. Author's interviews with high-ranking DPRK diplomats (names withheld by request), August 1996, January 1997, and October 1997, Moscow.
8. Ibid.
9. Author's interview with a high-ranking Russian diplomat stationed in the DPRK (name withheld by request), August 1996, Moscow.
10. In fall 1995, the United States donated $225,000 to the World Food Program (WFP) for humanitarian aid to the DPRK. On February 8, 1996, the Clinton administration decided to donate $2 million of humanitarian aid to the WFP to buy emergency food supplies for famine-stricken North Korean farmers. On June 14, 1996, Washington announced its decision to respond with a $3 million donation to the second United Nations appeal for humanitarian aid for the DPRK's flood victims. The KCNA periodically quoted a senior official from the DPRK Ministry of Foreign Affairs as stating that "humanitarian steps taken by the United States will remove distrust between the DPRK and the United States and create an atmosphere favorable for a smooth implementation of the U.S.–DPRK Agreed Framework." In 1997–98, Washington decided to

Notes

provide Pyongyang with more than 300,000 metric tons of grain in humanitarian aid. See *The People's Korea* (Tokyo), February 17, 1996, 1, February 24, 1996, 8, April 6, 1996, 1, and June 24, 1996, 1, December 24, 1997, 8.

11. For instance, such incidents in the past have included: ROK violations of North Korean airspace (December 1995), the North Korean submarine incursion into ROK territorial waters (September 1996), alleged North Korean interference with the 1997 ROK presidential elections, the DPRK satellite launch attempt (August 1998), and so on.
12. See *The People's Korea* (Tokyo), September 16, 1995, 8.
13. Author's interview with a senior DPRK diplomat at the United Nations (name withheld by request), June 1996, New York.
14. Ibid.
15. Author's interview with a senior DPRK diplomat at the United Nations (name withheld by request), June 1996, New York.
16. See *The People's Korea* (Tokyo), September 23, 1995, 3.
17. The DPRK seems to prefer such gifts as the 500 cows provided by Hyundai founder Chou Ju-yon during each of his two trips to Pyongyang in 1998.
18. Pyongyang's policy toward the ROK is overseen by the National Reunification Committee under WPK Central Committee Secretary Kim Yong Sun, whereas diplomacy with the United States falls under the Foreign Ministry's Department of North American Affairs.
19. Author's interview with Dr. John Merrill, May 31, 1996, New York.
20. Author's telephone interview with a senior official from the Russian Ministry of Foreign Affairs (name withheld by request), April 29, 1996, Moscow.
21. Author's interview with Dr. Tom Hubbard, April 1, 1996, Columbia University, New York.
22. KEDO Deputy Executive Director Itaru Umezu told the author that on many occasions he had witnessed "very cordial, friendly atmosphere between South and North Korean negotiators within the KEDO framework."
23. Author's interview with Dr. Tom Hubbard, April 1, 1996, Columbia University, New York.
24. KEDO's membership now includes Japan, South Korea, the United States, Finland, New Zealand, Australia, Canada, Indonesia, Chile, Argentina, Poland, and the European Atomic Energy Commission. See *KEDO Annual Report 1997/1998*, 13.
25. Author's interview with a senior DRPK diplomat at the United Nations (name withheld by request), New York, May 1996. Besides KEDO members, other countries that have made various contributions in cash or in kind to KEDO funds in 1996–98 include Germany, the United Kingdom, Netherlands, Norway, Singapore, Brunei, Malaysia, Thailand, the Philippines, Switzerland, the Czech Republic, Hungary, Oman, and Greece.
26. Author's interview with South Korean Ambassador Choi Young-jin, then-KEDO deputy director, April 1996, New York.
27. After all, in the wake of the devastation suffered by the North as a result of the Korean War, its economy and society were rebuilt in a mere five years by a million-men strong army of foreigners (i.e., Chinese People's Volunteers) and generous grants and aid from the Soviet Union. This reconstruction took place without any major encroachments on national sovereignty or debilitating effects on political unity and social cohesion. As his father did in the 1950s and again in the 1970s, Kim Jong Il too hopes to withstand politically and socially destabilizing aftershocks of the country's gradual economic opening to international market forces.
28. KEPCO is the ROK government-owned electric utility that was selected by KEDO to become the primary contractor in the implementation of the LWR construction project.
29. Author's interview with a senior North Korean diplomat at the DPRK's U.N. mission in New York (name withheld by request), June 5, 1996, New York.
30. See *The People's Korea* (Tokyo), September 16, 1995, 8.
31. Ibid.
32. Author's interview with a senior North Korean diplomat at the DPRK's U.N. mission (name withheld by request), June 5, 1996, New York.

Contributors

VLADIMIR D. ANDRIANOV is an economist and a full Member of the International Higher Education Academy of Sciences and the International Informatization Academy in Moscow.

NATALYA BAZHANOVA is a senior researcher at the Institute of Oriental Studies, Russian Academy of Sciences, Moscow. She is the author of many articles on North Korean economics, and a book on North Korea's foreign economic relations.

EVGENIY P. BAZHANOV is deputy director of the Russian Diplomatic Academy (Russian Foreign Ministry) and director of the Institute for Contemporary International Studies, Moscow. He served previously as a Soviet diplomat in Beijing, San Francisco, and Singapore. Professor Bazhanov is the author of many articles and several books on Russian relations with East Asia, especially China.

VALERY I. DENISOV is Russia's current ambassador to North Korea. He served previously for over two decades in the Soviet embassy in Pyongyang and has held senior positions in the Soviet/Russian Foreign Ministry in the area of East Asian policy.

GEORGIY KAUROV heads the Press Service of the Russian state nuclear company, Rosenergoatom. He formerly led the Information Directorate at the Russian Ministry of Atomic Energy (Minatom). Dr. Kaurov is the recipient of a USSR State Prize in physics.

VLADIMIR F. LI is director of the Center for East Asian Affairs at the Russian Diplomatic Academy and deputy director of the Institute for Contemporary International Studies, Moscow. He is the author of many studies on Russian policy in South Asia and East Asia.

ALEXANDRE Y. MANSOUROV is a research associate in residence at the Korea Institute at Harvard University. In the late 1980s, Dr. Mansourov served in the Soviet embassy in Pyongyang as a young foreign service office. He received his Ph.D. in political science from Columbia University in 1997.

VALENTIN I. MOISEYEV was formerly head of the Korean Division in the First Asia Department of the Russian Ministry of Foreign Affairs. He served in the Soviet/Russian embassies in both Seoul and Pyongyang.

JAMES CLAY MOLTZ is assistant director and research professor at the Center for Nonproliferation Studies (CNS) of the Monterey Institute of International Studies. He also directs the Newly Independent States Nonproliferation Project at CNS. From 1993–98 (vols. 1–5), Dr. Moltz served as founding editor of *The Nonproliferation Review* and, since 1996, has edited the web-based *DPRK Report*.

ALEXANDER PLATKOVSKIY is a journalist who has written on East Asia for many years. His foreign postings have included service as a correspondent for *Komsomolskaya Pravda* in North Korea from 1987 to 1990 and as the bureau chief for *Izvestiya* in Beijing from 1996 to 1998.

MITCHELL B. REISS is dean of international studies, director of the Reves Center, and professor of law at the College of William & Mary in Williamsburg, Virginia. Until June 1999, he served as assistant executive director and senior policy advisor at the Korean Peninsula Energy Development Organization (KEDO) in New York. He was previously KEDO's first general counsel. He is the author of the book *Bridled Ambition: Why Countries Constrain Their Nuclear Capabilities* (1995). Dr. Reiss holds a Ph.D in international relations from Oxford University and a J.D. from Columbia University's School of Law.

ROALD V. SAVEL'YEV served in the Soviet embassy in North Korea for over a decade in the 1970s and 1980s. He is now a senior fellow in the Korean Center of the Institute of Far Eastern Studies in Moscow.

LARISA V. ZABROVSKAYA is a senior researcher at the Institute of History, Archeology, and Ethnography of the Peoples of the Far East in Vladivostok, part of the Far Eastern Branch of the Russian Academy of Sciences. She is the author of a recent book on Russian policy toward South Korea.

ALEXANDER ZARUBIN was formerly a staff member of the Russian Security Council. He retired from the Russian military in the early 1990s with the rank of general.

ALEXANDER ZHEBIN was formerly the North Korean correspondent for the newspaper *Izvestiya*. After a term as a research fellow at the Institute of Far Eastern Studies in Moscow, he joined the Russian Foreign Ministry and is now serving as a first secretary in the Russian embassy in Pyongyang.

Index

Academy of Sciences (DPRK), 30
Academy of Sciences (Russia), x
Adamov, Evgeniy, 20
Agreed Framework, xi, 1, 2, 3, 4, 9, 20, 21, 26, 68, 76–85, 87, 93, 101, 108, 137, 138, 152, 156, 218, 233, 239, 244
   China and, 172
   DPRK's stake in, 50, 236–244
   implementation of, 8, 10, 11–12, 154, 169–172, 237
   natural disaster effects on, 76–90
   prospects for, 108–109, 112, 122, 153
   Russian views of, 219–235
   signing of, 99
   terms of, 47
Agreement between the Government of the USSR and the Government of the DPRK on the Education of Citizens of the DPRK, 28
Agreement on Scientific Cooperation between the Academy of Sciences of the USSR and the Academy of Sciences of the DPRK, 28
agricultural production, in DPRK, 43, 71, 77
Amnok River basin, flood damage to, 77
Andrianov, Vladimir, 6
Andropov, Yuri, 37, 145
Anju, flood damage to, 77
ASEAN countries, 228, 229
Association of Koreans of Russia, 183
atomic bomb
   Chinese test of (1964), 141
   DPRK and, 8, 9, 26, 35, 102
Atomic Energy Institute (DPRK), 32
Atomic Energy Institute (ROK), 159
Atomic Energy Safety Institute (ROK), 159

ballistic missile tests, of DPRK, ix, 102–05, 184
Barannikov, Victor, 36
Bauman Higher Technical School (*VTU imeni Baumana*), 17
Belle Wood, in naval exercises, 185
Belous, Vladimir, 35
Blix, Hans, 78, 148, 150
Bosworth, Stephen, 88, 157, 158, 159, 160, 161
Brezhnev, Leonid I., 37, 61, 63, 181

Burns & Roe, 166
Bush, George, 222, 233

Carnegie Endowment, 3
Carter, Jimmy, 3, 233
Central Intelligence Agency (CIA), 145, 152
Central People's Committee (CPC), 115
Central Transportation Authority, 82
Chang Song T'aek, 116
Chang Song U, 116
Chanjin River, power plant on, 54, 59
Chemical Weapons Convention, 109, 234
Chernenko, Konstantin, 37, 63
China, x, 46
   collective security system of, 217
   disagreements and policy problems of DPRK and, 131–132
   as DPRK ally, 9, 108, 163, 174, 175
   Korean Peninsula and, 171–178
   as nuclear shield, 140, 150
   nuclear strategy of, 140–143
   trade with ROK, 174
Cho Chen Nam, 23
Cho Kuk-kin, murder of, 192–193
Cho Myong Nok, 118, 119
Cho Myong-rok, 86, 104
Choe Byong Gwan, 161
Choe Gwang, 86
Choe Hak Kyun, 25, 29
Choe In Hwa, 259(n38)
Choe T'ye Bok, 116
Choe Yong Rim, 117
Choe Young Gyong, 119
Choi Hee Cheng, 25
Choi In Ha, 163
Choi Jung Sun, 150–151
Choi Mun Song, 117
Choi Young-jin, 157, 167
Ch'onch'on-gang thermal power plant, 56, 59
Chong Byong Ho, 26, 116
Chongchong River, flooding of, 77
Chonggye-ri, flood damage to, 77
Chongjin
   as DPRK port, 44, 45

271

272 Index

Chongjin (*cont.*)
　thermal power plant of, 57, 59
Cho Son-u, 200
Christensen, Thomas J., 172
Chung Ju-yung, 71
Churchill, Winston, ix
Clinton, Bill, 178, 223
coal, as DPRK energy source, 46–47, 51, 52
Cobalt-60, 17
Cold War, xi, 139, 140, 172, 194, 210, 216
Committee on Atomic Energy (CAE), 120
Constellation [U.S. aircraft carrier], 146
CSS-2 missile, of China, 103
CSS-4 missile, DPRK program for, 102

De Gaulle, Charles, 8
demilitarized zone (DMZ), 11, 87, 111, 146, 163, 192, 232
Democratic People's Republic of Korea (DPRK), ix, 1–2, 26, 29, 32, 36, 61, 127, 130, 142
　collective security system for, 217
　constitution of, 113
　cooperation with ROK, 70–71, 109
　cultural relations with Russia, 204–206
　disagreements and policy problems of, 131–132
　economic crises and difficulties of, 41–46, 60–75, 202–204
　economic reform in, 90
　electric energy sector of, 52–58
　energy sector of, 46–59
　foreign investments in, 45, 46, 70, 74, 88, 197, 238
　foreign trade of, 7, 45, 64, 66, 163, 236
　illegal access to nuclear secrets by, 28, 36–37
　international context of, 125–194
　Japanese policy on, 135
　leadership politics in, 110–124
　negotiations with KEDO, 156–170
　North Korean laborers in Russia, 181–183, 193, 203, 206–207
　nuclear cooperation with Russia, 15, 27–37
　nuclear crisis in, 1–12, 49
　nuclear institutions and organizations of, 21–37
　nuclear potential of, 143–146
　nuclear program of, ix, 16–38, 49, 125–171
　political liberalization of, 90
　power plants of, 59
　reform in, 69–70
　Russian arms supplies to, 130
　Russian political attitudes toward, 199–202, 219–222
　Russian specialists in, 1, 17, 21–22
　trade with Russia, 197–198, 202–203
Disarming Strangers: Nuclear Diplomacy with North Korea, 3
Doctors without Borders, suspended operations of, 88
"Double-headed Eagle," as U.S.-South Korean military game, 134
Downs, Chuck, 245(n4)
DPRK. *See* Democratic People's Republic of Korea (DPRK)
drug trafficking (DPRK in Russian Far East), 36, 182, 192, 206
Dudnik, Sergey, 203
Duke Engineering & Services (DE&S), 160

Eastern Pyongyang tide-power plant, 59
economic infrastructure, of DRPK, 44

electric energy sector, in DPRK, 51–59
environmental report (ER), for light-water reactor, 168, 169

"Fall Eagle" military exercise, 154
Federal Security Service (FSB), 200
Fitin, V., 190
Five-Year Plan, for electric power sector, 55
Flood Damage Rehabilitation Committee (FDRC), 82–84, 88
floods (DPRK), 76–85, 174, 206, 239
　economic implications of, 43, 81–85
　effects on health and school systems, 251(n1)
　military and security implications of, 85–89
"Focus Lens" military exercise, 134, 185
food, DPRK production decline in, 42, 77
food supplies
　crises in DPRK, 42, 67, 77, 83, 84, 99
　donations of, 82
Foreign Intelligence Service (FIS; Russia), 35
foreign investment, DPRK eased restrictions on, 70
four-party talks, Russian attitudes toward, 199, 219–235
Free Economic Zone (FEZ), in DPRK, 45–46, 68, 69
fuel oil, delivery to DPRK by KEDO, 267(n5)
Fyodorov, Valentin, 202

Gadhafi, Mu'ammar, 136
Gallucci, Robert, 20
Germany, reunification/convergence in, 212
glasnost (USSR), 94, 95, 182
gold-198, 17
Golden Triangle, 45
Golf II-class submarine, 104
Gorbachev, Mikhail, 94, 95, 97, 117, 128, 130, 139, 182, 197, 219
　foreign policy of, 129
　visit to Beijing, 132
grain production, decline in DPRK, 43
Gramenitsky, I. M., 30
graphite reactors, of DPRK, 107, 151
"Great Leader," Kim Il Sung as, 110, 112, 162
Gregg, Donald P., ix
gross domestic product (GDP)
　of DPRK, 41, 42
　of ROK, 42
gross national product (GNP)
　of DPRK, 41, 44, 45, 63, 74, 237
　of ROK, 43, 44, 45

Han Song Ryong, 116
Harrison, Selig, 3
Hayes, Peter, 3
helicopters, illegal purchase from Russia, 207
Hoch'on River, power plant on, 54, 59
Ho Gon, 24
Ho Jong, 157, 158, 159, 161
Hong Bom-do, 180
Hong Song Nam, 117, 189
House, K., 145
Hussein, Saddam, 137
Hwan-gang hydropower plant, 59
Hwanghaebuk-do, flood damage to, 77
Hwanghaenam-do, flood damage to, 77
Hwang Jang Yop, 89
Hwang Yong Yap, 175, 177
hydropower (DPRK), 46, 47, 51–52, 58
　plants for, 6, 59

Index

Hyundai Corporation, 49, 71

Ignatenko, B., 189
Independence [U.S.aircraft carrier], 146
Institute of Atomic Energy (DPRK), 29
Institute of Atomic Physics (DPRK), 24
Institute of Electronic Control Machines (DPRK), 24–25
Institute of Electronics (DPRK), 25
Institute of Machine Studies (DPRK), 24
Institute of Mathematics (DPRK), 24
Institute of Nuclear Physics (DPRK), 29
Institute of Oriental Studies (Moscow), 7
Institute of Peace and Disarmament (DPRK), 83
Institute of Physics (DPRK), 23, 24
intercontinental ballistic missile (ICBM), 105, 184
International Atomic Energy Agency (IAEA), 3, 8, 11, 16, 31, 32, 49, 78, 79, 80, 99, 105, 106, 160, 184, 222, 223, 229, 233
 8th conference, 1994, 142
 inspections by, 106, 148, 149, 225
 safeguards agreement with the DPRK, 17, 140
International Convention on Nuclear Safety, 142
International Federation of Red Cross and Red Crescent (IFRC), 77
International Monetary Fund, 174
Iodine-131, 34
Iran, 137
Iraq, 137, 146, 148
Itaru Umezu, 157, 167
Ito Hirobumi, 180
Izvestiya, 6, 149

Japan, x, xi, 45
 bilateral relations with United States, 235
 collective security system of, 217
 Korean policy of, 135–136
 nuclear program of, 155
 views on Korean reunification, 213
Jiang Zemin, 173
Ji Gil Song, 117
Joint Nuclear Control Commission, 109
Joint Venture Law of 1984, 68, 72
Jon Muu Sop, 115
juch'e (self-reliance) ideology, 2, 7, 31, 45, 52, 53, 65, 68, 69, 84, 89, 95, 117, 162, 236, 237, 241, 243, 244

Kaesong City, flood damage to, 77
Kangae hydropower plant, 55, 56, 59
Kang Song San, 117–118
Kangwon, 77
Kantor, Arnold, 145
Karasin, Grigory, 191
Kaurov, Georgiy, 5–6
KGB, 8, 35, 36, 37, 145, 152
Khabarovskiy Kray, 181, 183, 186, 192
Khromov, Gennadiy, 104
Khrushchev, Nikita, 7, 94, 95
Kim Byong Sik, 115
Kim Ch'aek, 119
Kimch'aek Polytechnic University, 23, 29, 30, 119
Kim Dae-jung, 177, 229, 243
 "sunshine diplomacy" of, 11, 243
Kim Dar Hyon, 117
Kim Gi Nam, 116
Kim Guk T'ye, 116
Kim Ha Gyu, 118
Kim Il Chol, 86

Kim Il Sung, 3, 8, 30, 33, 52, 60, 61, 68, 75, 181, 197
 death of, 89, 93, 110, 112, 119, 166, 174, 236, 237
 energy directives of, 53
 foreign policy of, 129
 interview with, 132
 nuclear weapons program of, 22–23, 96–97, 128, 136, 148
 regime of, 94, 98, 111, 187
 relations with Russia, 94
 son and heir of. *See* Kim Jong Il
 strategy of survival of, 136
 views on isolationism, 99
Kim Il Sung State University (Pyongyang), 23, 30, 32, 119, 120
Kim Jong Il, 7, 10, 32, 35, 110, 136, 162, 188, 189, 215, 221
 implementation of Agreed Framework by, 241
 light-water reactor project and, 238
 nuclear blackmail and, 93, 94, 100
 regime of, 61–63, 67–68, 75, 86, 87, 112–115, 118–119, 175–176, 199
 strategic decisions of, 69, 82, 84, 88–89, 237–243
 as Supreme Commander, 85, 86, 87
 World Youth and Students Festival of, 95–96
Kim Myong Guk, 118
Kim P'yong Il, 118
Kim Song Ae, 118
Kim Yong Chun, 86, 118
Kim Yong Ju, 113, 115, 118
Kim Yong Nam, 97, 115, 116, 131, 190
Kim Yong-sam, 156
Kim Yong Sun, 116
Kin Soh Yin, 24
Kipling, Rudyard, 215
Kolesnikov, M., 152–153
kolkhozy (collective farms), Korean-run, 205
Kommersant, 37
Komsomolskaya Pravda, 7
Korean Central News Agency (KCNA), 77, 78, 161
Korean College, 204
Korean Electric Power Company (KEPCO), 166, 167, 168, 243, 268(n28)
Korean-Japanese Kanghwa Treaty of 1876, 162
Korean Machine Import and Export General Society, 187
Korean Military Armistice Commission, 135
Korean Peninsula
 antinuclear declaration, 140
 China and, 171–178
 constitution for a unified, 232
 in cooperative security, 210–218
 denuclearization of, 119, 132, 139–140, 170, 234
 impasse in, x
 liberation from Japanese colonial yoke, 237
 military balance in, 214–216
 natural disasters of (mid-1990s), 76–90
 Primorskiy Kray and, 179–194
 reunification of, 122, 143, 173, 184, 210–211, 212, 213, 231–232
 Russia's plan for a settlement on, 229–235
 U.S. policy on, 133–135
 U.S. troops on, 133, 232
Korean Peninsula Energy Development Organization (KEDO), xi, 4, 5, 6, 9, 10, 15, 47, 80, 88, 99, 119, 120, 121, 122, 123, 124, 153, 170, 219, 226, 234, 240
 construction sites of, 25
 financial problems of, 11

KEDO (*cont.*)
  formation of, 2, 19
  membership of, 268(n24)
  mission of, 242
  negotiations with North Korea, 156–170
  projects of, 58
  Russia and, 20, 34
Korean People's Army (KPA), 85, 86–88, 98, 102, 103, 118, 123, 239
Korean Power Engineering Co. (KOPECO), 166–167
Koreans, in China, 204–205
Korean War, 9, 135, 148, 157, 176, 237
  economic and social effects of, 268(n27)
Kosovo, xi
Kozyrev, Andrei, 149, 224
Kryuchkov, Victor, 143–144
Kuala Lumpur Joint Declaration of June 12, 1995, 26, 78, 121, 238
Kumganggang cascade hydropower plant, 59
Kumho. *See* Sinp'o
Kunadze, Georgiy, 199
Kurapato Island, nuclear waste storage on, 154–155
Kuryong River, flooding of, 16, 78, 251(n12)
Kuzenkov, O., 186
Kuznetsov, Vladimir, 202
Kye Un Tye, 115

Labor Regulations for Foreign-Invested Enterprises in Free Economic and Trade Zone of 1993, 72
Leasing of Land Law of 1993, 72
Lee Bo-myun, 180
Lee Cha Gon, 24
Lee Gun-mo, 33
Lee Ha Il, 118
Lee In Gyu, 186, 190
Lee Jong Ok, 115
Lee Jong Rin, 24
Lee Song Dae, 187
Lee Ul Sol, 119
Levchenko, G., 189
Levchenko, S., 144
Li, Vladimir, 8–9
Liberal Democratic Party of Russia (LDPR), 199
Libya, 137, 146, 147
light-water reactors (LWRs), 20, 81, 88, 121, 122, 151, 157, 158, 161, 164, 166, 167, 170, 186, 225, 226, 237, 239, 240
  model of, 160
  site surveys for, 167, 168
  transfer project of, 122, 124, 160, 165, 242
Li Jong Hua, 88
Lim Ch'un-ch'u, 119
Lomakin, V. P., 181

MacArthur, Douglas, 185
Mangyongdae School, 119
Mansourov, Alexandre Y., ix, 5, 7, 9, 10
Mao Zedong, 136, 172
Mazarr, Michael J., 3, 4
Mikhailov, Viktor N., 35, 37
Mikheyev, V. V., 182
"Milgon" military exercise, 134
mineral deposits (DPRK), 44
Minju Choson, 119
Mirim hydropower plant, 59
missile development program, of DPRK, 101–105, 177, 184

Missile Technology Control Regime, 109, 234
Moiseyev, Leonid, 173, 174
Moiseyev, Valentin, 6–7
Moscow Energy Institute, 17
Moscow Engineering Physics Institute (MEPhI), 17
Moscow State Institute of International Relations (MGIMO), 221
Mount Kumgang, tourist development in, 71

NAC International of Norcross, Georgia, DPRK nuclear cleanup by, 80
Nam-gang hydropower plant, 59
Nam Gye-Bok, 104
Nam Hong Woo, 24
Namkung, Tony, 3
Nampo, flood damage to, 77
Nam San Nak, 118
National Defense Commission (NDC), 86
National Security Law (NSL), of South Korea, 154, 242
natural disasters
  effect on Agreed Framework, 76–90
  military implications of, 85–89
  security implications of, 85–89
Nautilus Institute, 3
Nazdratenko, Evgeniy, 183, 186, 187, 201, 202, 203, 206
Neutral Nations Supervisory Commission (NNSC), 232
Nikolayev, Konstantin, 199
Ninth Academy, in the province of Qinghai, 141
Nissan trucks, as *Scud* mobile launchers, 105
Nitsch, Larry, 171
Nixon Doctrine, 233
Nodong missiles, 103, 104, 108, 143, 184
nomenklatura, of DPRK, 96, 116, 117, 119
Non-Proliferation of Nuclear Weapons (NPT) Treaty, 1, 11, 17, 19, 32, 33, 34, 49, 99, 106, 107, 128, 138, 140, 142, 148, 150, 153, 154, 184, 223
  DPRK and, 138–155, 245(n5)
North Hwanghae province, 77, 181
North Korea. *See* Democratic People's Republic of Korea (DPRK)
North Korea and the Bomb: A Case Study in Nonproliferation, 3
North Korean Academy of Sciences, 23
North Korean Central Telegraph Agency, 191
North Korean Committee on the Environment, 154–155
North Pyongan province, flood damage to, 77
Notary Public Law of 1995, 72
nuclear blackmail, 93–100, 215
Nuclear Energy Research Institute (DPRK), 23
nuclear freeze, flood-induced (DPRK), 78–81
nuclear fuel
  storage on Kurapato Island, 154
  USSR supplies to DPRK, 17
nuclear nonproliferation regime, 138–155
nuclear power plants. *See* DPRK, nuclear program
Nuclear Regulatory Authority (DPRK), 168
Nuclear Research Complex (Yongbyon). *See* Yongbyon Nuclear Research Complex
nuclear waste
  disposal of, 80, 154–155, 223
  in Korean Peninsula waters, 154–155
nuclear weapons, ix
  as means of dealing with changing world, 96
nuclear weapons program (DPRK), 2, 41–42, 91–124
  economic aspects of, 39–90

Index

history of, 13–38
military and political factors of, 91–124
Russian reaction to, 222–225
studies on, 3–5

Oberdorfer, Don, 4
oil, ROK shipments to DPRK of, 47
oil refinery, lag in DPRK, 42
oil shipments, 98
O Jin U, 25, 86, 89
Old Choson, 237
Ounbong hydropower plant, 55, 56, 59
Ounggi thermal power plant, 56, 59

Paek Nam Sun, 117
Page, Trevor, 83, 88
Pakch'on, 48, 77
　nuclear infrastructure of, 78
Pakistan, exports of missile technologies to, 4
Pak Kwang O, 120
Pak Nam Gi, 116
Pak Song Chol, 115, 192
Panov, Alexander, 186, 188, 189, 190, 200
Park Chae Gyong, 118
Park Huongyu, 31
"Pasex" military exercise, 135
People's Republic of China. See China
perestroika (USSR), 95, 127, 128, 182
Phosphorus-32, 17, 34
Platkovskiy, Alexander, 7
plutonium, 16, 35, 37, 49, 50, 79, 102, 106, 122, 150
　storage of, 102, 160
Plutonium-239, 26
Pongaen, flood damage to, 77
Ponhawa hydropower plant, 59
preliminary safety analysis report (PSAR), on light-water reactor, 168, 169
Preliminary Works Contract (PWC), 160, 258(n15;n16)
Primakov, Evgeniy, 200
Primorskiy Kray (Maritime Province)
　Korean Peninsula and, 179–194, 201–206
　Koreans in, 36, 180–184, 204–207
Propaganda and Agitation Department, of Workers' Party of Korea, 115
Puch'on-gang hydropower plant, 59
Pukchang thermal power plant, 55, 56, 59
Puren cascade hydropower plant, 59
Pyongsan, 48
P'yongsong, flood damage to, 77
Pyongyang, ix, x, 1, 2, 16
　flood damage to, 77
　Soviet-DPRK Agreement on, 19
　thermal power plant of, 59
Pyongyang State University, 23, 48

Qian Qichen, 171

radioactive materials, smuggling from Russia to North Korea, 37
Radiochemistry Institute (Yongbyon), 48
Radiological Institute (DPRK), 23
Rajin, as DPRK port, 44, 45, 187
Rajin Business Institute, 71
Rajin-Sonbong Free Trade Economic Zone (FEZ), 45, 69, 186, 201, 203, 238
Reiss, Mitchell B., 161, 163
Republic of Korea (ROK), ix, xi, 131, 142

attempt to expand economic ties with DPRK, 70–71
collective security system for, 217
cooperation with DPRK, 70–71, 109
cultural issues with Russia, 204–206
economic growth of, 42
energy sources of, 47
gross national product of, 67
rural population in, 43
trade with China, 131
reunification
　of North and South Korea, 122, 143, 173, 184, 206, 210–211, 212, 213, 231–232
　North–South views on, 211–212
rice production (DPRK), 43, 77
Richardson, Bill, 90
"Rimpac" military exercise, 135
Ri Myong Sik, 161
Rinsan, flood damage to, 77
road transport (DPRK), 44
Rockefeller Brothers Fund, 3
Rockefeller Foundation, 3
Roh Tae-woo, 97, 130, 139, 231
ROK. See Republic of Korea (ROK)
rural population (DPRK), 43
Russia (Russian Federation),
　China relations with, 129
　exports of crude oil, 47
　foreign policy of, xi
　Korean minority in, 179–184, 204–206
　North Korea Treaty with, 185–192
　North Korean laborers in, 181–183, 193, 203, 206–207
　relations with DPRK, 9, 10, 197–209
　Security Council of, x
　interception of scientists from, 206
　stake in Korean Peninsula, 12
　State Duma of, 143
　views of Agreed Framework, 219–235
　views on reunification, 213
Russian Diplomatic Academy, 7, 8
Russian Foreign Ministry, 5, 6
Russian Liberal-Democratic Party (LDPR), 186
Russian Ministry of Atomic Energy (Minatom), 5, 6, 15, 18, 19, 20, 31, 34, 36, 102, 105, 199
Ryo Yin Gan, 23
Ryuksolbi firm (DPRK), nuclear fuel supplies to, 18

Sakhalin Island, 203, 205
Samkyong Technical Services Corporation, 166
Savel'yev, Roald, 8
Scalapino, Robert, 3
Scientific and Technological Commission (China), 141
Scientific Research Center on Atomic Energy, 30
Scowcroft, Brent, 145
Scud missiles, 102, 103, 184
Sea of Japan, nuclear waste dumping in, 223
Second Ministry of Machine-Building, 141
Sekai, 132
Selenium-75, 34
Seleznyev, Gennady, 189, 199
semiconductors, lag in DPRK, 42
Seoul Institute of Defense Research, 154
Series 9559 contracts, of USSR and DPRK, 15
services and construction, in DPRK, 44–46
Seven-Year Plan, for electric power sector (DPRK), 55, 58
Shahroud, as Iranian missile test site, 104

Shevardnadze, Eduard, 97, 131
Sigal, Leon V., 3, 4
Sinp'o, 11, 20, 25, 123, 159, 242, 243
Sinuiju, flood damage to, 77
Site Survey Report, for LWR project, 168, 260(n60)
So Ch'ol, 119
Socialist Party of Japan, 132
Sodium-24, 17
Sodusu hydropower plants, 55, 56, 59
soktochon, 156
Sonbong, as DPRK port, 44, 45
Sondok airport, 163
Song Chang Ho, 24
South Hamgyong province, flood damage to, 77
South Hwanghae province, flood damage to, 77
South Korea. *See* Republic of Korea (ROK)
South Korean Ministry of National Reunification, 143
Soviet Union, x, 216. *See also* Russia
  collapse of socialism in, 98
  democratization of, 130
  dismantling of, 35, 133
  joint nuclear activities with North Korea, 15–20, 27–37
  military aid to DPRK, 60
  North Korea treaty with, 185–192
  policy toward nuclear crisis, 1–12
So Yun Sok, 117
Special Economic Zone (SEZ), of Nakhodka, 203
Stalin, Joseph, ix, 65, 94, 130, 136, 205, 219
START I and II, 139
State Committee on Science and Technology (DPRK), 25
steel and steel-based products, DPRK production decline in, 42
Sunc'hon thermal power plant, 59
Sungni Oil Refinery, 186
"sunshine policy," of Kim Dae-jung government, 11, 243
Supreme National Confederate Assembly, 231
Supreme People's Assembly (SPA), 25, 26, 86, 113, 115, 140
Sup'ung River, power plant on, 54, 59

Taedong hydropower plant, 57, 59
Taek Kwan-oh, 30
Taepodong missiles, 50, 101, 103–104, 108, 155, 184
T'aep'yonman hydropower plant, 57, 59
Taiwan, 178, 228
  sinking of freight vessel from, 89
Tangun, 237
"Team Spirit" military exercise, 106, 134, 135, 140, 185, 238
"Tekhsnabexport" Foreign Trade Association (FTA), 17, 18
Thatcher, Margaret, 136
Theater Missile Defense (TMD), 108
thermal power plants, in DPRK, 52, 53, 55–56, 58
tidal-power plants, 52, 59
"Toksuri" military exercise, 134
Tong-A Consultants Company, 167
Tonhwa hydropower plant, 59
Tonno-gang hydropower plant, 55, 59
Torkunov, Alexander, 221
tourism, in DPRK, 45, 71, 73
Treaty of Peking, 180

Treaty on Friendship, Cooperation, and Mutual Assistance, between DPRK and China, 141, 173
Treaty on Friendship, Cooperation, and Mutual Assistance between the Soviet Union and the DPRK, 153
Tsutomi Hata, 155
Tumen River Basin Development Program, 46
The Two Koreas: A Contemporary History, 4

United Institute for Nuclear Research (UINR; USSR), 5, 21, 28–29, 30, 141
United Nations, 11, 146, 227
United Nations Environmental Protection, 223
United Nations Register of Conventional Arms, 109
United Nations Security Council, 8, 37, 217, 229
U.S. Congress, 12, 107
U.S. Department of Energy (DOE), 80, 169
United States, x, xi, 45, 93
  bilateral relations with Japan, 235
  collective security system of, 217
  estimates of DPRK nuclear power by, 145–146
  food aid to DPRK by, 267(n10)
  Korean policy of, 1–12, 133–135
  troops on Korean Peninsula, 133, 232
  views on reunification, 213
uranium, DPRK deposits of, 22, 25, 36, 48, 206
Uranium-235, 18, 24, 31
Uranium-238, 24

Vasileyev, Alexander, 183–184
Vengerovsky, A., 186

W. Alton Jones Foundation, 3
Wan Yongxiang, 173
Wiwon hydropower plant, 55, 59
Woolsey, R. James, 145
Workers' Party of Korea (WPK), 22, 25, 30, 53, 61, 81, 89, 110, 119, 132, 135
  energy proposals of, 57–58
World Food Program (WFP), 83, 88
World Youth and Students Festival (DPRK), 7, 95–96
World Youth and Students Festival (USSR), 94

Yang Hyong Sob, 115
Yeltsin, Boris, 10, 188, 189, 191, 197, 199, 201, 219, 223, 224
Yevstafyev, Gennady, 35
Yongbyon, 16, 80
  flood damage to, 78
Yongbyon Nuclear Research Complex, 23, 25, 35, 102
  facilities of, 16–17, 22, 78, 149, 151
  flood effects on, 79–80
  U.S. nuclear specialists at, 81
Yongbyon Special Administrative District, 78, 79
Yonhab firm (DPRK), nuclear fuel supplies to, 17, 18
Yon Hyong Muk, 117
Younbaek plains, flood effects on, 82

Zabrovskaya, Larisa, 9–10
Zanegin, B., 189
Zarubin, Alexander, x, 10
Zhang Qiyue, 173
Zhebin, Alexander, 6
Zhirinovsky, Vladimir, 199
Ziang Zemin, 141